THE
MAN–MADE
SUN

OTHER BOOKS BY T.A. HEPPENHEIMER

Colonies in Space

Toward Distant Suns

The Real Future

THE MAN-MADE SUN

THE QUEST FOR FUSION POWER

T. A. Heppenheimer

An Omni Press Book

LITTLE, BROWN AND COMPANY • BOSTON • TORONTO

FIRST EDITION

———

Excerpt from "Ambitious Energy Project Loses Luster," from
Science, Vol. 212, pp. 517–519, May 1, 1981. Copyright © 1981 by
W.D. Metz. Reprinted by permission of *Science,* American Asso-
ciation for the Advancement of Science.

"Santa Claus Comes to Fusion," by Paul J. Reardon. Reprinted by
permission of the author.

Excerpt from "Three Bad Signs" from *Other Things and the
Aardvark* by Eugene McCarthy. Copyright © 1970 by Eugene
McCarthy. Reprinted by permission of Doubleday & Co., Inc.

Library of Congress Cataloging in Publication Data

Heppenheimer, T. A., 1947–
   The man-made sun.

   "An Omni Press book."
   Bibliography: p.
   Includes index.
   1. Fusion reactors. I. Title.
TK9204.H47   1984   621.48´4   83–22240
ISBN 0–316–35793–6

VB

Designed by Dede Cummings

*Published simultaneously in Canada
by Little, Brown & Company (Canada) Limited*

PRINTED IN THE UNITED STATES OF AMERICA

To Angela, my love

Contents

Dramatis Personae   ix

Acknowledgments   xiii

1   This New Fire   3
2   Robert Hirsch   33
3   Quest for Power   55
4   Doing It with Mirrors   74
5   Lasers and Microspheres   107
6   Arms and the Man   138
7   The Entrepreneurs   163
8   The Breakthrough   188
9   Plasmas on the Potomac   217
10   The Face-Off   240
11   Fusion's Highest Potential   269
12   Fusion in the World   293

Bibliography   324

Notes   327

Credits for Illustrations   332

Glossary   333

Index   339

Dramatis Personae

ARTSIMOVICH, LEV.   *Director of the Soviet fusion program, 1950–1973.*

ASHWORTH, CLINTON.   *Fusion-power expert at Pacific Gas and Electric, San Francisco.*

BISHOP, AMASA.   *Director of the U.S. fusion program, 1964–1970.*

BODNER, STEVE.   *Head of the laser-fusion program at Naval Research Laboratory, on temporary loan to the OMB as a budget examiner during the fall of 1981.*

BRUECKNER, KEITH.   *Research director for KMS Industries, who was active in laser fusion and who persuaded Kip Siegel to put company money into its development.*

BUCHSBAUM, SOL.   *Executive vice-president at Bell Labs, with strong ties to Princeton University. A leading consultant in the field of fusion energy.*

BUSSARD, ROBERT.   *An assistant director in the fusion program, 1973–1974, and inventor of the Riggatron tokamak design. Founder and chairman of INESCO, Inc.*

CLARKE, JOHN.   *Ed Kintner's deputy when Kintner headed the fusion program; director of this program since early in 1982.*

COENSGEN, FRED.   *Experimental director in the fusion program at Livermore.*

CONN, ROBERT.   *Professor of engineering at UCLA; expert on the design of fusion reactors.*

COPPI, BRUNO.   *Plasma physicist from MIT; inventor of the Alcator and collaborator with Bussard in inventing the Riggatron.*

DEAN, STEPHEN.   *Ed Kintner's director in charge of tokamak research in the DOE.*

DEUTCH, JOHN.   *Director of Energy Research in the U.S. Department of Energy, 1977–1979, and Kintner's boss. Also Undersecretary of Energy, 1979–1980, and Ed Frieman's boss.*

DUNCAN, CHARLES.   *Secretary of the U.S. Department of Energy, 1979–1981, under Jimmy Carter.*

EDWARDS, JAMES.   *Secretary of the U.S. Department of Energy, 1981–1982, under Ronald Reagan.*

EMMETT, JOHN.   *Head of the laser-fusion program at Livermore, since 1972.*

EUBANK, HAROLD.   *Princeton physicist in charge of neutral beams on PLT.*

FOSTER, JOHN.   *Director of Lawrence Livermore Lab, 1961–1965, and one of the nation's leading experts in assessing advanced technologies. Chairman of the Foster Committee, which reviewed the fusion program in 1978. Also chairman of a different Foster Committee, which reviewed laser fusion in 1979.*

FOWLER, KEN.   *Director of the magnetic fusion program at Livermore since 1970, and Coensgen's boss.*

FRIEMAN, EDWARD.   *Director of Energy Research in the U.S. Department of Energy, 1979–1981, and Kintner's boss.*

FURTH, HAROLD.   *Experimental director of the Princeton fusion program prior to 1981; then Director of the Princeton Plasma Physics Lab.*

GOTTLIEB, MEL.   *Director of the Princeton Plasma Physics Lab, 1966–1981.*

HIRSCH, ROBERT.   *Director of the U.S. fusion program, 1972–1976.*

HYDE, RODERICK.   *Livermore physicist who has worked on using laser fusion for rocket propulsion.*

KEYWORTH, GEORGE.   *President Reagan's science advisor and head of the OSTP.*

KINTNER, EDWIN.   *Robert Hirsch's deputy, 1975–1976; subsequently the director of the U.S. fusion program, 1976–1981, and John Clarke's boss while director.*

LOGAN, GRANT. *Livermore plasma physicist who played key roles in inventing the tandem mirror and the thermal barrier.*

McCORMACK, MIKE. *Democratic congressman from Washington state, 1971–1981, and author of the Magnetic Fusion Act of 1980.*

MEADE, DALE. *Experimental director in the Princeton fusion program since 1981.*

NUCKOLLS, JOHN. *Livermore physicist; expert on pellets for laser fusion.*

PALMIERI, TOM. *Official in the OMB; Don Repici's and Steve Bodner's boss.*

PEWITT, NELSON DOUGLAS (DOUG). *Acting Director of Energy Research in the Department of Energy during part of 1981, and Kintner's boss while Acting Director. Subsequently an OSTP staff member under George Keyworth.*

REARDON, PAUL. *The chief engineer in charge of building the TFTR, 1975–1982.*

REPICI, DON. *Budget examiner at the OMB who was in charge of the fusion budget, 1977–1981.*

ROSENBLUTH, MARSHALL. *Director, Institute for Fusion Studies at the University of Texas, 1980–, often called "the pope of plasma physics" and regarded as the world's leading authority.*

SCHLESINGER, JAMES. *Chairman of the AEC, 1971–1973, and Secretary of U.S. Department of Energy, 1977–1979.*

SHANNY, RAMY. *Executive vice-president at INESCO, Inc., under Bussard; currently president.*

SIEGEL, KEEVE M. (KIP). *Founder and president of KMS Industries and KMS Fusion.*

SPITZER, LYMAN. *Inventor of the stellarator, founder of fusion research at Princeton, and director of the Princeton Plasma Physics Lab, 1961–1966.*

TRIVELPIECE, ALVIN. *Assistant director for plasma physics research under Hirsch, 1973–1975; Director of Energy Research under President Reagan, and Kintner's boss.*

WOOD, LOWELL. *Livermore physicist; laser-fusion expert.*

ZIMMERMANN, GEORGE. *Livermore physicist; author of the LASNEX computer program.*

Acknowledgments

In May 1981, I was at the annual meeting of the American Institute of Aeronautics and Astronautics, held in Long Beach. There I met Ed Kintner, the director of the fusion program within the United States Department of Energy. He was there to give a talk on this program. I told him I was hoping to write a book on fusion, and he was immediately interested. At his suggestion, that September I attended a meeting of his Fusion Power Coordinating Committee, held in Los Alamos, New Mexico. At that conference, he mentioned my name and my project to his lab directors and senior managers, and suggested that they help me. In the ensuing months he sent me a large amount of valuable source material, met with me for six hours of interviews, steered me to people who were in a position to help me, and reviewed drafts of my text. In aiding my entry into the fusion community and in smoothing my way, he did me a tremendous service, and it is a pleasure to thank him.

So it was that during the summer and fall of 1982, I loaded up my Pinto with my luggage and with a Panasonic cassette recorder, and set out in search of the fusion program. The ensuing odyssey covered some 11,000 miles of driving and took me to the major fusion centers of Princeton, New Jersey; Washington, D.C.; Massachusetts Institute of Technology; and Lawrence Livermore National Laboratory, east of San Francisco. Also I attended the week-long fusion conference of the International Atomic Energy Agency, held in Baltimore during the first week of September. Along the way, a large number of fusion leaders were kind enough to grant interviews. In particular, there were the following.

Princeton: Kees Bol, Anthony DeMeo, Harold Eubank, James

French, Harold Furth, Mel Gottlieb, Donald Grove, Dale Meade, Robert Papsco, Paul Reardon, Lyman Spitzer, Theodore Taylor.

Livermore, magnetic fusion program: Frederic Coensgen, T. Kenneth Fowler, Theodore Kozman, Grant Logan, Ralph Moir, Richard F. Post, Thomas Simonen, Keith Thomasson, Robert Wyman.

Livermore, laser fusion program: Harlow Ahlstrom, George Chapline, Victor George, Roderick Hyde, Michael Monsler, John Nuckolls, Mike Ross, Lowell Wood.

KMS Fusion: Judy Francis, Alexander Glass, David Solomon.

INESCO, Inc.: Robert Bussard, Ramy Shanny, Carl Weggel.

MIT: Bruno Coppi, Ronald Parker, C.L. Fiore, Stephen Wolfe.

Washington, D.C.: Stephen Bodner, Gail Bradshaw, John Clarke, Stephen Dean, Robert Hirsch, George Keyworth, Edwin E. Kintner, Mike McCormack, N. Douglas Pewitt, Don Repici, Alvin Trivelpiece.

Other fusion experts: Clinton Ashworth, Charles Baker, Hans Bethe, Robert Conn, Arthur Kantrowitz, Shigeru Mori, Sebastian Pease, Marshall Rosenbluth, Hans-Otto Wüster.

In addition to granting interviews, a number of these people were kind enough to review drafts of the text and to point out errors in facts or interpretation. The people who did this were Dale Meade, Mel Gottlieb, Don Grove, and Paul Reardon; Fred Coensgen and Ken Fowler; John Nuckolls, John Holzrichter, and Rod Hyde; David Solomon; Robert Bussard; Steve Bodner, John Clarke, Robert Hirsch, Ed Kintner, Mike McCormack, and Don Repici; Robert Conn and Arthur Kantrowitz; and Ed Frieman.

I owe particular thanks to a number of people who helped me in other ways. To John Nuckolls, for arguing with me till I understood the merits of laser fusion. To Steve Bodner and Don Repici, for guiding me through the mysteries of Washington, and particularly of the Office of Management and Budget. To Bruce Abell of the White House staff, who arranged my interview with George Keyworth. To Gail Bradshaw of the Department of Energy, who arranged my interviews with John Clarke and Al Trivelpiece. To Mike Ross and Diane DeHollander at Livermore, who compiled a fine collection of artwork and illustrations for me. To Ray Kidder and John Holzrichter, for letters and correspondence that provided valuable background material concerning laser fusion. To Mike Ross and Janet Sitzberger, who arranged my visit and my schedule at Livermore. To Dominique Surel, who did the same at MIT. To Tony DeMeo and Diane Carroll, who did the same at Princeton. To Betty Graydon and Jane Holmquist, librarians at the Princeton Plasma Physics Lab, who gave me office space and a great deal of help.

Two people have been invaluable in providing me with background information. Joan Bromberg of Cambridge, Massachusetts, sent me a manuscript copy of her book *Fusion: Science, Policies, and the Invention of a New Energy Source,* since published by the MIT Press. It is a history of the fusion program from 1951 to 1978, and an essential reference work for anyone with a serious interest in fusion power.

Also, Stephen Dean of Fusion Power Associates, Gaithersburg, Maryland, provided me with a number of publications from his office. Fusion Power Associates is an industry-sponsored group advocating rapid development of fusion energy; its monthly newsletter has proven quite useful. In addition, this group sponsored an excellent conference held in San Diego early in 1983.

The cassette tapes of my interviews will be deposited permanently with the Center for the History of Physics, American Institute of Physics, 335 East Forty-fifth Street, New York, New York 10017.

A note is in order as to the disclosure of classified information. Page 118 contains what purports to be a description of an early laser-fusion pellet design invented by John Nuckolls. This description is based upon the disclosures that that design involved a radiation-filled cavity, what physicists call a hohlraum, and that pellet compression was to be achieved using soft X rays. These disclosures are consistent with classification guidelines of the U.S. Department of Energy. With this, and with an understanding of Planck's radiation law and of related pertinent physics, I have inferred the pellet description as presented. The actual description remains classified and has not been disclosed to me. Similarly, the pellet-design concepts in Figure 7 are based on the unclassified literature.

Finally, there are acknowledgments of a more personal nature. Thanks go to my literary agents Neil McAleer and Bob Weil, who got this book started as a project. Bob Weil in particular offered some very good literary criticism while I was preparing the proposal for this book. At Little, Brown, my editor, Bill Phillips, has from the start offered a great deal of encouragment as well as of excellent criticism. Whatever quality this book may possess is due in large measure to his active involvement. Thanks also go to his assistant, Colleen Mohyde, and to two other key people, Elisabeth Humez and Char Lappan.

It is a pleasure to show my appreciation to George Hazelrigg, who made me welcome in his Princeton home and extended to me his hospitality for the duration of my stay in Princeton.

Thanks are also due to my Angela, who held the fort at home during my lengthy absences, who all along has put up with my late-night sessions with cassette recorder and typewriter, who has frequently been more patient than me, and who throughout has brought me love, warmth, and affection.

And finally, it is appropriate to remember a wonderful old professor of mine with the unlikely name of Maria Zbigniew v. Krzywoblocki, who harks back to my undergraduate days at Michigan State University in the mid-1960s. He introduced me to plasma physics and controlled fusion in a course he taught in 1967, some fifteen years before I would seriously consider how I might be able to involve myself in the fusion program and make a contribution.

Fountain Valley, California
November 10, 1983

THE
MAN-MADE
SUN

1

This New Fire

THE ivy grows thick at Princeton University. We think of ivy as thin tendrils making their way across brickwork and walls, but on Nassau Hall, ivy planted a century ago has developed virtual tree trunks thicker than a man's arm. Old Nassau itself stands in serenity amid shade trees, as it has since George Washington was four years old. Even in late autumn and winter, the nearby lawns and buildings share a similar serenity. Then in the springtime the overcoats are packed away, the greenery springs forth from every branch and bush, and the campus fills up with young people on their bicycles.

A short walk to the south are the residence halls with their walls of gray stonework, two and three stories tall. Built early in this century, they show steeply gabled roofs and English-style chimneys, with leaded windows looking out onto courtyards, and ornamental wrought-iron lamps to light the way at night. Now head for the southbound Washington Road, past the fountains of the Woodrow Wilson conference center with its walls gleaming white, past that geological Guyot Hall that gave its name to the Pacific's curiously flat-topped undersea mountains. Across Lake Carnegie with its boathouse, the road narrows and is lined with rows of trees. It is not hard to imagine horse-drawn carriages using it in days when Princeton was younger. A little farther on is a traffic circle, with Gulf, Exxon, and Arco stations surrounding it. Here the road connects with the northbound Highway 1.

Once you are past a cluster of buildings near the traffic circle, the country opens up. To the right stand the research labs of an electronics firm, across some hundreds of yards of sunlit lawn. To the left is an extensive open field; the Gothic tower of a campus building juts above the distant trees and recalls the countryside around Stratford-upon-Avon. A mile farther and a Holiday Inn brings us back to the States; then a low hill, and a turnoff from the road. There a sign announces the destination, the Forrestal Campus. Here are the brick and stucco buildings of the Princeton Plasma Physics Lab, interspersed among lawns and parking lots, set amid thick, leafy woods. Inside these buildings, over a thousand people are working to gain one of the great achievements that may be left to us in the waning years of this century, and that can illuminate the next with bright hope. They are working to develop controlled fusion energy. To achieve this, they have built a mammoth machine that will soon be producing the hottest temperatures in the solar system—a hundred million Celsius degrees, seven times hotter than in the center of the sun that shines on the ivy of the campus buildings back up the road.

Fusion, indeed, is the energy source of the sun. Our sun, small and cool as stars go, has a temperature at its center of 15 million Celsius degrees. The pressure there, with the entire weight of its outer layers pressing downward, is nearly a billion times greater than at the bottom of the Pacific Ocean. The material at the solar center is mostly a mixture of hydrogen and helium, which we encounter on earth as light gases; but under that extreme pressure these gases are compressed to nearly fifteen times the density of lead. Here is the true solar furnace, the place where fusion reactions produce the sun's energy. This happens when the nuclei of hydrogen or other light atoms collide violently and combine to form heavier nuclei, or at least different ones. Always when this happens the mass of the combined nuclei is less than that of the original ones; the difference appears as energy.

Yet the surprising thing is not that this furnace is overwhelmingly powerful, but that in any local region of the sun's interior, it is so tenuous. This sun-stuff, under conditions found at the center, produces energy at a rate of only two watts per ton. That is not a very intense rate. A bathtub cooling releases energy thirty thousand times faster. One might think that if we could build a starship powered by solar material, at the conditions of its deep interior, then we would have the very model of an advanced design. Yet in power-to-weight ratio, a Roman galley full of slaves would be much more efficient. In fact, even when the slaves were asleep their body heat alone would yield more energy than this sun-stuff.

What makes the sun work is that it is so large. Its outer layers, hundreds of thousands of miles deep, trap the interior heat so effectively that even this feeble furnace can keep the center at its 15 million degrees. Viewed another way, the sun's surface is only a few times hotter than the flame of a blowtorch, but whereas the blowtorch flame is only an inch or so across, the sun's diameter approaches a million miles. The sun's energy source, therefore, is almost unimaginably weak, and if this were the best that fusion could offer us, we might do better to contemplate seeking a new energy source based upon squirrels running around in cages. Fortunately, fusion can occur in many forms, some of which are extremely energetic.

In physicists' fusion experiments such as those conducted at Princeton, the work revolves around one simple fact: to produce fusion at a usefully energetic rate, the fusion fuels must be heated to the temperature that will be reached inside Princeton's big machine: a hundred million degrees Celsius. This is a temperature so high, so far beyond our ordinary experience or comprehension, as to represent a concept that we name rather than grasp, as if it were the national debt or the population of China. If a pinhead could be raised to that temperature and kept there, it would radiate enough heat to boil water ten thousand miles away. A fifteen-foot sphere, raised to that temperature, would radiate as much energy as the entire sun itself.

At temperatures of only a few thousand degrees, even the hardest forms of steel flash into gas. At temperatures far below a hundred million, even a gas ceases to behave as a gas. Like everything else, a gas is made up of atoms, and an atom features a cloud of electrons whirling in orbits around a central atomic nucleus. At very high temperatures—a hundred thousand degrees, say—the electrons begin to escape. They fly away from their atoms, like little planets escaping from their solar systems. As the temperature rises still higher, more and more electrons escape in this way. The atoms left behind, bereft of some or all of their electrons, are called by a new name: ions. An ionized gas, made up of ions and electrons, is also called by a new name: plasma. Plasmas are different from ordinary gases. Unlike such gases, they conduct electricity, and respond to the influence of a magnet. This is a fortunate fact, for it provides the key to producing plasmas at a hundred million degrees, and at even hotter temperatures. The way we achieve and control such temperatures is by having very good insulation. With plasmas, however, the insulation cannot be provided by rock wool. Instead, it is produced with the aid of powerful electromagnets.

An electromagnet is a coil of wire or cable wound around a central form; an arrangement of copper bars will also do quite nicely. When electric current flows through the cable or copper, a powerful magnetism is set up nearby. A piece of iron, set loose in the vicinity, will fly toward the cabling or the copper bars and strike them with great force. The property of space near such a current-carrying coil, which causes the iron to behave in this way, is called a magnetic field. To learn more about this field, take a plate of opaque glass or other nonmagnetic material, and place it on top of the coil. Sprinkle iron filings on it—before the magnet is turned on, not after, if you don't want the filings to fly away. Turn on the current in the magnet. The filings will arrange themselves in patterns of curving lines, what are called lines of force.

Physicists visualize magnetic fields in terms of such lines of force. The lines play the same role in a physicist's thinking as contour lines play in the thinking of a geographer, when he studies relief maps of mountains and flatter terrain. Where contour lines are close together, it means the land is steep; where they are far apart, slopes are gentle. Similarly, where magnetic field lines are close together, the field is strong; the magnet attracts iron very powerfully. Where the field lines are far apart, the field is weak. On a contour map, crossing contour lines means that we are climbing or descending a hill. Similarly, crossing magnetic field lines means that we are moving between stronger and weaker regions in the field.

The key to achieving a hundred million degrees, then, is that a plasma responds to a magnetic field rather as those iron filings do. Each ion or electron rides on its own field line, and continues to stay there unless it is disturbed. If a plasma's particles are trapped in this way, its ions or electrons, the plasma can be penned up and kept from reaching the walls of the chamber in which it is to be confined. When we trap or confine the plasma in this way, we also trap its heat, even at a hundred million degrees. The plasma could still cool off by radiating its heat, by glowing like the sun. But by our taking care that the plasma is clean, free of impurities, this glow is kept from getting too bright. We heat the plasma by feeding energy into it, but a bright plasma means that the energy comes right back out again, without making it any hotter. A dim plasma is best; it will absorb the energy we feed in, and heat up even further.

In fusion experiments like those at Princeton, the plasma is often composed of ordinary hydrogen. This common industrial gas is cheap, convenient, and easy to handle. It is also the gas of which most of the sun is made. A hydrogen plasma, then, is truly, in every sense, like a little piece of the sun's deep interior. Of course, compared to that interior, the plasma is small, rarefied, and evanescent, and hence it produces virtually no fusion energy. Tomorrow's fusion reactors will rely on more energetic reactions. The most interesting of these employ deuterium and tritium,

Motion of Charged Particles

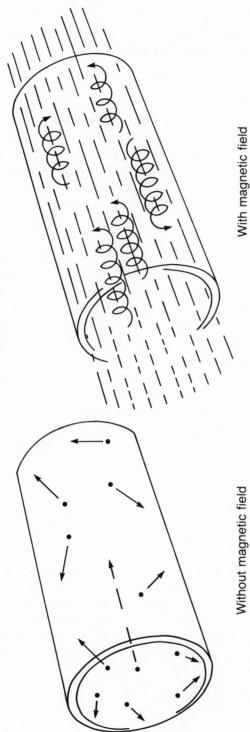

Without magnetic field

With magnetic field

FIGURE 1. *Confining plasma with a magnetic field. Without such a field, ions and electrons, the charged particles in a plasma, fly in all directions. With a magnetic field, these particles tend to spiral around field lines and can be kept from hitting the surrounding metal chamber.*

the heavy isotopes of hydrogen. These have nearly the same chemical properties as ordinary hydrogen, but differ in the structure of their atomic nuclei. The substitution of deuterium and tritium for the sun's hydrogen then raises the reaction rate from virtually nil to a useful level.

Deuterium is present in seawater at a proportion of about an ounce per ton. For many years it has been extracted in a straightforward way, in quantities measured in tons, for use in certain types of nuclear reactors. In principle it is possible to build a fusion reactor that will run on deuterium alone. If all the electricity in the U.S. were generated by such reactors, the deuterium needed would amount to about a hundred pounds per hour, which would be like a flow from a kitchen faucet.* This deuterium could be had from a small separation plant, whose intake would simply be a five-inch pipe feeding in ordinary seawater. During the 1960s, when Glenn Seaborg was chairman of the Atomic Energy Commission, he used to go up to Capitol Hill and present an argument that would whet congressional interest. He would tell the legislators that deuterium in the world's oceans amounts to the energy equivalent of five hundred Pacific Oceans brimfull with the highest grade of fuel oil. Put another way, every time you drink a cup of coffee you are downing enough deuterium to provide the energy to run your car for a week.

The catch is that the all-deuterium reactions are difficult to ignite. The reaction employing deuterium plus tritium is also difficult, but is notably less so, so it has received the most attention. However, tritium does not exist in nature, at least not in useful quantities, but it can be produced from the element lithium in a nuclear reactor, or in a fusion reactor. Fortunately, there is plenty of lithium, particularly in seawater and in certain heavy brines.

At the Princeton Plasma Physics Lab, perhaps by now you have managed to find a parking space, over by the trees. Walk along the roadway, toward the entrance of the main building. Outside on the lawn, enclosed within a fence, are two large white dish antennas, slanted upward. They point at a communications satellite orbiting 22,300 miles up, and link this Princeton lab to the powerful computer facilities at Lawrence Livermore Laboratory, near San Francisco. Livermore has the main computers for the nation's entire program in research on controlled fusion. That lab has two additional fusion programs, very active and

---

*The corresponding figure for coal is about 300,000 tons per hour. This is like an endless train of coal cars on the railroad tracks, continually roaring by at eighty miles per hour.

vigorous, separate and distinct from the Princeton one; their leaders often compete with Princeton's for attention and for funds. The name Livermore seldom crops up in conversations within the Princeton lab, but that is only natural; since when have Ivy Leaguers deigned to notice a bunch of brash upstarts from the West Coast? Still, those white dish antennas are there, and remind the thoughtful that not all of the leading fusion-research teams wear Princeton's orange and black.

Go through the doorway into the lobby of the main building of the Princeton plasma lab. It looks like the lobby of a fairly nice hotel, a Holiday Inn or whatever, that happens to have been taken over by a group of people holding a scientific convention. There is the reception desk, where guests check in and out. Visa and American Express cards will not be honored, but that desk is there because the Department of Energy requires that visitors sign in and wear badges. There isn't much for the receptionist to do, particularly at midday, so she spends a lot of her time reading magazines. Still, the regulations are observed.

The lobby has rust-red carpeting and is full of elegant gray sectional sofas. One entire wall is picture windows facing on a courtyard full of picnic tables; when the weather is nice, people like to have lunch outside there. Along with the couches, there are various pieces of what look like abstract sculpture. The creator of most of them is Harold Furth, the lab director. These "works of art" are actually a few of his ideas for fusion devices of the future. For example, on a stand between two sectional end-sofas and next to an ashtray is a light-blue form, about eighteen inches high. In cross-section it is triangular with rounded edges, and is twisted along its length. It looks vaguely feminine; it could almost be the torso of the Venus de Milo doing aerobic exercises. Actually, it is the shape of part of the plasma in a device called a stellarator.

A little farther along is another abstract sculpture on a plastic base. A marine biologist might take it for a greatly enlarged model of the calcium skeleton or shell of a radiolarian, a microscopic bit of plankton. It could also be a novel design for a five-bladed ship propeller, its blades helically twisted; but then why would the blades be hollow, with only their peripheries filled in? This too is a shape for physicists rather than mariners. It shows the main magnetic coils of a tokatron, still another of Furth's ideas.

At Princeton, the focus of attention is on a class of devices known as tokamaks. The tokamak features a large plasma chamber in the shape of a doughnut, or torus. Tokamaks were invented in the Soviet Union; the name is actually an acronym for "toroidal magnetic chamber" in Russian.

Princeton's prize fusion machine is the TFTR, the Tokamak Fusion Test Reactor, and a scale model of its entire system is prominently displayed in the middle of the lobby. The model is about as big as a large desk; on the same scale, a man would be somewhat the size of a little toy soldier. The model features not only the tokamak itself, surrounded by auxiliary equipment; also shown is the TFTR basement, full of instruments used in studying the plasma during a test. All these intricate details are made of plastic in different colors, and if there are children who dream of someday working on fusion, a model like this would be the ultimate hobby kit.

Then, along a wall are large color blow-ups showing recent scenes of TFTR construction and operation. The pièce de résistance is a color TV set on a table next to the exit to the courtyard. It is linked by closed circuit to a camera that continuously and with unblinking eye looks at the TFTR. It is not as though the view changes from minute to minute, or from day to day. A TV view of this tokamak is not full of drama and excitement like one of a rocket lifting off. It is much more like Andy Warhol's eight-hour movie showing a continuous view of the Empire State Building, from one single point. There is good reason for this closed-circuit TV system; in case of an accident someday, it would be valuable for people to see what's going on. Within the lobby, though, one would think a recent color photo of the scene would suffice, but no; there is that TV monitor. It says a lot about how important the TFTR is to Princeton.

Two corridors lead away from this lobby. One, carpeted in the lobby's rust-red, takes you toward the TFTR control room. Just now, however, there is nothing to see. The TFTR is shut down for three months, while physicists and workmen install two neutral beams. These are boxlike structures the size of a railroad freightcar, welded together from thick plates of stainless steel. They house the powerful electrical equipment that shoots energy into the heart of a plasma, to heat it by tens of millions of degrees.

The other corridor proceeds from the opposite side of the lobby. With its linoleum flooring, its bulletin boards, and the offices lining its length, it has the look of an academic department. There are several such corridors, all connected, and if you don't know the way, it is easy to get lost. If you find your way past the library, you are doing OK, but if you pass the Coke machines, you have to go back and make a right turn. After several hundred feet of this, you can walk through a set of red double doors. Now the corridor has changed; it looks industrial rather than academic. The floor is concrete; there are pipes along the walls, and there is a steel staircase. On the second floor nearby is another, shorter set of stairs, and at its top is a set of double doors marked WARNING, KEEP OUT. Behind them are the people who work with the second of Princeton's major tokamaks, the Princeton Large Torus, or PLT.

Why is fusion energy important and worth pursuing? Is it just one more hi-tech future energy source, or is there more to it than that? Authors who have pet alternative-energy sources to write about often take the approach of throwing large rocks at their competition while lavishing praise upon their own subject. Anybody who has read a few books or articles of this nature thus can well suspect that there are plenty of problems with any energy source that anyone is likely to put forward. This suspicion is entirely correct.

What we are living with is not an energy crisis; rather, it is a kind of economic Paradise Lost. Everyone today would love to have fuels that are clean, cheap, environmentally acceptable, unlimited in supply, and available from domestic sources. That is precisely what we had during the middle decades of this century, with oil and natural gas. In 1925, these premium fuels accounted for only 20 percent of the nation's total energy needs. By 1970 they provided 75 percent.

Those fuels were all that anyone could ever want. They were versatile, powering everything from ocean liners to lawnmowers. They were easy to get. You didn't need a dangerous underground mine or a big pit in the ground stripped wide open; just a pipe drilled into the soil of Texas would do. They were low in sulfur, low in smoke and soot or ash, easy to refine and process into a host of products. They were so abundant that their prices stayed level for decades at a time. Even the rationing of gasoline during World War II was not intended to save scarce petroleum for the war effort. We had plenty of the stuff; instead, the measure was imposed to conserve rubber for the tires of military vehicles.

During these middle decades, our economy leaped and leaped again. Tens of millions of families became homeowners, tens of millions of young people went off to college. Shopping malls and high-rise clusters sprouted on what had recently been farmland, while interstate highways laced the nation. Our use of electric power increased tenfold and more, much of this growth being due to the increased use of air conditioning. (Air conditioning is one of the serious inventions of this century; it made possible the Sunbelt.) For decades, our land was merry with the sound of hammers, the sight of construction cranes, the smell of wet cement; and it all was founded on cheap energy. The nation we know today was built on oil at $3 a barrel, coal at $4 a ton, and natural gas at $0.16 per thousand cubic feet.

Today, we find that geology and economics have put sharp restrictions on how much oil and gas we can produce within our own borders, and at what price. Still, this is a far cry from saying that we are running out of energy, or even of fossil fuels. The good news is that we have hundreds of years' worth of fossil fuels readily available, as coal, as oil shale, or as tar sands. The bad news is that we may have to use them.

If these fuels were as cheap and convenient as Texas oil and gas, we would not today be talking of energy crisis. The fact that they will last for "only" a few hundred years would be a mere matter of academic quibbling, for these few centuries still will be a very long time. When Queen Elizabeth was preparing to fight the Spanish Armada, she was not framing policy initiatives to address needs that would come to the fore during the administration of Margaret Thatcher. The problem with these alternate sources of fossil fuels is not they they will someday run out. The problem is that if we are to use them, it will take heroic efforts to keep them from pushing us backward to an earlier industrial age, a Dickensian world of coal and soot, of energy available only at very high prices and with considerable inconvenience.

Tar sands contain vast quantities of a gooey black mess somewhat resembling petroleum, but much thicker. It is too thick to flow into wells; like coal, it must be strip-mined. Oil shale is worse; it calls for vast underground mines or open pits, and stands to fill whole canyons in the Rocky Mountains with the rocky rubble from which the oil has been extracted.

Petroleum can be pumped directly from the well to the refinery, then processed in a straightforward way and sent out into the world as gasoline and other products. Coal, tar, and shale oil require industrial processing before they or their products can be pumped at all. To turn them into fuels suitable for our homes and cars then will call for vaster industries yet, which will not only be very expensive and add to the cost of these fuels, but also will use up a fair part of their energy just in the processing. With enough chemical effort, petroleum can be turned into plastics, fertilizers, synthetic fibers. The cost and complexity of turning petroleum into synthetic fibers is rather on a par with that required to turn coal or tar into synthetic fuels.

Synfuels, as they are called collectively, will not only be expensive in themselves; the creation of such industries will also be quite risky to the investors. For decades, oilmen have been saying that shale oil would come into its own as soon as the price was right. In 1972, the conventional wisdom was that shale would become profitable if oil ever hit $5 a barrel. By 1980 oil cost $30 a barrel and was heading higher. By then, finally, companies like Exxon and Union Oil were showing serious interest in synfuels. But it wasn't because shale had become cheaper; it was because President Carter was priming the oil pump with his $20 billion Synthetic Fuels Corporation. A couple of years later, oil prices had stopped rising. They weren't dropping; they simply were flat. Nevertheless, that was enough to kill nearly all interest in synfuels, which would still have been more expensive than even the highest-priced oil. Exxon, which had been building a large shale center near Parachute, Colorado, bailed out. By then, no one could say whether synfuels would become profitable even if the price of gasoline were to rise to $2 a gallon.

So the synfuels are there, but we can't expect to turn to them except in dire need. Also, if we are to use them, we will have to pay the price. It thus is a little wonder that people should be thinking seriously about the "permanent" or "inexhaustible" energy sources, sufficient to sustain the world for millions or even billions of years. These alternatives to fossil fuels fall into three categories: solar energy, nuclear energy, and fusion.

Solar energy is everyone's favorite, and no one can doubt that its use will grow. Unfortunately, solar energy offers nothing remotely like the versatility we are accustomed to expecting from our present sources. There are a few things it can do very well, such as heating water on rooftop panels; there are others it does very poorly, such as generating electricity. Even so, we are beginning to see the first experimental solar-electricity plants. They feature large arrays of photovoltaics, solar cells made of silicon, which turn sunshine into direct current (DC). However, these plants have been heavily subsidized, with tax write-offs and other benefits. Today the homeowner can buy a complete photovoltaic system, from ARCO Solar. It features a 108-square-foot array for your rooftop, together with all the equipment needed to provide power at 110 volts. The cost is $15,000. Its power is three kilowatt-hours per day, which is worth thirty cents.

In 1980, photovoltaics accounted for 4.2 megawatts of electric generating capacity, about one thousandth of one percent (.00001) of the nation's total. Suppose this capacity were to grow at a rate of 60 percent a year, which is almost unbelievably fast; by 1990 it would increase a hundred-fold. By then, solar energy still would account for no more of the country's electric capacity than a single medium-size nuclear reactor.

There is little doubt, then, that in looking to the future of electricity, we have to find something better than solar power.

Both fusion and nuclear power, by contrast, can be electricity generators par excellence. Both of these will rely on reactors that are compact, powerful, complex, and quite expensive, but that exist for a very simple purpose: to boil water, which in turn will supply large quantities of hot, high-pressure steam, to turn turbines. The turbines are attached to generators, the spinning generators produce the electricity, and the steam then is cycled through those large cooling towers that people picture when they think of nuclear power. The cooling towers condense the steam back into water, so it can be reboiled and sent through the cycle anew.

Despite this usefulness, nuclear power poses problems even more severe than those of a strip-mine or a costly synfuels plant, and it is worthwhile to understand why. The reason lies in radioactivity. Radioactive materials, which are invariably produced by nuclear plants, give off gamma rays. These are similar to X rays, but are much more energetic and penetrating. When gamma rays strike cells in the body, they produce a great deal of damage, including mutations, birth defects, and cancer. There are radioactive substances, such as strontium-90, that are readily absorbed by the body and that lodge in the bones or other tissues, where they stay a long time. Deep within the body's inner recesses, then, they deliver up their radiation just where it will do the most harm.

There are two ways of producing radioactive materials: by splitting atoms, and by irradiating nonradioactive substances with neutrons. The first of these is by far the more dangerous, and it takes place in a nuclear plant. A nuclear plant produces energy by splitting uranium atoms. The trouble is that there is no way to control the products of these splittings. The split atoms do not just disappear; they stay within the reactor as a mix of a large number of different atoms, producing what we call nuclear waste. Within this waste are strontium-90 and other substances that are as bad or worse. It is impossible to keep a nuclear plant from producing them; they can only be managed after the fact. The whole issue of nuclear safety then hangs on absolute, positive guarantees that never, ever will the nuclear waste be able to escape into the outside world.

Fusion, by contrast, does not rely on splitting atoms apart; it works by fusing them together. This might be a distinction without a difference, except that what comes out of these fused-together atoms is nothing remotely resembling nuclear waste, ablaze with its radioactivity. Instead, the product is the gas helium, the same helium found in the sun and in children's balloons, and it cannot be made radioactive at all. Fusion reactions also produce vast swarms of energetic neutrons. The neutron is a constituent of the atomic nucleus; it gets its name from having no electric charge, negative or positive. The electric charges in atomic nuclei are very intense, but, being electrically neutral, a neutron can freely approach and enter such a nucleus. When it does, it may make that atom radioactive, by changing the nuclear structure. That is the type of radioactivity that will be produced by a fusion plant, the radioactivity from irradiated reactor parts.

But this type of radioactivity is much less dangerous, much more readily handled and dealt with, than that from the waste of nuclear plants. Rather than accept the presence of strontium-90 and then try to deal with it after the fact, the designer of a fusion plant can control from the start the types of radioactive materials that will be produced. Thus the production of strontium-90 can be avoided altogether. In addition it is possible, right at the outset, to avoid using materials that are readily made radioactive—that activate easily, in nuclear parlance. Just as the designer chooses metals like steel for their strength, or like copper, for their electrical conductivity, so he also can choose to build the fusion reactor using metals with safe activation properties. There even are materials, closely resembling the heat-resistant tiles used on the Space Shuttle, that do not activate at all. Fusion, then, offers the hope of benefiting from nuclear power's energy, without having to worry about its radioactivity.

Through the coming decades, we will not see oil and gas replaced by some new fuel of comparable cheapness, cleanliness, and abundance. Having seen paradise lost, we will not soon return to that vanished energy Eden. Instead, we will be muddling through, doing the best we can with a combination of all available energy sources. Still, our future will be very different if we have to turn to synfuels on a massive scale, compared to the prospects that we will face if we can develop them slowly, modestly, and amid no compelling desperation or national need. That, in turn, will depend on how well we can proceed with the development of our nonfossil alternatives. Yet a nuclear nation might well be even less desirable than one founded on coal and oil shale. Solar energy, however useful, will be too weak a reed to support our economy.

An energy future that might well be worth seeking would rely on a combination of solar plus fusion energy. The two complement each other nicely and can work together effectively. This will be particularly true when we seek to produce gasoline. Solar energy can grow sugar cane on huge plantations in the tropics, with the cane being crushed and fermented. The alcohol distilled from this ferment could then be put through a catalytic chemical process and re-formed into gasoline. But energy is needed for the distillation, and for the catalysis. Fusion can supply it, once we understand how to build fusion reactors and make them work.

Whatever is to be our energy future, however, it is already high time to begin pursuing it vigorously. Suppose it could be shown that there is enough oil and gas in the ground to last us for fifty years; no more. To replace these fuels, then, this coming half-century would have to see us build a large number of alternate-energy facilities. Suppose that on average each such facility had the energy of a nuclear power plant; how many would we need, and how often must one of them be installed? A typical nuclear plant is rated at a million kilowatts. Thus, we would need some 1,500 of these alternate-energy plants to replace our oil and gas, so we would need to install a new one about every twelve days throughout each of these coming fifty years. Evidently, then, it is none too soon to get started.

Back at Princeton in the 1980s, there is that KEEP OUT sign on the doors leading to the PLT control room, but that sign isn't meant for you; you have your visitor's badge. On the opposite side of the doors bright fluorescent lights shine on linoleum flooring. Lift up a floor panel; under it is a snaky maze of blue cables. The computer center is on your right, behind a set of large windows, with its blinking lights and its solemn high priesthood attending to its data in the manner of a long-established ritual. Just ahead and on your left is the control room for the PLT, where the tokamak people work.

There are about two dozen of them, mostly young college types, casually dressed, sitting in office-type swivel chairs. Some of them wear yellow bowling shirts with a statement of their accomplishments silk-screened on the back: PLT NEUTRAL BEAMS, with a diagram of this tokamak appearing just below, and then 75 MILLION DEGREES. This has been their temperature record for a plasma, in degrees Celsius.

Most of them sit in front of racks of electronic equipment, which feature green TV screens and more blue cables. The cables run to overhead racks leading back to the PLT. Many of the instruments closely resemble stereo receivers as seen from the back. A receiver's back panel has none of the attractive, elegant design of the front; it is no more than a functional array of dials, switches, and plug-in jacks. So it is with the instruments on the control-room racks, except that many of them have those small TV screens off to one side, to display plots of data. Other consoles resemble digital pinball machines, with red numeric displays and bright multicolored lights. Within this room are over three hundred such instruments, stacked one atop the other and filling rows of large blue cabinets. Strategically interspersed among them are computer terminals, with keyboards and larger green TV screens. These bring convenience to the complexity. Simply by typing in commands, any of these physicists can operate the equipment and get the data. The electronics, cabling, and consoles all exist to allow these people to think about what they are going to do, not about how they are going to do it.

Another feature of this control room shows that physicists, too, have their traditions. The room is quite dimly lit; you would get eyestrain if you tried to read a book there. This harks back to the days before computers, when physicists also had their experiments and electronic consoles, as well as those small green TV screens known as oscilloscopes. During a test, data appeared on those screens as bright green wiggly curves. Today all the data is stored in the computer, but in the old days, the only way to record what was happening was to photograph directly the green curves on those screens. The dimness of the light improved the photos' contrast, making the oscilloscope curves stand out clearly. Today all the data goes into the computer, which will print out plots of it for anyone who types in the proper commands. Still, the habit of keeping the room dim has persisted. There is something oddly comforting in this. The next time we wonder why law courts and real-estate offices stick to their hidebound ways rather than adapting to the modern age, it may be worth remembering that in some ways, even the most up-to-date physicists haven't quite adapted to it either.

A day's run of experiments on the PLT starts early, at 7:30 in the morning, with a senior technician at the control panel. His official title is Responsible Person; his name on a display board even announces that he is so. He is the actual tokamak operator, twiddling the dials and adjusting the controls. Sitting close by is the Physicist in Charge, who tells the Responsible Person what kind of plasma he wants to study, what kind of tokamak operation he needs. The day begins with a bit of housekeeping. Someone has to go look around the PLT and make sure that no one has gone in and left a wrench or anything else lying around loose. When the PLT is turned on, its magnetic fields are so strong that they can pick up a wrench from twenty feet away and send it crashing with great force into the side. Indeed, a few years ago such a mishap broke a glass observation window in the PLT's main chamber, which had been pumped down to vacuum. The resulting rush of air not only wrecked the vacuum; it drew in dust and dirt, contaminating the clean metal walls of the tokamak's interior. It took quite a cleanup job before things could once again be set right.

Tokamaks produce neutrons, which are a form of radiation that can be dangerous to people. The PLT doesn't produce many, but it makes enough so that Department of Energy regulations forbid anyone from being inside with it during the actual experiments. Thus, after the checks for loose tools and whatnot, the PLT is sealed up within its room, and a red warning light goes on outside the door. Then the Responsible Person starts his countdown clock, which beeps every ten seconds. At eighteen seconds to go, a buzzer sounds; at four seconds, a bell. When the clock reaches zero, green dots quickly trace curves across the TV screens, taking less than a second from start to finish. That is all the time during which the PLT is actually operating; in fact, these tests are so short that they are referred to as "shots." Still, during these million microseconds the instruments are busily soaking up data and feeding it to the computer. The Physicist in Charge, meanwhile, closely watches these green dots in their paths. If he likes what he sees he may say, simply, "Good." Or he may just nod; these brief gestures reflect his years of experience and the many thousands of times he has watched similar dots. Then it's time to begin getting ready for the next shot.

The PLT operates entirely automatically, firing every three to five minutes, as controlled by the countdown clock. Everyone there in the room can count on a continuous, regular run of these shots, to be observed and measured. After each shot, everyone in turn will call up his data and display it as graphs on a TV screen. The people who work there have their lab notebooks, resembling the composition books we all once used at school, but whose pages are quadrile-ruled rather than lined. After each shot, each physicist will carefully note the shot number, recorded on a wall counter, and write whatever comments are appropriate. If someone's instruments and equipment are working right and giving good data, he then can spend a happy afternoon observing his bit of the PLT's activity. If the data aren't so good, it isn't necessarily a problem; he'll simply adjust the amplifier or turn to a different channel on the instrument. The next shot will come in just a couple of minutes and maybe the data will be better. And if there's really a problem, the PLT shuts down promptly at six in the evening, and this physicist can go in to fiddle more directly with his equipment.

The power for the shots comes from motor-generators, which are among the more dramatic features of the facilities. They are housed in big green steel casings resembling sawed-off Quonset huts, five and ten feet long. Within the casings are rapidly rotating flywheels weighing 96 tons. Their power comes directly from the 138,000-volt line from the power station. The flywheels soak up energy and run faster and faster during the few minutes between shots. Then at the proper time they shift from taking up electricity to producing it. As the countdown clock nears zero, they give off a screech, evidently the sound of brakes being applied. In three seconds they feed back their stored energy, slowing their spin as they briefly shoot two hundred megawatts, enough power for a fair-size city, through the tokamak's magnetic coils. In the Soviet Union, where similar work is under way, there are no such motor-generators; the experimenters tap directly the power line voltage from the power plants. If the Princeton experimenters did that, the lights of the town would dim with every shot. But instead there is only that low screech.

During the interval between shots, the motor-generators recharge, while cooling water surges through the tokamak coils, heated by their huge current flows. In addition, the tokamak's interior walls are prepared for the next test. A sphere of titanium is inserted into the interior and a heavy electric current flows through it. With this current, and in vacuum, the titanium ball acts like a mothball. It gives off vapor, which coats the walls with a layer a few atoms thick. This titanium then acts as a sponge, soaking up atoms of impurities that would otherwise quench the plasma and keep it from getting hot enough. But none of this activity is apparent in the control room. There, the people have a welcome few minutes to look over their data and think about what to do next.

In addition to the Responsible Person, the Physicist in Charge, and the various people sitting at their consoles, there is another important person, with no title whatever. He is known simply by his name. He is the real person in charge, but no title reflects that fact; he might best be known as "the guy whose experiment we're running right now." If you ask people what they are doing, they'll say, "Oh, we're running Bill Hooke's experiment today," or whoever's it is. Even a graduate student might be the person in charge, albeit briefly, if he can persuade the more senior types that he deserves a crack at having the tokamak run for his benefit, producing plasmas that his instrument can study most effectively. Whoever he is, the holder of this exalted station is indeed a busy person, talking to one after another of his colleagues about how the equipment is running, or what the data are saying. It is for his benefit that the Responsible Person and the Physicist in Charge will change the tokamak's operating conditions, shot by shot. For the next one, they may change the rate at which the equipment brings up the strength of the electric currents flowing in the plasma, or they may agree to increase the power of the neutral beams. Another screech, and that shot is done. This will go on for days, even for several weeks.

The PLT shuts down early these days, but often enough it has run till midnight, for a total of sixteen hours. Such late-night sessions have frequently featured arguments over raising this neutral-beam power, to heat the plasma more vigorously. When the PLT shuts down for the night, the neutral beams also shut down, and don't get turned on again till mid-morning of the next day. Then they take several hours to warm up properly. If the beam operator were to turn up the power too quickly, he would risk a short circuit within the neutral-beam systems, which could burn out key parts. Everyone knows that, but the fact remains that the people running the experiments want the greatest beam power they can get, as quickly as they can get it. On the PLT, the Physicist in Charge is forever saying to the Responsible Person, who operates the beams, "Come on, let's get this voltage up." But the Responsible Person knows that the beams work best if they are powered up slowly and then kept below full power. He is likely to turn up his dial a lot less than the others would like him to. Traditionally, the best data come at midnight, because that's when the beams are finally up. The beam people are always a little antsy about working at full power, so if they do it at all, they give it only at the end of the night.

Just now it is late afternoon, and it will be possible to go in to look at the PLT itself.

Since a torus is a thick ring, like a doughnut or life preserver, you might expect, entering the PLT room, to see something like the tire of some colossal truck. What is actually there is more nearly reminiscent of the inside of your car's engine compartment, blown up to the size of a house. An auto engine is basically a set of pistons within cylinders, a rather simple mechanical arrangement. But this basic simplicity is surrounded by so many auxiliary mechanisms and equipment that only a trained mechanic can dig down through the complexity to the pistons and cylinders at the core.

That is how it is with the PLT. The central life-preserver shape is covered first with green and yellow-painted arrays of thick curving copper bars, to conduct the megawatts of current that generate its magnetic fields. Then there are trussworks formed from steel bars, to hold these conductors in place against the forces of their powerful magnetism. Crowding around the periphery are optical instruments and equipment used to study and observe the plasma in the PLT interior, hung with signs like DELICATE, EXPENSIVE EQUIPMENT: DO NOT BUMP. Some of these instruments are as large as a walk-in closet. Still, the periphery must leave room for the neutral beams, whose boxy structures may be nearly as large as the tokamak chamber itself. Also there are vacuum pumps of considerable power. Nature may abhor a vacuum, but plasmas love them; it would be impossible to produce a useful plasma if the PLT could not be pumped down to produce a very high degree of evacuation.

Across the top of all this equipment are racks holding pipes, tubes, and blue electrical cables, as well as a thick steel beam traversing the center. A heavy-duty crane, painted yellow, dangles from another beam, just below the ceiling. Also there are catwalks with ladders or steep, narrow flights of stairs; one thinks of the engine room of a ship. With all this intricacy of pipes and instruments there is little hope of glimpsing the central tire-shape of the PLT's plasma chamber itself. Indeed, all the casual observer may observe is that its basic shape is round.

Within the stark windowless walls of painted cinder block that house the PLT, there are rarely more than a few people. Every year or so the PLT, like its big brother the TFTR, will see crews of construction workers installing some major modification, some change in the basic equipment. Also, between tests a few physicists will be in here to work on their instruments or make adjustments. The purpose of these instruments is to allow these physicists, so far as possible, to observe the behavior of the plasma as if they were right there inside watching it. A shot proceeds very quickly, of course, and we would need to slow time down a thousandfold to follow what happens the way we experience events in the world around us. Still, if you could follow the shot millisecond by millisecond from a seat inside the PLT, you would see a most remarkable dance of the atoms.

It would begin with heavy electrical currents in surrounding copper coils, producing magnetic fields a hundred thousand times stronger than the field of the earth that deflects a compass needle. With the currents at full strength, a wisp of hydrogen gas puffs into the curving interior of the vacuum chamber. Almost immediately another powerful current, carrying 400,000 amperes, flashes through the gas. This is a hundred times more current than is carried by those overhead transmission lines whose steel towers march across the countryside. This electric flash strips the electrons from the atomic nuclei and raises the temperature to ten million Celsius degrees. The hydrogen, now turned to plasma, glows a bright reddish-purple, with the same intense color as a lightning stroke caught by color photography.

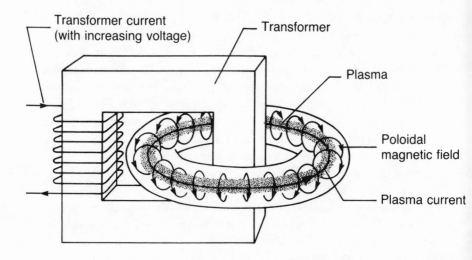

FIGURE 2. *How a tokamak works. The transformer, passing through the tokamak's "doughnut hole," is fed with a current at steadily increasing voltage. By the laws of electromagnetism, this increasing voltage drives a powerful current through the plasma, which fills the interior of the "doughnut." This plasma current, in turn, sets up a magnetic field known as a poloidal field, which helps keep the plasma confined. The direction of the magnetic field lines is given by the "right-hand rule": if you point your right thumb in the direction of the plasma current, your fingers will curl in the direction of the field lines.*

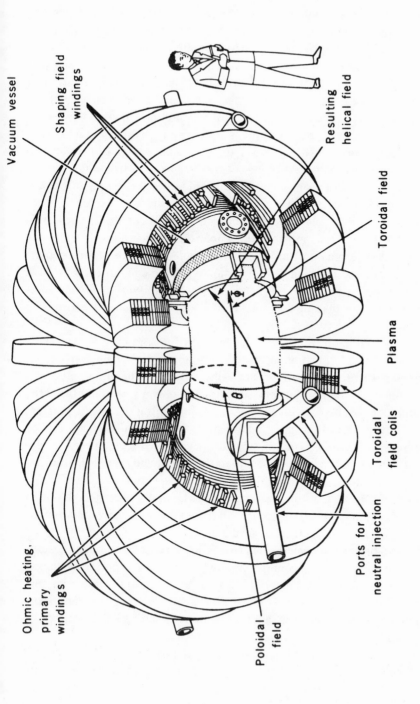

Vacuum vessel

Shaping field windings

Resulting helical field

Toroidal field

Plasma

Toroidal field coils

Ports for neutral injection

Poloidal field

Ohmic heating, primary windings

FIGURE 3. *How the PLT works. The poloidal field is in the θ direction. The large ring-shaped objects are toroidal field magnets, which set up a magnetic field in the Φ direction. The toroidal field is much stronger than the poloidal field. Together they combine to give a magnetic field, whose lines of force spiral around the plasma in helices. This helical field gives very good plasma confinement.*

Within the first milliseconds this channeled lightning bolt writhes violently, resisting entrapment. But the confining magnetic fields quickly drive it back, forcing it to flow and to circulate around the chamber. Then it is time for the neutral beams. They produce intense streams of hydrogen atoms, shooting them into the heart of the plasma with high energy, to heat it still further. Carrying well over two megawatts of electric power, enough to serve several thousand households, these neutral beams are the closest things now existing to the ray-gun cannons of science fiction. The collision of an atomic ray-gun beam with a lightning bolt then would represent the essence of a fusion test.

The plasma writhes again, and you would expect its purple brightness to increase many times over, as if it were a star suddenly exploding. Yet, paradoxically, the plasma turns nearly dark in its center. The reason is that the glow comes mostly from impurities, which give off light only where their atoms are not completely ionized. In the plasma center, temperatures are so high that ionization is total; the impurity atoms have lost all their electrons, and glow only dimly. Only in the cooler outer regions of the plasma do these atoms still retain a few electrons, to produce a more vivid purple.

As the atoms in the neutral beam slam into the bare nuclei of the plasma, the plasma temperature soars above eighty million Celsius degrees. Its pressure also increases dramatically. When that puff of gas first filled the chamber, its pressure stood at only a few millionths of the normal atmosphere. With the heating from neutral beams, its pressure rises higher than that in your auto tires: so powerful is eighty million degrees. Again, then, the plasma surges outward, seeking escape. But again the magnetic fields are equal to their task, and there is no escape.

Now the neutral beams shut off, a half-second after they pulsed on. If this were an actual fusion reactor and not a laboratory test, the plasma by now would have been made to ignite. Its atomic nuclei would now be smashing into one another, quadrillions of times each second. In some of these collisions the nuclei would fuse, stick together, and release energy copiously. The plasma would continue to glow with its purple brilliance, and within its vacuum chamber this new fire would stand as a miniature and man-made sun. The physical processes producing that laboratory glow would be the same as those that produce a similar light, deep within the sun itself.

For now, however, it is enough merely to kindle this fire; sustaining it will be work for another day. Today's work has tested the efficiency with which the magnetic fields contain the plasma, and has shown that, indeed, there has been no escape for the plasma as a whole. Now, however, a half-second into the test, the plasma is emulating the sun in still another way.

Within the outer regions of the sun, strong shock waves continually cross and recross, stirring the material of the sun's outer layers into turbulence. These waves grow especially violent as they pass into the rarefied regions above the solar surface. When such waves advance into these regions, they blast the sun-stuff outward, so the sun is always surrounded by escaping plasma. This material forms the silvery, shimmering, pearly-white corona that we see during a solar eclipse. And beyond the corona this plasma continues to flow outward, forming the solar wind, the interplanetary wind that blows comets' tails about like so many flags in the breeze.

Within the plasma of Princeton's fusion experiment there are similar waves, which also drive material at the outer regions to flow outward and away. But these waves cause the plasma to lose its substance far more rapidly than happens in the sun. During the half-second while the neutral beams were on, ions and electrons were leaking away, and the test operators were puffing small amounts of gas into the chamber, to replace what was being lost. Now, however, when the beams shut off, the gas-puffing also stops.

In the next one-fortieth of a second, then, the plasma will all leak away, in what amounts to a rapid diffusion. We think of diffusion as a slow process; an opened bottle of perfume will spread its fragrance throughout a room, but only after some minutes. However, at eighty million degrees, conditions are very different from those we know. Still, it must not be thought that the magnetic fields have somehow failed to do their job. If they had not been properly designed—and design was a problem that plagued the early days of fusion research—the plasma would escape in bulk and in a few millionths of a second, or ten thousand times more rapidly.

So the shot comes to an end. The currents are made to die away, the fields lose their intensity, the plasma cools—and, ironically, for a brief instant it glows more brightly, as electrons begin to reattach to their ions. Soon the vacuum pumps will go to work, pumping away that wisp of rarefied hydrogen that will be all that is left of the plasma from its moment of glory. All will be still, all will be quiet. Yet one is left with a thought: if this is what a plasma is like within the PLT, what will it be like in the TFTR, twice as large and with ten times the power?

It is the ensemble of all such shots, run successfully with many different settings, that together make up a tokamak experiment, and there will be many such experiments during the ten or more years in which a big tokamak is active. Since such experiments may run for up to sixteen hours a day, with a shot every few minutes, the Princeton plasma lab has an electric bill of a third of a million dollars per month, even though it gets the cheapest rates in town. The connection between fusion experiments and large electric bills, however, is no new thing, but dates back more than fifty years. The first suggestion that fusion reactions power the sun was published in Germany by R. Atkinson and F. Houtermans in 1928. The Russian physicist George Gamow, who had worked with Houtermans on early fusion studies, reported on the paper at a meeting in Leningrad. Following the meeting, Gamow was approached by Nikolai Bukharin, a member of the Politburo in Moscow, who made a proposal: Gamow could have the use of all of Leningrad's electric power for one hour every night if he would try to duplicate in the laboratory what was going on in the sun. But the time was not yet ripe to begin such research, and Gamow declined the offer.*

In all this, it might seem there is a paradox. Everyone hopes that fusion is to be developed as a power source, capable of producing energy to light a city. But these tokamaks soak up the power of a city, at least during their short tests. One is quite justified in asking where the fusion energy is going to come from. The answer is that these experiments are concerned with achieving the conditions for fusion, rather than fusion itself. Fusion per se would actually get in the experimenters' way. It would produce copious floods of neutrons, making heavy shielding necessary and turning parts of the tokamak sufficiently radioactive that they would be harder to work with. But to achieve *conditions* for fusion, albeit in a nonfusing plasma—that has been the difficult problem in this research. Once the physicists learn enough about how to produce such conditions routinely, they will be ready for the next challenge: actually producing fusion energy and using it.

The fact that we produce these conditions today for only a second or less is reminiscent of the early phases in the development of a new rocket engine, such as the SSME (Space Shuttle Main Engine), which powers the Space Shuttle. During this development, the initial experimental versions of the rocket were bolted to a test stand and fired for no more

*Gamow later emigrated to the United States. Had he accepted Bukharin's offer, it might have cost him his life, for a few years later Bukharin was purged by Stalin: expelled from the Communist Party, condemned at a show trial, and shot.

than a half-second. The final rocket that emerged from development runs today for several minutes at a time, but in the early phases the tests were no more than these short bursts. They served to measure performance, while keeping the thrust chamber from getting too hot or the turbopumps from having to stand up to the stress of extended use. It took several years to develop this new rocket, and during those years the difficult problems included keeping the thrust chamber continuously cooled and assuring long operating life for the turbopumps. The early half-second tests allowed the rocket engineers to avoid these problems, while still learning a great deal about the basic performance.

In fusion, the step from second-long pulses to nearly continuous operation will not be easy. Still, many people will declare that its problems stand to be less severe than those they have experienced to date. Thus far, scientists have mostly been studying the basic plasma physics—and already more than thirty years have been spent. The next step will be a matter of installing systems to inject new fuel continuously while removing the spent products of reacted fuel. Special vacuum pumps will be added, to pump away impurities that would otherwise radiate away the plasma's energy. There will be a need for powerful microwave heaters, somewhat like those in microwave ovens, to heat the plasma for many seconds at a time, far longer than can be done with neutral beams. Other microwave sources, tuned to different frequencies, will not heat the plasma but instead will drive strong electric currents through it: so varied are the uses of microwaves. The power supplies, those big motor-generators with their flywheels, will be built to store more energy. The magnet coils will be replaced with new designs, of a type known as superconducting; they will produce their magnetism without needing flows of current from an electric supply. Still, these represent added complexities that are not needed right away. That is why, even though the tokamaks run for only a second at a time, everyone is happy.

It is easy enough to explain how a magnetic field can confine a plasma, if everything works right. However, controlled fusion has been the most complex and intractable problem yet faced in physics. In seeking to confine a plasma, the very first problem is one that would not even exist anywhere else. The plasma must be made to stay put. However, the plasma resists confinement. When penned up by surrounding magnetic fields it tries with all the tricks at its disposal—and it has quite a few—to writhe, wiggle, leak, or push its way out. Confining plasma with magnetic fields is a little like trying to hold a blob of liquid suspended in midair and formed to some desired shape, entirely by blowing on it with jets of air.

The plasma physicists' goal has been to develop magnetic bottles that can hold their plasma routinely and readily. The term *magnetic bottle* was coined in the mid-1950s by Edward Teller, one of the inventors of the hydrogen bomb. Early examples of such bottles included the cyclotrons and other particle accelerators built in the 1930s and 1940s. These used simple arrangements of magnetic fields to confine small numbers of electrons or protons—say, a billion or so. These particles were spaced so far apart that they interfered very little with each other, behaving for the most part as isolated particles. Such bottles indeed could confine their contents almost indefinitely, or at least for as long as physicists might wish. Their modern descendants are the storage rings, large ring-shaped magnetic channels within which up to a trillion protons or other particles circulate for hours on end. But in a plasma, which is denser, the particles do not behave as in isolation. They influence each other's motion, in ways that help one another escape.

Thus, whenever physicists confine a blob of plasma behind magnetic fields, they find it indeed behaving like a man-made sun—though not in the way they would prefer. Fusion, the production of artificial sunshine, does not happen at all readily or easily. What does happen, what has proven very difficult to avoid, is that waves within the plasma drive it to leak away at its edges, forming an artificial corona and solar wind. Even the sun as a plasma blob is leaky, prone to lose material; these laboratory-size blobs are much more so. Thus, no fusion device will ever amount to a thermonuclear Coleman lantern, to be filled with fusion fuel that will burn till exhausted. Any fusion device will always be continuously fed and continuously leaking. The challenge for plasma physicists, then, has been to understand the sources of these waves and of other causes of plasma loss, then to control them and to reduce the leakiness of their bottles to the point where it would cease to be a problem.

Plasma physics has a lot in common with summoning spirits from the vasty deep. Anyone can call them, but will they come? Any group of plasma physicists can build a fusion device, but it is another matter altogether to get it to work as advertised. The plasma, for its part, behaves like a Cheshire cat, fading away and disappearing. When it does this, however, it leaves no smile. Rather, it leaves worried frowns on the faces of the physicists, who would like to know how it got away and what it has been up to this time.

What sort of plasma-confinement devices can we build, then? Before answering this question, it is worthwhile to explain just what we need in the way of plasma confinement. The plasma must be dense enough, and must be confined long enough, to achieve fusion when heated to a hundred million degrees. Specifically, we have the confinement parameter, $n\tau$. This is the product of two numbers: n, the plasma density, and τ, the time it takes to leak away or escape confinement. The density of air as we breathe it is about 3×10^{19} molecules per cubic centimeter; that is, 3 followed by 19 zeroes. A good fusion plasma will be 100,000 times less dense; say, 3×10^{14} particles per cubic centimeter. To produce useful fusion energy, then, it must be confined for about one-third of a second. The reason is given by the "Lawson criterion," named for the British physicist J.D. Lawson: useful fusion energy requires $n\tau = 10^{14}$. In this example the criterion is satisfied, because (1/3 second) $\times$ (3×10^{14}) = 10^{14}. Higher plasma density allows shorter confinement time, but lower density requires a longer confinement time.

This very low density makes it easier to understand how a fusion plasma can reach temperatures of a hundred million degrees. The reason hangs on the difference between temperature and heat. In our ordinary experience the two are very closely related, but this is not always true. For example, the water in the Gulf of Mexico is at a rather tepid or moderate temperature, but the Gulf itself contains a great deal of heat; this heat energy is what produces hurricanes. With a plasma it is the other way around. The temperature is high beyond comprehension, but the amount of hot material in it is so small, so tenuous and rarefied, that its heat is quite insufficient to melt or damage the walls of the plasma chamber. In fact, the reason we must confine the plasma is not to keep it from heating the walls, but to keep the walls from cooling the plasma. If the plasma ever touches the walls, then its heat is gone in an instant, and so is its high temperature.

The Lawson criterion can easily be used to show how close a given device is to being able to produce fusion energy. Thus, high-temperature experiments on the PLT typically feature $n = 1.5 \times 10^{13}$, $\tau = 0.025$ seconds; hence $n\tau = 4 \times 10^{11}$, very nearly. This is quite far from what we need. By contrast, the Princeton tokamak, the TFTR, is to have $n = 3 \times 10^{13}$, $\tau = 1$ second; hence $n\tau$ will reach 3×10^{13}. Obviously, the TFTR will get much closer to being a real fusion reactor.

Now let us think about how to confine the plasma using magnetic fields. The basic approach is to start with a long tube surrounded by hoop-shaped electromagnets, closely spaced all along its length. These magnet coils produce a strong, even field that keeps the plasma away from the walls of the tube. However, they do absolutely nothing to keep the plasma from leaking out the ends of the tube, and so the problem becomes what to do about these ends. There are two ways to deal with them. The first is to bend the tube around to form a torus, joining the ends together. That gets rid of them; a doughnut has no ends. Making a torus does not in itself solve the problem of confining a plasma, but it stands as a good start. By adding auxiliary magnet coils and other equipment, one can turn this basic torus into a tokamak. That is Princeton's approach.

The other possibility is to keep that original straight tube with its magnetic hoops, but to look for ways to plug the ends. The plugging must be done with magnetic fields, of course. Further, it won't do simply to wind electric cable around the ends, or to weave such cabling tightly over these ends. Magnets follow their own rules, and such simpleminded approaches won't work. Still, by being very clever, it is possible to invent magnet arrays that in fact do seal these ends, at least to a large degree. That is the approach being pursued at Lawrence Livermore Laboratory.

There is still a third approach, which is also being pursued at Livermore. It seeks to avoid the need for magnets altogether. In this approach, tiny pellets of fusion fuel are zapped with the energy from a large and very powerful laser. The laser beam then acts as an immense match, lighting the fusion fuel. Also, the laser energy produces such high temperatures on the outside of the pellet that these outer layers blow off explosively, compressing the pellet interior strongly. This interior then is squeezed to very high densities, so that it need be confined for no more than vanishingly short times. This approach is known as laser fusion. To be specific, the fusion fuel will hold together as a pellet for only about ten trillionths of a second, 10^{-11} seconds, before it blasts apart. By the Lawson criterion, then, the density must be about 10^{25} particles per cubic centimeter. This is actually more than ten times the density of water.

All this might look as though fusion research amounts to a debate between gentlemen over how best to build magnets, or perhaps whether to use them at all. Such a statement is on a par with saying that the relationship among Ford, Chrysler, and General Motors amounts to a genteel debate as to how best to build a motor car. In fact, the relationship between Princeton and Livermore is often highly competitive and deeply political. Fusion research is no mere academic ivory tower. Talented physicists have devoted their entire careers to wrestling with its problems. Large sums of money have been spent, and more is on the way; the federal budget for the fiscal year 1984 shows that over $600 million is to be spent on fusion research. With such sums at stake, physicists are often under enormous pressure to succeed in their experiments. Jobs go on the line; powerful people have their egos engaged; congressmen become interested.

The politics of fusion extend to Washington itself, but they take a very different form from what appears in the civics books. All of us know about the three branches of government: legislative, executive, judicial. Their relationships are fixed by the Constitution. Thus, for example, the President submits his budget requests to Congress, which debates them and decides whether to pass them. This is the stuff of much of the Washington news in our daily papers.

However, Congress usually challenges the President on only a few budget issues, and even these are often largely symbolic. Most of the items in a presidential budget are dealt with in Congress as routine business, and passed with no more than modest change. Energy programs generally are among these items. Fusion politics does not mean a Congress and President at highly visible loggerheads, with a clear-cut victory or defeat in store. It is much more subtle than that. Fusion politics is the art of influencing the President's budget *before* he sends his request to Congress.

When a fusion scientist goes to Washington, he also sees three independent centers of power, but these are not the ones of the civics books. They are the Office of Science and Technology Policy, the Office of Management and Budget, and the Department of Energy. The first of these is the least familiar; it is headed by the President's science advisor. The relationship between these power centers is not fixed by constitutional provisions but is highly fluid and variable, changing often from month to month, and from issue to issue. What determines the relationship is influence. This has nothing to do with mistresses or with suitcases of money laundered in Mexico. It has a great deal to do with controlling the development of information. When budget decisions are to be made, the side that has the most information and can make the best case is usually the side that will win.

Imagine that some Powerful Official is seeking to develop such a case to support his point of view. His usual method is to convene a high-level commission of scientists to review the question at issue, and to take care to keep the choice of members in his own hands. The general public occasionally sees something of this procedure, when this Official is the President. The question of Social Security reform was handled by this sort of commission, as was the sticky question of how to deploy the MX missile. More commonly, the process takes place very quietly.

There is a certain art to this. When it is done right, our Official will have a panel of thoroughly distinguished people with unassailably expert credentials. Their report will show an impeccable independence of judgment, which by sheer coincidence will agree exactly with what the Official wants. His opponents in the controversy then will stand confounded, since their views will not be backed by similar high-level and independent experts.

It doesn't always work this way, however. Such commissions, after all, are legally independent, free to follow where the evidence leads them, and have been known at times to get out of control. When this happens, they may recommend actions very different from those that our Official had in mind, issuing recommendations that he has little choice but to accept in good grace. Other things can go wrong as well. Sometimes the commission isn't quite as high-level as the Official would like. Then his adversary can convene his own commission, staffing it with people of even higher caliber. He then can win the battle by making the Official's panel appear as no more than a group of partisans, whose views could be known in advance and therefore discounted. Then again, a commission's recommendations—at least as presented in their final report, which may be edited by the Official—may be so blatantly one-sided that they fail to gain credibility. When that happens, a new review by a new set of experts, convened by the adversaries of the Official, is not far off.

In all this activity, there are plenty of opportunities for good scientists to become known in Washington and to work as consultants. There is a continuous flow of top people among government, universities, and industry. Often by doing good work on a review commission a scientist gains the visibility that will land him his next job in government.

The theme of this book, then, is not one of otherworldly savants in academic departments. It is one of pressure, controversy, and politics. The people who have carried fusion forward have done so by winning battles not only against technical problems in the laboratory, but also against budget problems in Washington. Today, fusion is fast approaching the day when it will graduate from being a matter for physics research and will advance to being a real energy option. When that day comes, the nation's policymakers will make a choice. They will not go on indefinitely supporting three different approaches to fusion power. Instead, they will decide that one of these three has won the right to stand as the basis for fusion's future. The other approaches, by contrast, will stand as losers. The theme of this book is the shaping of this competition and this choice, in Princeton, at Livermore, and in Washington.

2

Robert Hirsch

EARLY in the 1970s, the fusion program in the United States fell under the leadership of a hard-driving manager who knew a good thing when he saw it. This was Robert Hirsch, and the main thing he wanted to do was to raise the fusion budgets. As he said at the time, "We're now at the point where a lot of money could do a lot of good." Before long, his aggressive leadership had resulted in approval for a number of large new fusion facilities. Nor was he one to be put off easily. At a conference in Albuquerque in 1974, one physicist asked if the Princeton Large Torus, then under construction, might develop a bad case of plasma leakage. The session chairman answered, "Bob Hirsch wouldn't allow it."

It would be going too far to say that Hirsch wanted to build a fusion reactor while he was still in high school. However, he certainly wanted to do something for the good of the country. Civic interests came naturally to him. During the early 1950s, while he was in high school, his mother was president of the local chapter of the League of Women Voters in his hometown of Wilmette, Illinois. He spent the Eisenhower years getting bachelor's and master's degrees, then went to work for Atomics International, near Los Angeles. Then, in 1959, he was introduced to fusion.

He was taking additional courses on the side, at the University of California in Los Angeles. The one that particularly piqued his interest was titled "Foundations of Future Electronics." It was not devoted specifically to fusion; the course treated it instead as one of several topics that were covered. Still, to the young Hirsch, this brief introduction was sufficient: "I thought it was very interesting, that it would be an exciting thing to work on. It looked like it could be very valuable to mankind." In fact, it so caught his fancy that he decided he had found the next thing he would want to do in his career. He would build a working fusion reactor that would produce power.

First, though, he needed more education; a master's degree just wasn't enough. At the suggestion of a friend, he applied to the Atomic Energy Commission for a fellowship. He didn't think he'd get it. His college grades had been at the level of B+ and A−, and he stood in awe of anyone who had a Ph.D. When the fellowship came through, he says, "I was amazed. I had done all the work to fill out the forms and everything else, but I really didn't feel I was, well, worthy. That was my view of myself then. I held people with Ph.D.'s on such a high level that I had a hard time identifying myself with them. That was then; I'm smarter than that now." Fifteen years later, when President Gerald Ford named him director of energy research within the Energy Research and Development Administration, he would be managing the work of over a thousand Ph.D.'s. But in 1960, with his new-minted fellowship grant, he was not looking that far ahead.

What he had was more than enough: "It was a whole new dimension. It was exciting, shocking, and scary as hell." Eventually he would see that getting the Ph.D. didn't stop people from being human beings; "they get more education in facts and technical problem-solving, but they don't necessarily get smarter." At the moment, he had to decide where to go with his fellowship. He decided to go back to his home state, and picked the University of Illinois. It had strong programs in both physics and engineering, and he wanted to study both. He felt that was the best way to proceed toward his ambition of building a fusion power system.

He went on to get the first Ph.D. in nuclear engineering granted by that university, and soon afterward went to work for Philo T. Farnsworth in Fort Wayne, Indiana. Fort Wayne was the kind of town Senator Eugene McCarthy would later write about in a poem:

> This is a clean, safe town.
> No one can just come around
> With ribbons and bright thread,
> Or new books to be read. . . .

Farnsworth, for his part, was one of the last of America's great tinkerer-inventors, in the tradition of Edison, Eli Whitney, and the Wright brothers. He was one of the two men who had invented television. He had made many of the key inventions, which included key contributions to the TV camera and picture tube, and had defended his patent rights in court battles. In the mid-1960s he was interested in fusion, and proposed to invent his way toward a reactor.

Hirsch and Farnsworth didn't have much in the way of diagnostics, instruments used to study the condition of their plasmas. Thus they weren't in a position to learn in great detail how their plasmas were behaving. Their sparsely equipped laboratory was only one step up from the earliest and most primitive days of plasma physics, when one of the main diagnostic instruments had been the human eyeball. In those days people used to study the plasma through a porthole; if the plasma got brighter, they figured they must be doing something right. Hirsch and Farnsworth, however, had something a little more advanced: a neutron counter. Fusion reactions give off neutrons, so this counter gave them an indication of how well they were proceeding.

They did much of their work using plasmas made from mixtures of deuterium and tritium, the best fusion fuels available. This was no simple matter. Tritium is both expensive and radioactive; its use calls for great care and plenty of paperwork. Nevertheless, using tritium meant that if they got more neutrons than before they could be sure they were on the right track. Certainly they never got close to producing impressive amounts of fusion power; the neutrons simply made a handy diagnostic. Still, this work left a lasting impression on Hirsch. It gave him a strong belief that one of the most important features of fusion research would be to work with tritium. It was part of his attitude: "Don't play around with idealized systems any longer than you absolutely have to. Get to work on the real problems as fast as you can."

Late in 1966 Farnsworth's health began to fail, and more and more of the responsibility for the research fell on Hirsch. International Telephone and Telegraph had been funding this work, but in 1967 the money started to get tight, and their management told Hirsch to look to the Atomic Energy Commission for funding. Hirsch prepared a proposal and sent it off to Amasa Bishop, head of the AEC's fusion program.

A proposal for the funding of a new project is quite a formal matter. In Hirsch's words, "We spent about a year preparing the proposal, getting it reviewed, and dancing around it." In the end, Bishop decided that Farnsworth's ideas couldn't be fitted into his existing structure of projects and federally funded laboratories, so he declined to fund them.

But Hirsch had made an impression on the people at the AEC. He went back to Indiana, tried unsuccessfully to raise the half-million dollars a year he knew was needed, and decided that his road to fusion would have to lead away from Fort Wayne. One day in 1968 he phoned Bishop, explained his situation, and asked if there was a possibility of coming to work for him in Washington. Bishop said they could talk about it. Not long afterward they both were in Washington for a physics conference. They met in a bar one evening to talk, and Bishop soon hired Hirsch as a staff physicist. At that time, there were only two other technical people working for Bishop in his Washington office, so Hirsch had found a position where his talents could be noticed quickly. He was thirty-three years old.

Just then, fusion was in the doldrums. The 1968 budget was $26.6 million, a sizable cut from the 1960 peak of $33.7 million. Even this 1968 level had been hard won; in 1966 it had been lower still, $23.1 million. What was more, these modest sums were being split among five different labs. The reason why fusion research lay dead in the water was that during most of the 1960s, the best that anyone's machines were doing was to reach a very poor level of performance known as Bohm diffusion, named for the physicist David Bohm. This was the best that Princeton was getting, for instance; other labs were not doing even this well. Bohm diffusion was a peculiarly intractable form of plasma leakage. It was many times faster than could be tolerated, and at that time no cure was in sight. To quite a few physicists, it looked like a barrier or limit forever barring them from success in fusion. Certainly there were very few data to indicate that this barrier might be breached, and a lot to say it couldn't.

Nevertheless, by 1968 the suspicion was growing that Bohm diffusion might not be an intractable limit after all. There were at least a few experiments, straws in the wind, that had succeeded in confining the plasma significantly longer than was predicted by Bohm's equation. Most of these experiments involved plasmas far cooler and more tenuous than the physicists would have liked, but still the fact that they could break through this Bohm barrier at all, reducing the plasma leakage below that predicted level, was potentially quite significant. Also, at an international fusion conference in 1965, the Russians had reported that they were beating Bohm's predicted confinement time, ten times over, with a kind of machine they called the "tokamak." The Russians had not had very good diagnostic instruments, however, so this report had drawn little attention in the West. It was easy to believe that these Moscow scientists had simply made incorrect measurements.

Now, however, the world's fusion researchers were preparing for a new conference, which would in fact be held in the Soviet Union, near the

Central Asian city of NovosibirsK. Robert Hirsch went to work in Washington in the late spring of 1968, just in time to be included in the U.S. delegation attending this conference, which was scheduled for August. He flew to London and then on to Moscow, where he stayed overnight. Once in the Soviet Union, he later said, "those were the first times that I had to stand in long lines that seemed relatively senseless, and one began to feel that the Soviet culture really is dramatically different." He had a diplomatic passport and was traveling as an official of the U.S. government, within a group of American scientists all of whom were known to the Soviets. Nevertheless, for trivial things like getting his ticket stamped, he had to wait in line while a bureaucrat sat there, fondling papers before she would finally stamp them and let him go. There were lines and lines, but finally he could fly on to Novosibirsk. His airliner was very cramped, and he flew at night. Anyone who looks out from an airliner at 30,000 feet knows how little can be seen, but the Soviets didn't want the Americans to fly during the day. Presumably they might look out the window and see something forbidden.

Novosibirsk, literally "new Siberian city," is a dreary industrial center with a population of about a million. The conference actually took place in the science center of Akademgorodok, "academic town," which was much more pleasant. It was set among lakes and pine forests, with footpaths winding through the woods. The conference hall, to Hirsch, was "an attempt at stylish architecture that missed the mark because of very poor workmanship. Everything was pre-cast concrete and the corners didn't fit together." However, the summertime walks to the conference from the hotel, a half-mile away, were very pleasant.

Hirsch soon found there were unanticipated hazards, though not from the KGB. One evening, after the conference sessions were over and after dinner, some of the Soviet scientists invited some of the Americans to share some 80 proof vodka. They brought a bunch of bottles, and everyone sat down by a stairway. Soon they began offering toasts: "Here's to our comrades from the United States, whom we love to have here. We welcome you with open arms and we drink to your good health." Following the Russian custom, everyone drank bottoms-up. Then, as guests, the Americans had to respond: "Here's to our Soviet colleagues, who are fine fellows and who are such good people. We salute your good health and happiness." Then again it was bottoms-up. After about an hour of this, Hirsch was in trouble: "I can hold my liquor well, but I had gotten to the point where I could not feel the end of my nose, I could not feel my cheeks, my ears had long since left me. But they weren't

letting up for a minute, and I knew I had to beat a hasty retreat."
Suddenly, Amasa Bishop walked by, and Hirsch saw his opportunity:
"Excuse me, I've got to talk to my boss." The two of them left together
and walked to the hotel. Then, outside the hotel, Bishop sat down with
Hirsch on a park bench and began to talk physics. Hirsch was struggling to
stay alive, let alone talk intelligently to Bishop, but somehow he managed
to do both.

There were rumors floating around that the Soviets had done something special; the word *tokamak* was in the air. The host for the conference, Lev Artsimovich, was a very genial sandy-haired man who was
widely liked and admired. He was also the director of the fusion research
group in Moscow, which had done nearly all the work on tokamaks.
Artsimovich had not invented the tokamak; its inventor had been Andrei
Sakharov, later to win the Nobel Peace Prize for his work as a Soviet
dissident. But Artsimovich had perceived its possibilities early on, had
nurtured the concept from infancy, and had worked to demonstrate its
potential in his own laboratory. Now he was about to step to the podium
and deliver the talk that would begin the process of convincing the rest of
the fusion world.

He claimed that his T–3 and TM–3 tokamaks had achieved plasma
confinement fifty times better than would have been allowed by Bohm
diffusion. But this wasn't all: he also claimed a new temperature record
for his plasmas, close to ten million Celsius degrees. As Hirsch listened to
Artsimovich, he heard no single statement or display of data that was
immediately electrifying. Nevertheless, as he recalls, "here was this
Russian man who was clearly brilliant and very sincere, and very humble,
standing up and describing results he had achieved. He had put a lot of his
time and effort and soul into this work, for years, and now he was
describing what had come of it, in a very matter-of-fact way.

"Listening to him, even discounting for some of the potential over-
statements, discounting for the difficulty of doing some of the diagnostic
experiments and so forth, it seemed to me that what they had done was
clearly significant. The way they had done their experiments seemed to
me to be very credible. I was so impressed by what he said at that time,
and by what I could then glean from the Russian scientists that I talked to,
that I expected when I got back in my office the phone would start ringing
immediately, with physicists wanting to build tokamaks in the United
States. It was so clear that this was a significant step forward. If I had been
in a laboratory, I would have gone back and started building a tokamak

even before I got the money for it. This was a machine that could lead to a practical reactor."

Back in his office, however, his phone was silent. Nobody wanted to go out on a limb by committing himself to believing the Soviet results. On the contrary, many physicists remained reluctant. Perhaps the most reluctant was Harold Furth, who led the experimental group at Princeton. In 1968, Furth was already one of the graybeards of fusion.

In his case that description was literally true. In addition to a thoroughly expressive face and a sardonic outlook that has long made him a joy to interview and to quote, for many years he has sported an unusually bushy beard and shock of hair. When the Denver *Post* featured an article on fusion, it ran with a cartoon showing him in full bushiness, looking for all the world like a Rocky Mountain prospector who had happened to become a professor of physics. In the cartoon, he was holding a wine bottle with a miniature sun inside. A colleague, flashing a slide of this cartoon on a screen at a conference, remarked, "That's supposed to be a magnetic bottle—I think."

Then, as now, Princeton had the nation's strongest program in fusion research. It was built around a machine called the Stellarator Model C, which was closely similar to the tokamaks in important respects. Yet this stellarator had been mired in Bohm diffusion and had never achieved a breakthrough remotely comparable to what Artsimovich was claiming. Thus, Furth questioned the Soviet claims heatedly, using all the physics arguments at his disposal. No one denied that the Soviet work was open to question, for as in 1965, the Soviet diagnostics were still not very good. As Furth said, "It looked like they were being way too optimistic. I said that what they were seeing instead were effects due to hot electrons, which could fool them." He did more than raise arguments; he backed them with cold cash. He made five-dollar bets with Hirsch and a number of other fusion people, that Artsimovich was wrong.

Furth wasn't the only one with doubts. At the Culham Laboratory in the United Kingdom, a group led by Sebastian Pease had developed a technique for measuring plasma properties by shining laser light through the plasma. Pease, too, questioned whether the Soviet claims could be believed, but Artsimovich was not miffed. Instead, he saw Pease's questions as an opportunity to convince the fusion world at large. He invited Pease to send a team of experts to study his tokamaks, using their lasers. Early in 1969 the British scientists loaded five tons of laser equipment aboard a Pakistani Airlines Boeing 707, which had a scheduled flight to Moscow, and then flew off themselves. While the fusion world waited, the British set up their instruments around the T–3 tokamak and went to work.

Hirsch was quite unhappy with Furth's attitude: "The Princeton people brought forth only arguments to downplay the Soviet results. There is a matter of raising valid questions, but there is also a matter of going beyond that to where it is abundantly clear that people don't want to believe and therefore don't believe." Not all the Princeton leaders were so recalcitrant. Furth's boss, Mel Gottlieb, the director of that lab, felt that the tokamak was worth taking seriously. But Hirsch and his own boss, Amasa Bishop, particularly wanted to bring the Princeton people on board. They wanted to build a tokamak in the United States and, despite their limited budget, to do so as quickly and as inexpensively as possible. The easiest way to do this would be to get the Princeton leaders to volunteer their crown jewel, the Stellarator Model C, to be converted into a tokamak.

Meanwhile, during the winter and spring of 1968–1969, people were asking more questions about tokamaks, and the answers were coming out right. Then, that spring, Artsimovich himself came to the United States to deliver a series of lectures at Massachusetts Institute of Technology. What was more, he held seminars. Soon a number of leading fusion physicists were beating a path to Cambridge, Massachusetts, to sit at a long table opposite the tokamak guru and talk with him at first hand. Soon a group of MIT physicists would be at work setting up their own fusion-research lab, expressly for the purpose of studying tokamaks. Several other groups of physicists were also proposing to build tokamaks. Evidently, they were the coming thing. Furth and his fellow critics had to think carefully about what they would do next, now that so many of their colleagues had become interested in tokamaks.

Furth had initially shown a very grudging attitude toward the idea of converting the C-Stellarator. He felt that if Princeton did it at all, it would only be to show the foolishness of this enthusiasm for tokamaks, to prove once and for all that the Soviet work wasn't right. By late spring, however, it was becoming clear that by making this conversion, Furth would quickly put Princeton in the forefront of tokamak research, which was a field his colleagues were very eager to pursue. Late in June, Bishop convened a meeting of his lab directors in Albuquerque. The directors of the other labs all urged Mel Gottlieb to convert the C-Stellarator. Gottlieb decided he would do it. At lunchtime, Gottlieb and Furth went downstairs together for a swim in the pool. Gottlieb told Furth of his decision, expecting a big fight. But Furth said, "Maybe you're right."

Two months later, the British group in Moscow had the data they were seeking. Their leader, Nigel Peacock, phoned his boss, Sebastian Pease, back in England. Pease in turn telephoned the fusion program office in Washington. Artsimovich was right. The T–3 was really not only as good as he claimed, but better. The temperatures and plasma confinement indeed were setting and even exceeding the records he had reported in Novosibirsk. When the phone call was finished, the Washington people looked at each other with big grins. One of the staff members got up and danced on the table. They had reason to cheer. For nearly a decade, the worldwide fusion community had been stuck in the mire of Bohm diffusion. Now, beyond a doubt, they had broken free, with a breakthrough that improved plasma confinement nearly a hundredfold.

With all this, tokamaks soon were bursting out all over. By October 1969, Amasa Bishop had approved funding for five of them: at Princeton, MIT, Oak Ridge in Tennessee, the General Atomic Company in San Diego, and the University of Texas in Austin. In Congress, the powerful Joint Committee on Atomic Energy was astonished. At its hearings in March 1970, its chairman, Chet Holifield, questioned Hirsch closely: "We want to know why you are going to five devices simultaneously. Is this a new romance of these laboratories? Is there a duplication involved? What is going on here?" But to Hirsch, these five tokamaks were no more than a start. After all, there were many questions to be answered. Moreover, none of these tokamaks would burn deuterium and tritium, a combination known as DT. And DT burning would be an essential step toward building a fusion reactor.

In the meantime, Hirsch was in a position to move up within the fusion program. Even before hiring Hirsch in 1968, Bishop had remarked that he would be leaving the government in a few years, and Hirsch would then be a candidate to replace him. However, Hirsch had a problem: he disagreed with his boss. He saw Bishop not as a forceful leader pushing the fusion program toward a goal, but as a man dominated by the laboratories he was funding. Indeed, he felt that not only were the labs doing whatever they wanted, but that they were flaunting their independence. One instance particularly galled him: "They had built the Spherator at Princeton, and then Livermore wanted to build one also. We did a study and our committee said that one is enough, so Bishop said, 'No, Livermore, you can't do it.' Six months later we read a progress report that said they were building one. I went to Bishop and said, 'You've got to stop this. You told them not to do it, and you can't maintain control if they can go ahead against your orders.' Bishop's approach to the problem was 'Let's go back and see if we can't justify two machines.' He allowed the second machine to be built. This was beyond what I could tolerate, and I spoke up about it." Criticizing his boss's decisions was not a good way to succeed his boss, and the results were predictable: "If I was a good boy, and didn't speak up, and kept my nose clean, I could have had his job. But I could not stand by when my strong feelings were that the program was not run the way it ought to be. In doing that, I was putting myself at risk. But I did it anyway." When Bishop left in 1970, he passed over Hirsch, naming as his replacement Roy Gould, a professor from California Institute of Technology.

Hirsch was still very much the man who wanted to build a fusion reactor. Gould was very much Bishop's successor; like Bishop, he felt he should defer to the lab directors. These directors were the heads of the fusion programs at Princeton, MIT, Oak Ridge, General Atomic, Livermore, and Los Alamos. They were very interested in studying plasma physics, but had only a secondary interest in building a fusion reactor. Fortunately, Gould was willing to defer not only to them but also to Hirsch. He was uncomfortable with Hirsch's aggressiveness, but gave him a free hand. In November 1971, Hirsch told a congressional hearing that, with vigorous funding, a demonstration fusion reactor could be ready by 1995.

Hirsch by now was doing a good deal of missionary work on behalf of fusion, meeting with people in Washington, some of them quite important. He was articulate and well prepared, and he found that others felt comfortable with him. He came across as having reasonable ideas as well as the drive to make them happen: "I really believed in fusion power. While I was being aggressive, imaginative, and expansive, I believe that I was being generally realistic, and nobody I ever talked to felt I was a nut." As 1971 and 1972 went by, more and more influential people were getting to know him and to feel that his judgment was good. These included people on Capitol Hill as well as senior deputies to James Schlesinger, chairman of the Atomic Energy Commission.

In mid-1972, Gould left the AEC to return to Caltech. Like Bishop two years earlier, he recommended against Hirsch's being his successor. However, this time the decision was not in Gould's hands. Spofford English, Schlesinger's assistant general manager, called Hirsch into his office and said that he was recommending him. Then Robert Hollingsworth, the general manager, called Hirsch in and talked about the post. Then Schlesinger himself called him in. When Hirsch did get the job, he recalls, "I felt thankful, and I felt that now I could go do something really significant."

He would do just that, during the next five years. Under his direction the budget for magnetic fusion leaped and leaped again. From an anemic $32.2 million in 1971 it reached $63 million in 1974 and $118 million in 1975, then rose by an *additional* $100 million during each of the next two years, to $316 million in 1977. The rise for laser fusion was even more dramatic: $3.5 million in 1970, $20 million in 1972, $112 million in 1977. Those were years when energy was in the forefront of the nation's concerns, and Hirsch took full advantage of that.

He also had an influential ally in Capitol Hill, Congressman Mike McCormack. McCormack, a chemist by background, was one of the few congressmen to hold a technical degree. He was a member of the Joint Committee on Atomic Energy, and also was chairman of the Subcommittee on Energy Research and Production of the House Committee on Science and Technology. Better still, from the first he was a strong partisan of fusion. Thus he was able to fight the good fight not only in Congress, but also in exchanges with the Office of Management and Budget. He could, and did, intervene directly with Roy Ash, director of the OMB. Hirsch, by contrast, could not approach Ash so directly; he had to go through channels. Moreover, McCormack's membership on the Joint Committee was very helpful, since that committee controlled all congressional action concerning the Atomic Energy Commission.

In 1972, though, all this lay in the future. To do something significant, Hirsch first had to come to grips with his lab directors, whose independence he was determined to reduce: "I had inherited a collection of very bright people, very good physicists, who in many respects were not aggressive but who were very conservative, and who wanted to do a lot more basic physics than I thought was appropriate. They wanted to solve all the problems before they would take the next step. I had a bunch of people who were generally timid and only weakly committed to making practical fusion power." The general view was that DT burning would happen someday, but that it was somewhere off in the hazy future and should not concern anyone just then.

Hirsch, by contrast, made no bones about what he wanted, which was to build a big tokamak and have it burn DT. Everyone appreciated that while the subtleties of plasma physics might in time permit a wonderfully clever solution to the problem of confinement, for the time being the road to success demanded a direct approach: build larger tokamaks. Increasing the size would move the walls farther away from the plasma, which then would take longer to leak away; it was as simple as that. But Hirsch insisted that he wanted more than just another machine for studying plasma physics. He wanted it to stand as a major step toward the fusion reactors of the future, and to do this, he insisted it burn DT. So bold a step would bring the complexities of handling radioactive tritium, of coping with floods of neutrons from the fusion reactions, and of dealing with reactor parts that would become radioactive from the neutrons. But to Hirsch, these features were problems to be faced up to and to be delayed no longer: "It was very obvious for some time that we were playing hydrogen plasma physics," without using tritium and without actually getting to the stage of fusion reactions. "Every part of me and the way I think says you've got to get into the dirt and slop of making the thing happen before you know what your real problems are."

Still, there was a widespread expectation that the program indeed would some day get around to burning DT, and that when it did, the event would take place at Oak Ridge. This Tennessee laboratory had a lot of advantages for a DT experiment, which would be both big and radioactive. Oak Ridge had a long history of dealing with radioactive reactors, dating back to the Manhattan Project. It was a big government center with lots of open space around it. The Tennessee Valley Authority was nearby, with all the electric power a big tokamak could ever need, and, what was more, Oak Ridge already had an active fusion program. In particular, two managers of Oak Ridge's tokamak design group, John Clarke and Mike Roberts, were unusually interested in DT burning. Hirsch started by feeding them money and telling them to go design the DT tokamak he wanted.

Princeton, by contrast, was a stumbling block. Its lab managers would be more than glad to build a big new tokamak, but on one point they were adamant. They were not about to allow tritium and DT burning within their laboratory. After all, Princeton was an Ivy League university, not a federal nuclear lab, and the feeling was that radioactivity would never have a place there. On the day when its laboratory had to get into the radioactivity business, Princeton would bow out. The Princeton lab saw itself as a plasma physics lab pure and simple, and took the position that DT burning would get in the way of doing good plasma physics, while contributing nothing worth having. Indeed, the view at Princeton was that DT burning would be no more than a publicity stunt.

Hirsch could not simply go ahead and allow the Oak Ridge people to start construction. Even the Oak Ridge director, Herman Postma, was no more than lukewarm toward his plans. Moreover, Hirsch had to build a consensus among his lab directors before he could proceed. But as 1973 wore on, Hirsch's reasons for seeking that consensus increased. There was a new chairman at the AEC, Dixy Lee Ray, and she was quite impressed with the fusion work. What was more, she was in a position to provide more money to Hirsch. In June 1973, President Nixon announced that he would be dramatically stepping up spending on energy research, and gave Dixy Ray the job of deciding where the money should go. Hirsch felt that if he could show that his plans were proceeding at a pace faster than anticipated, he could win some of this money for fusion.

He thus drew up a strategy. Just then the Oak Ridge physicists were doing experiments which for the first time would use neutral beams to heat a plasma in a tokamak. Anticipating success, Hirsch proposed that these results be described as a "major breakthrough." He would then follow up this breakthrough by requesting funds for his big DT burner, to be started in 1976 and completed in 1979. This would be eight years earlier than the prevailing technical opinion called for. Hirsch then convened a meeting of his lab directors, along with the program's senior consultants, in July 1973 at Key Biscayne, Florida. Hirsch controlled his lab directors' budgets and was determined to take control of the program. Still, he did not see himself as a boss cracking the whip. He saw himself as a conductor presiding over a chorus of prima donnas. He knew what he wanted, and he expected that Key Biscayne would be the place for full discussion and democratic acceptance of the course of action he was seeking.

"We could have had it in the salt mines of Kansas as far as I was concerned," said one participant. "It was a nice place to stay, but God, it was a stressful time." There were four lab directors present: from Princeton, Oak Ridge, Livermore, and Los Alamos. Not one really wanted to build a DT machine. Nevertheless, by then the Princeton director, Mel Gottlieb, was having second thoughts. Hirsch was obviously bent on having his way, and if Princeton stuck to its guns, it might be left with little to do in the future. Already some of the Princeton physicists, sensing which way the wind was blowing, were nervously sounding out their Oak Ridge colleagues about job prospects in Tennessee.

The man who turned the meeting around was Solomon Buchsbaum, head of research at the Bell Laboratories. He was not a lab director, but like Princeton, the Bell Labs were in New Jersey, and Buchsbaum had quite a bit of influence at Princeton. Buchsbaum argued that DT burning was acceptable and logical, not a public relations stunt but an important step involving issues that should be faced. With this support, Hirsch was able to win assent for at least a cautious endorsement of his plans. "The program should seriously plan for DT burning experiments at a time earlier than previously anticipated," were the words of the report from the meeting. This was what Hirsch needed. He went back to Washington and passed the word to Dixy Lee Ray: "The decision should result in a saving of a number of years in the time required to develop commercial fusion power."

As the summer and fall progressed, however, Hirsch began to have doubts about the design Clarke and Roberts were preparing at Oak Ridge. Their concept stood to cost several times what Hirsch knew would be available. It was very large and featured technology more advanced than it really needed. Indeed, rather than simply being a tokamak that would burn DT, in many respects it was a leap ahead to a prototype reactor. Hirsch wanted to be bold, but he didn't want to be too bold. He kept trying to tell Clarke and Roberts that it was too big, too advanced, but he couldn't be sure they were getting the word.

These doubts came to the fore when he called his next major meeting, in December at AEC headquarters outside Washington. At that meeting the Princeton delegates soon were taking potshots at the Oak Ridge scientists, sniping at the cost of their design concept, the advanced performance it would require, and the need to freeze its detailed design before data would be available from their own intermediate-size Princeton Large Torus. On the second morning, Ken Fowler, director at Livermore, asked Herman Postma of Oak Ridge if Postma really thought his lab could begin detailed design and construction in 1976. Postma replied that he wasn't sure.

Hirsch felt he was being sandbagged. He had still been carrying the ball for DT burning, talking it up at high levels within the AEC, in Congress, even at the White House. Now here was the director of the lab he expected to rely on, saying he wasn't sure he could proceed according to Hirsch's schedule. Hirsch got very upset. It was obvious to everyone there that he was angry, but he kept his cool. Rather than giving vent to an outburst, he simply said, "Let's break for lunch." Then, when they returned for their afternoon session, Harold Furth of Princeton stepped up to the blackboard and gave a talk. This talk ultimately would define the nature of the DT-burning tokamak, win it as a project for Princeton, and, upon Mel Gottlieb's resignation a few years later, make Furth director of the Princeton lab.

Furth pointed out that the emerging technology of neutral beams offered a quicker route to a successful DT experiment than Hirsch had thought could be followed. Up to then, the idea was that a bulk plasma of DT would have to be heated as a whole to the proper temperature, then confined long enough to produce net power from fusion reactions. Furth argued that the plasma should consist entirely of tritium and should be hit with energetic neutral beams made entirely of deuterium. These beams, striking the plasma, then would produce abundant DT reactions. The plasma could be less dense than in the Oak Ridge design, the required plasma confinement less severe, the tokamak smaller, less technically advanced, and a great deal less costly.

Hirsch was extremely interested, and asked the Princeton people to prepare a detailed proposal, as an alternative to that of Oak Ridge. Fortunately for Princeton, about eight engineers from Westinghouse were already working there. They had been invited to Princeton to serve as the nucleus of a project group, in the belief that a new fusion project was in the offing. They did indeed serve as that nucleus, within a design team that was very hastily put together, with most of its members being back at Westinghouse's Pittsburgh facilities. In less than six months they came up with a pile of documents a foot and a half high, as a preliminary design for the new machine. Meanwhile, back at Oak Ridge, Clarke and Roberts were trying to modify some of the more objectionable features of the behemoth they had been unable to sell to Hirsch. However, their final effort still would cost three times what Princeton was proposing. In July 1974, Hirsch gave the nod to Princeton, christening their machine the Tokamak Fusion Test Reactor, or TFTR.

Now that Mel Gottlieb had won, and had placed his lab on the road to a major expansion, he needed a man to manage the project. He had his eye on Paul Reardon, who had recently finished building the accelerator at Fermilab, a large new center for high-energy physics just west of Chicago. In September Gottlieb phoned Reardon and invited him to come and talk about how to manage TFTR. In December, Gottlieb decided to hire him. At that moment Reardon was on the ski slopes at Devils Head, Wisconsin. His kids came up, found him at the ski lodge, and told him to go back and call Gottlieb. He tried to call after lunch, but the line was busy. It later turned out that they were on a party line, and a cat in a woman's nearby house had knocked over a telephone, while she was out. As

Reardon put it, "Poor Mel had to sit there stewing for about ten hours, wondering why I wouldn't return his phone call."

When Reardon got to Princeton in April 1975, he found that the TFTR had been approved, but there was no staff, and he had to start from scratch. The Westinghouse engineers had been working on their conceptual design, and Reardon had to begin by worrying whether it was complete. It wasn't. They had been emphasizing the tokamak in their work, without worrying as much as they should have about the neutral beams, and Reardon had to tell them to slow down and think about designing the tokamak and beams together. Reardon in his first months had both to hire his staff (he picked quite a few of them from Fermilab) and get a workable conceptual design. Then, to add to his burdens, he was told that going sole-source with Westinghouse just wouldn't do; he had to put the TFTR design out for open bid.

The Atomic Energy Commission had just been reorganized into the Energy Research and Development Administration, and its officials concluded that it wouldn't be legal to give the main design contract to Westinghouse. Reardon had foreseen that doing this might be a problem, but he had been assured, no, Bob Hirsch had it all squared away; Westinghouse could do the final design. When this turned out not to be so, it meant a sizable change in what Reardon had to do. Instead of continuing with the Westinghouse people, he had to backtrack and write a very detailed request for proposals, wait while the various interested companies prepared their proposals from scratch and got them in, then set up a board to evaluate them. All this activity, plus the need for the eventually selected contractors to take the time to get up to speed, delayed the project by up to a year.

The original letter from Princeton, announcing the request for proposals, went to a variety of potential bidders, including aerospace firms as well as companies that had done engineering on nuclear power plants. One recipient was Joseph Gavin, president of Grumman, a large aerospace company on Long Island. Gavin was interested and asked one of his associates, Robert Papsco, to go down to Princeton and see what TFTR was all about. Gavin had led Grumman in its work on the Lunar Module during the Apollo program; that moon landing had been the principal frontier of the 1960s. Now he saw fusion as potentially the next frontier for years to come. Once in Princeton, however, Papsco saw that the TFTR would be too big for Grumman alone to take on. So he called up a friend of his, Len Reichle, a vice-president at Ebasco, which was a major firm designing nuclear reactors. Then they bid as the team of Ebasco-

Grumman. They put in their bid, which included a conceptual design, in November 1975. In January 1976, they got the contract.

By then Hirsch was preparing for his next promotion, in which President Ford would appoint him director of energy research, and had turned over day-to-day management to his deputy, Ed Kintner. Reardon and Kintner worked closely together and got the contract approved in six weeks. That was probably the fastest contractor selection ever for so large a contract; it was worth over $100 million. As to why it went so fast, in Reardon's words, "Bob Hirsch wanted to get it done."

The people at Princeton already had their hands full with two other projects, the PLT and another experimental tokamak, the PDX. That was why Reardon put most of the TFTR work in the hands of outside contractors. He had about forty Princeton scientists working for him after a while, to manage what eventually became four hundred engineers and designers sent down from Ebasco-Grumman. Reardon insisted that everyone work at Princeton rather than back in the company offices. He took over some warehouses about a mile away and converted them into office space. As he tells it, "We had this great learning process. We had a brand-new staff of people we had hired in-house, from Princeton, from Fermilab, and a few stable citizens like Ken Wakefield who had been in the fusion business for a long time. We were just sort of getting them up to speed when we had to absorb another four hundred people from Ebasco-Grumman, who weren't up to speed. So 1976 was a very trying year."

One of the advantages of going with Ebasco was that it was a New York firm; many of its employees lived in New Jersey and could commute. For those who couldn't, particularly the ones from Grumman, the Grumman management leased a number of apartments in nearby Plainsboro and filled them with rented furniture. Thus, either commuting or living in these leased apartments and working in Reardon's converted warehouses, the engineers could go ahead. They were doing preliminary design, making calculations and tradeoff studies, weighing and balancing among a large number of desirable features that would be built into the TFTR. Preliminary design would define the tokamak and its surrounding complex sufficiently for people to know that it would work and what it would need for electric power, water, and physical space. But it couldn't be built on the basis of the preliminary design. That wouldn't show the details, including all the needed stress analyses; that wouldn't include the working drawings and blueprints. All those would come out of final design. Then the working blueprints would go into the hands of still another set of contractors, including heavy industrial construction firms like Rockwell International and Chicago Bridge and Iron. That would be when machinists would start to cut metal.

Designers love to keep on massaging their creations, forever seeking perfection, and Reardon knew that eventually he would have to get them to wrap things up. By January 1978, he had a fairly reasonable preliminary design for the buildings that would house the TFTR. At that point, Reardon took a calculated risk. He decided to pick a contractor and start putting up these buildings. He was faced with what he called "the infamous six-page list" of holes in the design, of things he knew would have to be changed in the buildings. These faults eventually had to be fixed for more money than he had anticipated. Nevertheless, his decision gave the project the boost it needed, to get out of the design phase and to start building equipment. As he said, "You can make formal management decisions, but when people see the buildings go up they really take it seriously. The designers get nervous. They say, 'My God, I've got to finish this thing up.' Because it's awfully visible when one part of the building is empty and you're the guy who's responsible for the equipment. If it's tucked away in some piece of paper, it's not so visible." It turned out that there were more holes than just those on the six-page list, and some parts of the buildings had to be redesigned even while they were under construction. But none of this really held up the project, and Reardon's decision turned out to be basically right.

Because so much of what was being built was one-of-a-kind that had never been built before, Reardon and his senior managers had worries about quality control. The relations among the contractors were defined by legal documents, as formal as in a real-estate transaction, and all budgets were closely audited. Still, just as with an income-tax form, there was some leeway for fudging. Thus, one company was building casings for the tokamak magnet coils. The casings had to be machined and had to have lots of holes, all drilled within close tolerances. After their machinist had gone to work, someone discovered that the holes were out of position. Right there was a quality-control problem: why should the holes be out of position? On the drawing it said where the holes were to be, sixteen inches from the pin, plus or minus one thousandth. Why didn't they do it? And the answer came back: Yes, we knew it, but the machinist made a mistake; he measured from the wrong side of the pin. Well then, why didn't somebody check the setup before allowing that guy to start drilling? Well, it costs money to check everybody's setup. After that episode, the quality-control experts could identify machinists' setups as a problem area needing more attention.

In the meantime, there was the problem of what to do with the faulty magnet casing. Legally, the Princeton managers could toss the part away and say, "Sorry, fellows, you blew it; it's your money." There had been times when this sort of thing happened and the contractor had to absorb the cost, just because a machinist had goofed. But in this case, the metal part had been machined from a $70,000 forging and another one couldn't be acquired for nine months. The Princeton managers could have told the contractors to go fix it, but that wouldn't have done the project any good. So Princeton put a stress analyst to work, and he figured out how they could use the part after all. Still, this had cost more money, so Princeton nevertheless wound up charging that machine shop something extra.

Not long afterward, it turned out that Princeton and Ebasco had made a mistake. Many parts were to be machined from particularly hard forms of stainless steel, tougher and stronger than ordinary stainless. That particular machine shop had planned its costs and schedules on the basis of a vendor's brochure, furnished by Princeton and Ebasco, which said that this material could be worked and machined like ordinary stainless. But when the machinists got to doing the work, they found that their tool bits wore out quickly. The material was very hard to cut, and the job was taking four times longer than the machine-shop boss had planned. This boss then said to his managers at Princeton-Ebasco, "What about this, fellows?" He also put in a claim for more money. Then, while they were negotiating that claim, this machine-shop manager kept in mind how much he had had to pay because of the magnet casing. He tried to get a little reimbursement through his claim, in ways the auditors wouldn't spot. It didn't work, though; the accountants were too clever for that.

On the other hand, an incident involving one of the neutral beams showed that the quality control could be too good. These beams were housed in huge boxes welded together from four-inch-thick plates of stainless steel. They had to stand up against not only the forces of powerful magnetic fields but also against the pressure of the atmosphere, since they were to be pumped down to vacuum. The first one was built in 1978 at Berkeley, California, and was transported by flatbed truck along a carefully planned route to Denver, to be machined. When this work was completed, it was loaded back aboard the flatbed to go back to Berkeley by the same route.

Somewhere in Utah the driver missed his turnoff from the freeway and got lost. But he soon found his way again, or so he thought, and went on tooling merrily along with his sixteen-and-a-half-foot-tall load. Up ahead was an overpass with a clearance of sixteen feet. But he didn't worry, because he knew the route had been carefully planned. He later said he'd been doing only forty miles an hour, but no one believed him, because he was out on a major highway in the middle of nowhere and had no reason to be doing less than sixty.

When he hit the overpass, his cab separated from the trailer and went a thousand feet farther down the road. This was fortunate for him, since otherwise he'd have crashed right through his windshield and flown on down the highway; as it was, he put quite a dent in the windshield. The beam-box went two and a half feet, all at the expense of the trailer and the overpass. The trailer broke its back and was smashed to the ground. The overpass was shifted on its abutments and sustained a quarter-million dollars' worth of damage. It was supported by I-beams six feet thick, braced with steel crossbeams; the impact drove the crossbraces through one of the steel I-beams. Eventually the beam-box got back to Berkeley. The people there pumped it down to vacuum. It was perfectly sound and entirely usable; the only damage was that a flange had been bent.

"We had too much quality control that time," Reardon chuckled. "The beam-box won and the bridge lost. I knew we had too much if we could knock down bridges, but that wasn't one of our standard tests." In its ultimate operation, its neutral beams would have much in common with the atomic ray-gun beams of science fiction, which could destroy buildings and knock down bridges. However, this was not quite what the science-fiction writers had in mind.

Another mishap, in December 1980, proved to be more serious. This one involved the TFTR power supply, two big motor-generator sets, each of which was nine times as big as the older motor-generators then in use for the PLT and PDX. The two together were to produce 950 megawatts of power for six seconds, enough for a city of half a million inhabitants during these brief bursts, every five minutes. Each was being built as a massive cylindrical rotor surrounded by a barrel-like stator, thirty-two feet across, and were to be housed in concrete pits resembling missile silos. Each rotor weighed six hundred tons. The stators, for their part, could not be shipped by rail or road freight when fully assembled, so they were brought in as three pieces, each weighing a hundred tons. To handle the heavy lifting, the Reliance Trucking Company of Phoenix, Arizona, brought in a gantry crane, with much greater lifting power than the crane permanently mounted to a beam just below the ceiling. The work crews put together the first set of stator sections, then the gantry picked up this 300-ton iron barrel and set it in the other pit. Then they assembled the first rotor, and with the gantry they lifted that stator out of its pit and lowered it down over its rotor. At that point they not only had one complete motor-generator; they also had lifted and maneuvered the stator twice.

Continuing with this work, they then assembled the number-two rotor in its pit, along with the second stator, which was built some distance away. When the gantry picked up this stator, several workmen were standing on the rotor to help guide the stator to its close fit. The stator was sixteen feet in the air when someone decided it should be rotated slightly on bearings, and passed the word to the gantry operator. The gantry bearings proved hard to rotate and started making funny noises, so the gantry man decided to back up. At that, the bearings made very loud noises. Someone yelled, "Get out! It's gonna go!" The riggers standing atop the rotor scrambled to get clear. They had just done so when the hoist let go and the entire stator fell sixteen feet to crash against the base of the rotor. The stator had a support ring at the bottom made of six-inch steel; it was crumpled. Someone was riding the gantry boom, an act certainly in violation of safety rules. When the stator let go, the gantry cables, stretched taut, rebounded and the boom bucked violently, flipping the rider. However, he somehow managed to hang on and got away with no more than a scratch on his hand. When Princeton started to look into its insurance coverage, it turned out that four or five companies were involved, so it took a year and a half to sort everything out. They had to buy a whole new stator as well as a number of parts for the rotor, but finally the insurance companies settled for $8 million.

In spite of these mishaps, by midsummer of 1982 the TFTR was coming together into a recognizable whole, with large subassemblies being welded on amid the flash of torches. All this was going on within a hall of Pharaonic proportions, a hundred and fifty feet long, over a hundred wide, and as tall as a five-story building. It featured walls of concrete, five feet thick, to keep in the neutrons. Within those walls the TFTR assembly showed structural beams the size of bridge girders, made entirely of stainless steel. Magnet coils sprawled across their supports, huge rings thirty feet across. The water mains, carrying cooling flows to these magnets, would be suitable for a city waterworks. Seeing all this, one was struck with the feeling: they are really doing it. This would not be just another physics lab; it would be a new industry being born.

In the comfort of the main lobby, with its sculptures and the large TFTR model, the TV monitor showed color views of the assembly of the plasma chamber, a vacuum-tight vessel built entirely of stainless steel. It resembled the classic toroidal space stations of science fiction, and with a width of twenty-five feet, it certainly looked as though it could serve as an experimental small-scale version. Each segment of the vessel had the shape that sections of an inner tube would have if one had been sliced like a pie. Each such segment featured two powerful ring-shaped magnet coils, over seven feet across, weighing nearly forty tons to the pair, and heavily braced to avoid deforming under their magnetic forces.

For a closer look, visitors could don yellow plastic hard hats and go on in. The noise in this industrial area was not overwhelming, but still people had to speak up to be heard. As a friend said to me, "If archaeologists dig up this site in fifty thousand years, they'll be convinced they've found the mother lode from which we got all our stainless steel." Off to one side of the mini–space station was a massive steel structure holding still another set of magnet coils, these to run around the periphery like bicycle tires fitted around an auto tire. A thirty-foot coil, the largest one, was lying there ready to be hoisted into place. At the appropriate time, this entire steel framework carrying its magnet coils would be hoisted, lifted, and lowered into position—very carefully. It then would entirely cover the mini–space station, hiding it from view.

With that, the TFTR tokamak would be complete and ready for use. The results of Robert Hirsch's resolve would be in operation as a complex of buildings and facilities, sprawling across a dozen acres of Princeton's Forrestal Campus. Hirsch's wish, which he had cherished for over twenty years, would finally stand realized in the finished TFTR, ready to be tested.

3

Quest for Power

To join the group that will be working with the TFTR, one of the main requirements is a Ph.D. in plasma physics. The way to get one is to spend several years wrestling with the esoteric mathematics that purports to describe how plasmas behave, or what is better yet, to spend those same years fiddling with instruments and making measurements on real live plasmas in the lab. The really good university physics departments have their own tokamaks and other fusion devices—not big ones like those at Princeton, of course, but still complicated and intricate enough to give their graduate students a proper preparation for the big time.

At Princeton, however, the big time starts early. Posted on a bulletin board down a corridor from the main lobby is a memo that tells the story:

> First-year graduate students will typically take graduate courses given by the Department of Physics in quantum mechanics, electricity and magnetism, and in statistical mechanics, together with the introductory plasma physics course (AS 551 and 552). AS 558 is a seminar—largely student-run—which meets one afternoon a week to discuss a wide variety of topics of current interest. Second-year students typically take the intermediate-level plasma courses (AS 553 and 554) [and] Professor Kruskal's stimulating mathematics course. . . . The study of plasma physics requires the folding together of knowledge from electricity and magnetism, atomic physics, hydrodynamics, statistical mechanics and kinetic theory, and applied mathematics. . . . In your first year, you will join one of the Plasma Physics Laboratory's experimental physics groups and participate, from the very beginning, in research which is at the forefront of knowledge in plasma

physics. . . . Typically, the first two years consist mainly of academic work
plus assistantship research. After passing General Examinations (usually at
the end of the second year), thesis research begins. Most thesis projects
take two to two and a half years.

When a new-minted physicist emerges, proudly holding his Ph.D. di-
ploma, it certifies that he is a dyed-in-the-wool, guaranteed, no-fooling
expert on plasma leaks. He has learned all their tricks, at least all that are
known up to now, and he has an appropriate vocabulary to describe
them. Thus, he is fully qualified to join a research group like those at
Princeton, where he can spend his career fighting the battle of those
leaks, struggling with the magnetic fields that he hopes will keep the
plasma in place, and manipulating the plasma in the hope of getting it to
do something useful. Like undergoing fusion.

Once his dissertation is complete and graduation day is fast approaching, it's time for job interviews. For the TFTR group, these are in the
hands of Dale Meade, head of the Experimental Division. A tall, robust
man with a soft voice, he is totally in love with his profession: "I'd pay
people to let me do what I do here." In his younger days he was a high-
school football quarterback; he lettered in track and basketball, too. Now
he works with a very different team, but one in which the camaraderie
and the spirit of winning can be just as strong.

What does Meade ask a young man, when they meet for their interview? "I try to find out from people exactly what they've done. That's one
of the difficult things to pin down—exactly what did you do? It's hard to
separate what the thesis advisor did, what others in the group did. How
much did the student do on his own, how much at the direction of the
thesis advisor? In many cases the thesis advisor has done the real work to
solve a problem, and the guy I'm interviewing has had to just go and do
routine work to carry out the details.

"As I get into that, I'm really trying to figure out whether or not this guy
has in him the ability to solve problems on his own. How does he react to
something he doesn't know the answer to? I try to see if he can figure out
things for himself, instead of just running to the thesis advisor.

"Secondly, I'm after people who have tremendous dedication. When
he gets stuck on something, how long does he pound away at it to get it
solved? I find out about that from discussions with his colleagues. What is
his personal dedication to the job? Has he worked in a group? Can he
sacrifice himself completely to the team? That's very tough; often these
qualities don't go together. Some people who work hard will work that

way for themselves and nobody else. That is, they'll really put a lot of effort if they're interested in a problem, but if they're not interested in it, even if their boss is or their colleagues are—they won't. Nobody is perfect in that respect, but I look for people who have enough of this group spirit to maintain their motivation. Because they'll be working with a team, a very dedicated, professional team.

"Then I ask a person straight out: 'What do you want to do?' At that point, I'm interested in finding out what he thinks about when he goes home at night. What I'm after is the kind of person who goes home at night and he thinks about physics. I'm interested in people who think about it on Saturday morning, on Sunday morning. That's the kind of guy that, when you get him, you've got a gem, because those will be the leaders. In fact, we do have a fair number of people who want to spend their weekends doing physics. If you look at this place on a Saturday, you'll find a lot of people here. They don't maintain this spirit forever. There's certainly a lot to be said about how as you get older, you tend to lose a lot of the adrenaline.

"The peak years for these people are from four to maybe fifteen years after getting the Ph.D. During those first few years they still have things to learn; they're not yet at the level where they'll be making the strongest contributions of which they'll be capable. Then after about fifteen years, age forty or so, they begin to get tired. They can't quite keep up with the younger people in the late-night sessions. But they do very good work during those years."

What about their salaries? "Not as high as at some other places. This place has always maintained an academic-type connection. As a result, there is something about being at Princeton that costs you ten to fifteen thousand a year, at least. Our competitors offer them forty-five when we pay them thirty. Livermore hires fresh graduate students for more than we pay people who have been here six years." Then why don't their people emulate the plasmas and quickly leak away? "It turns out we are fortunate. A lot of people still will stay here and work, work hard. I don't know how long that'll continue; we'll see. But a lot of these guys are very dedicated to this business. They've been working on it for many, many years. They could go elsewhere and make more money, but they chose to stay here, work on this project, work their tails off."

A lot of the problems they work on stem from the extreme conditions encountered within the equipment. To study the plasma, they use lasers so powerful that during the few billionths of a second that they are flashing, they are putting out the power of New York City, all in a sudden burst of light. Close by are photodetectors more sensitive to faint light than the photographic plates used on large astronomical telescopes. The plasma may have a temperature of a hundred million degrees. Three or four feet away is a magnetic coil cooled with liquid helium, at four degrees above absolute zero. In a neutral beam system, the beam may carry power of forty to sixty kilowatts per square inch, and if it were to strike metal this would be more than enough to turn hard steel to Silly Putty. Indeed, on the tokamak wall opposite the beam ports are slabs of heavy armor plate, to take the damage from the beams and then be replaced after a while. Eighteen inches from the beam is a pump handling more liquid helium, which is so sensitive it will flash into vapor if you do little more than look at it the wrong way.

Other extremes come from the electrical power. In the plasma are currents carrying a million amperes. Nearby, some of the instruments have sensitive detectors that generate billionths of an ampere. Then there are the plasma disruptions. When a plasma is fed with too much current too fast, it writhes violently, disrupts suddenly, and hurls itself against the tokamak wall. "This shakes the devil out of the machine, shakes the living daylights out of the vacuum vessel," said Meade. Indeed, if it were not braced solidly, the tokamak would break as if from an explosion within. Any instrument in the way of the disrupting plasma will be zapped, and some of them are. But those that aren't must faithfully follow the plasma's behavior, without reading improperly.

Princeton probably has the world's best diagnostic instruments. Meade and the other project leaders insist that when measurements are being made on plasmas, everything must be measured two different ways, preferably three, and the measurements must all agree. Some of their equipment exists nowhere else in the world. An example is their Thomson scattering system, an updated descendant of what Nigel Peacock took to Moscow in 1969 to study the Soviets' T-3. It works by detecting laser light scattered off the electrons in a plasma. The principle is somewhat the same as studying fog by shining an auto headlight through it, then observing how the light beam is made visible by the fog. To appreciate what is involved, think of a laser shining through the clear and pristine air of a mountaintop. The beam would probably be invisible, so little of its light would be scattered to mark its path. A fusion plasma is a million times clearer than the best mountain air, in terms of scattering laser light. Inevitably, then, the laser must flash with the power of New York City in order to produce a handful of scattered photons that will be picked up by the photodetectors.

"You asked what these guys are doing," Meade went on. "We have physicists who spend many long hours designing equipment, of a type that's never been built before. They spend many long hours constructing it, along with the engineering staff. Many, many frustrating hours getting it to work. The frustrations involve getting the circuits to work, getting them to perform for the first time. With most of these pieces of equipment, even though they have huge budgets—several million dollars—we never have enough time to do as well as we would like. These are not quite chewing gum and string. To some extent there is no margin on any of these instruments; they are all pushing the state of the art. Then, in using them, we push just as hard as we can to get the most advanced possible measurements. When you do that, you know you're asking for many difficult hours of getting things to work.

"It's not like building a Heathkit. When you buy your Heathkit, you're usually doing that for fun and enjoyment, like to build your own stereo. These guys work in a very high-pressure world. Things have been planned for many years, large amounts of money have been invested. A $500 million project needs this thing. It's maybe like building your Heathkit in your kitchen but having everyone watching over you, asking when it's going to work, why it's taking so long, why are you spending so much money, why don't you come back and work some more on it after midnight. A tremendous amount of pressure. And it's not just for a few weeks, either. It's for years."

If you're working on some instrument like these, your boss may tell you how badly he needs it, and you think things over and tell him when you'll have it. Then a glitch comes up. You go home and sleep till seven A.M., it's a rather uneasy sleep, then you're back there working again to find the glitch. Often glitches involve misalignment of equipment that you would have sworn was aligned properly. During a magnetic pulse, the structure of the tokamak may be deflected slightly. Then, during a shot, mirrors and lenses within the machine must be made to come into alignment at the proper time. But when it pulses, sometimes it shakes hard enough to shift one of these mirrors. Often they are inside the machine, so they can't be adjusted while the tokamak is running. All the operator can do is sit at his console and try every adjustment possible, then wait till he has a chance to get inside the machine. But the glitches do get found. In Meade's words, "We very seldom fail."

"We are living on the hairy edge," he went on. "The name of the game is to push parameters, to push the frontier. You can't sit back in a nice cozy little room and take data, because that's not where the action is. The action is always in pushing the plasma toward the very edges of disruption, pushing instruments toward the edges of their measuring capability." More than instruments get pushed. "Already we're pushing people to the limit. They're all being pushed right to the limit of how much they can work. These guys are working twelvish hours every damn day. We've got Don Grove up there, the head of the TFTR project, working Saturdays and Sundays to get this thing going. Sometimes the people in the Department of Energy say, 'Well, can't you work a little bit harder?' All I want is to have them walk behind us for a day. A month, maybe, give 'em a month to get tired."

Why do Meade's physicists do it, then? Why should they devote themselves so intensely, backing their commitment not only with hard effort but also with cash foregone out of their careers? The reason is that in fusion and plasma physics, Princeton is the center of the world. In *The Right Stuff*, Tom Wolfe wrote of the fraternity of aviators and test pilots, for whom the flight-test center at Edwards Air Force Base was the top of their professional pyramid. Princeton holds similar status among physicists. Livermore has an enormous construction project and is building what may be the next step beyond the tokamak. They will use it to demonstrate a potentially superior magnetic bottle, but will not burn DT to produce fusion energy. DT burning will happen first at Princeton. This is where the goals and timetables have been made explicit, where the people are working today who will actually do it.

Everyone who studies physics learns about atoms and their energy, and is reminded of their enormous impact on this century. But that is a far cry from having actually been there, been a part of the wartime Mahattan Project that unlocked this energy—and, incidentally, launched so many professional careers. Most physicists work at fairly routine research tasks, in industry or in universities. Few can deliberately set out to join a group that will be making history. Princeton is such a group. The people there stand to be present at the creation of fusion power, and they know it. They are living as a part of it all, from the latest plasma disruption to the next coffee break. They know that forever afterward, in the community of physicists, those who were there will be looked up to like that happy few who stood together upon St. Crispin's Day.

"They're out on the frontier, they're going to do something that is going to end up being, I think, compared to the Fermi experiment in 1942," Meade went on. "It'll be recorded as the first demonstration of significant thermonuclear fusion in the laboratory, it's really a significant historical event." The Fermi experiment was Enrico Fermi's first atomic pile, built in a squash court under the stands of Stagg Field at the University of Chicago. That was where atomic energy was first controlled and released; the Atomic Age began there. Today the stands are long gone. The place is marked by a plaque and a bronze sculpture that looks vaguely like the dome of a nuclear power plant and vaguely like a skull. But for decades, anyone who was there has not even had to say, "Back in 'forty-two, when I was with Fermi in Chicago . . . ," in order to receive the rapt attention of any physicists nearby. In such gatherings, his reputation would have preceded him, and the mere statement "in 'forty-two" would suffice. That is how it may be for today's Princeton people, in future years.

What's more, they have plenty of opportunity to strut their stuff, at professional meetings and conferences. At major conferences like that of the American Physical Society in 1982, held in New Orleans, over a hundred people from Princeton were present to give papers. In Meade's words, "We insist on it. It's important that they talk to those other guys, make sure that there are no stones unturned." It's a matter of being able to compare notes and to say, "Tell me about this feature on your Thomson scattering experiment; maybe I should have it on mine." Again, in Meade's words, "I may not be able to offer these guys a high salary, but I try very hard to offer them professional development, professional recognition. So when they do work that way, I try hard to make sure they get recognition for it. It isn't so easy; it takes a while. But we try hard to systematically make sure that people who work on developing these diagnostics, spending many, many years, get a chance to report not only on the hardware of their diagnostics, but some part of the big thing that those diagnostics were a part of. So they can give, for example, a talk on the main physics results."

At that moment, in mid-August 1982, we were less than a month away from the most significant conference of all. Held every two years, sponsored by the International Atomic Energy Agency, it would be a gathering of the worldwide fusion clan into a single week-long set of sessions held in Baltimore's newly built Harborplace. There the leaders of fusion work in Japan, Europe, and the Soviet Union would join their U.S. counterparts, in the 1982 meeting of the same conference where Artsimovich had told of his tokamaks near Novosibirsk in 1968. "At the meeting in Baltimore, the guy who built the Thomson scattering system, went through several years of agony on PDX, finally got this to work, it's the best one anywhere in the world—he is giving the paper on PDX." This is a tokamak as large and significant as the Princeton Large Torus.

"His paper has all the plasma physics and confinement results for the past year and a half. Dave Johnson is his name. He's the first guy with the first paper of the entire conference. That was his reward for sticking it out."

However, there is a large difference between the Fermi experiment and the TFTR. At Stagg Field, Fermi's pile generated half a watt of nuclear power. When TFTR has its big day, it will produce 30 million watts of fusion power, which is a power output in the same class as in the reactors of the first nuclear submarines. This output, the goal of Meade's people, is called $Q = 1$. Q is the ratio of fusion power produced to power fed into the plasma. $Q = 1$ means power out equals power fed in. But this power fed in is not in the least like the total power it takes to run the system. The bookkeeping used in calculating $Q = 1$ would never pass muster with any accountant, but still it does have its own peculiar logic.

In operating the TFTR, to begin with, there will be several hundred megawatts produced by the motor-generators and running through its magnet coils. None of this gets counted, because none of it goes into the plasma. When asked about it, physicists shrug and comment that real fusion reactors will be built with superconducting magnets, cooled to only a few degrees above absolute zero by using liquid helium. Superconductors have the nice property that a current once started within them will flow forever, needing no additional energy, as long as they are kept cold. Believing in superconducting magnets has been compared to believing in motherhood and that sunshine is good for you; but TFTR's magnets use ordinary copper, which is not superconducting. Anyway, the power in the magnets gets left out in the bookkeeping.

Then there is the power for the neutral beams. They demand 115 megawatts to run them, the power delivered by their wall sockets, as it were. Of this, 30 megawatts gets shot into the plasma to heat it. Someone who wanted to fudge could say that none of this should count, either. Real fusion reactors will probably operate with ignited plasmas, which produce so much energy they will keep themselves hot without neutral beams or other outside heating. Such a plasma, producing energy with no power fed in, would then have a Q of infinity. But such an argument would be taking things too far. From the start, the goal of TFTR has been to produce a "substantial" amount of fusion power, which has been taken to mean as much as will reach the plasma from the neutral beams. However, on PLT, the predecessor to TFTR, the best ever achieved was an "equivalent Q" of 0.03, which would have been reached if the PLT was run using tritium (which it was never designed to use). The next step after

TFTR will likely be about as big a jump. In any case, TFTR will stand as the base from which all else will follow.

For Meade's experimenters, though, while the hours may be long and the glitches frustrating, when they go to the TFTR control room they can enjoy a number of welcome changes when compared to the PLT. For one, it is much more brightly lighted. Also the TFTR control room is much livelier in decor, with carpeting instead of linoleum, and the electronics racks are in pastel-colored cabinets of orange and yellow, rather than the drab blue of the PLT. The lively colors cost no more than GI gray and make things much cheerier. The carpeting is supposedly there to keep down the dust, but somehow the computer center next door manages with the same old linoleum. However, the old-style racks of oscilloscopes and other instruments, stacked like so many stereo components, are conspicuous by their absence. Instead there are rows of computer terminals, and nothing else. All data and control commands go in and out through the TV screens of the terminals. The complete system is called CICADA, Central Instrumentation, Control, and Data Acquisition, and is actually the same kind of control system that will appear in the next generation of nuclear power plants. CICADA is completely digital, with as many as fourteen computers that talk back and forth to each other.

Another change makes the study of the data much more systematic. Everyone in a control room has a lab notebook, quadrile-ruled in light green. They write down in them comments about each shot, such as how their particular instruments performed and whether there was anything special in the data. However, this has proved to be a somewhat hit-or-miss procedure, since there has been no systematic way to know what were the best shots to study. People might think they had a good run, then later find out that someone's instrument wasn't working and they needed his data, too. With the TFTR there is a Scientific Coordinator with a clipboard. After each shot he walks up and down asking each person, "Did you get that shot?" That way he will know which shots had all the instruments working. He also notes comments about the shot itself, within the TFTR. Then it is easy to let everyone know which ones were the best, a day or so later.

Longtime aficionados of tokamak control rooms may miss one more little drama, which is the nightly argument over raising the neutral-beam power. On the PLT the beams are under manual control, and the experimenters can hope to get more power by leaning on the Responsible

Person, who operates the beams. But on the TFTR, the beams are under computer control, along with everything else. If anyone complains to the beam operator, he will reply that he doesn't want to break into the control arrangements and try to operate it manually. Still, not all is lost. During the most critical phases of work, the TFTR will be kept running all week long and around the clock, with three shifts. The beams will be kept warmed up and will always be available for full power.

"We will rush to do $Q = 1$ as fast as absolutely possible," said Meade. His people will not busy themselves unnecessarily with fascinating little side roads in plasma physics. Instead, they will concentrate on troubleshooting, solving problems as they arise, to meet their schedule. "Our main experiments will be to continually push toward the $Q = 1$ demonstration, at the earliest possible time. Nothing will stand in our way— unless the plasma or the machine starts to act up. We want to get as much power out of the machine, as fast as we can. If we run into difficulties with the plasma, we will have probably the world's best diagnostics system, costing about $50 million, to allow us to untangle these problems as they arise."

The work has already started, for TFTR achieved its first plasma at 3:06 A.M. on December 24, 1982. Since that was the day before Christmas, there was the hope that no one would have to work past two in the morning. That witching hour approached and they still weren't ready, so someone did a very logical thing: at 1:55, he unplugged the control-room clock. That first plasma was nothing special; it lasted for a twentieth of a second and its temperature was only a hundred thousand degrees. It merely served to prove that they could initiate a plasma in so large a machine. But when it happened, it came as a splendid Christmas present, and everyone was delighted. Harold Furth celebrated by hosting a champagne party in the lobby of the lab. Paul Reardon even wrote a poem:

Santa Claus Comes to Fusion

'Twas the night before Christmas and all through the cell
Not a creature was stirring. Just the warning bell.
The diagnostics were hung on the tokamak with care
In hopes that first plasma soon would be there.
With Harold in his coat and I in hard cap
We were looking around for a place to nap.
When out of the cell there arose such a clatter

Don Grove rushed in to see what was the matter.
With ladder in hand and quick as a flash
He jury-rigged a fix to keep on with the bash.
Minutes later, ahah, to our eyes did appear
Good vacuum, good control, nothing to fear.
On TF! On OH! With EF to steer,
First plasma, for sure, soon will be here.
Now we'll puff in the gas.
Does TFTR really have class?
Wait a moment, what's that glow?
First plasma, of course. Ho, ho, ho!

After installing more equipment, during much of 1983 they did tests in which the plasma was heated with electric currents only. The neutral beams were not yet in place. They started at low levels of current in the plasma, a hundred thousand amperes or so, and gradually worked their way up to a million amps. This gave them their first look at how well the plasma was confined, and how that confinement compared to other tokamaks. Their goal was 0.2 seconds for the confinement time, the time it takes for the plasma to leak away. In their first major run of experiments, they set a record by reaching 0.19 seconds. This is very short by ordinary standards, but physicists have already gotten very excited over achieving confinement times a third or a half as long, in other machines. As we've seen, a tokamak works by puffing in gas, having the gas turn to plasma and stay confined for a split second, then letting it leak away to be pumped off by vacuum pumps. In TFTR, a fifth of a second will be quite enough. These TFTR shots ran for one and a half to two seconds, or seven to ten times longer than their actual confinement time.

Also the scientists were looking at how well they can control those plasma impurities that cause energy to be lost from the plasma. Impurities also limit how much current a plasma can take; a clean plasma can take more current, though too much makes the plasma disrupt. In Meade's words, "In the initial startup phases the plasma is always a little dirty, and you try to always nuzzle up against this boundary," the limit at which the plasma disrupts. Thus they also were studying whether the disruptions were stronger than anticipated, knowing that later in the test program, the disruptions will be stronger still: "We want to make sure that we don't have any difficulty with the machine shaking itself to pieces." But the disruptions proved less severe than expected.

Meanwhile, other groups were working with the neutral beams, with the first two being installed during the first part of 1984. This gave them half of their neutral beams. By August 1984 they hope to have these beams working well, and between August and December 1984 they will repeat the previous year's experiments. Confinement, impurities, disruptions—these will all have to be studied anew, but this time with the beams. After December 1984, they plan to shut down again for three months, to install the other two beams. As Meade explained, that won't be all. "We will install the famous second motor-generator set, which we've been waiting for. We've been waiting ever since the accident, because we need that to go to really full power." They will also install a high-power limiter, "which," says Meade, "we are feverishly designing right now," a device that keeps the plasma from expanding against the walls. At that point, by the spring of 1985, they will have everything in place for the full-up, high-power experiments.

"Between April and December of 1985, in a nine-month period, we will gradually push every parameter we can. We will get the magnetic field up to full value, get the plasma current up to full value, get the neutral beams up to full value, push on the density, and hopefully then get the temperature we need." This will also be the time when the TFTR people will be able to set a new temperature record, and get new T-shirts. The old ones will be the ones reading PLT NEUTRAL BEAMS—75 MILLION DEGREES. The new ones will tell of the TFTR and will read 120 MILLION DEGREES. Physicists measure very high temperatures in kilovolts. A kilovolt is 11.6 million Celsius degrees, 20.9 million Fahrenheit. In Meade's words, "On PLT and PDX you have the only two machines in the world that have ever made six kilovolts, in a tokamak. But for TFTR, we must go for ten. We think actually the central temperature will be even higher, might even be fifteen. Might." They will set this record in December 1985, "if we're lucky," Meade added.

All this work will be done in plasmas of ordinary hydrogen, to avoid producing neutrons that would make the TFTR radioactive. But in December 1985, for the first time, they will face such neutrons. In going to higher temperatures still, they will switch to using deuterium in the beams, which will give them more power, and they will shoot those beams into a deuterium plasma. However, they won't do very many shots that way: "We will only do about six hundred, because when we use the deuterium beams we will make D-D reactions that are semi-fierce," Meade explained. When working with deuterium alone, let alone with DT, the TFTR will produce up to half a million watts of fusion power from the D-D reactions only, with much of this power coming off as neutrons. A half-megawatt of fusion power would ordinarily be an immense achievement, but on TFTR it will be no more than an incident along the way. Only eight months later, in August 1986, Meade's group will be ready to move on to the true goal, and to operate with the far more powerful fuel of DT.

At this point, everyone will be face to face with a contradiction: they will be ready to sprint ahead to $Q = 1$, but will not be ready to handle a radioactive machine. At its peak of radioactivity, following routine DT experiments to be made continuously a few years later, the TFTR will be so radioactive that a man standing next to it would receive as much radiation in only forty seconds as the Department of Energy will allow him to receive throughout an entire year. Working with DT is something everyone wants, but then there is the not-so-little matter of keeping the technical staff from getting fried. The solution will be to refuse to allow anyone near it, never ever, till it has the chance to cool down and let the radioactivity decay. In the meantime, it will be handled by robots.

There is a large port in the side of the vacuum vessel, which can be opened for a long articulated arm that snakes its way inward. The arm will be under computer control, preprogrammed to flex its way around the curving interior, guided by rails. Then, at the appropriate spot where work is to be done, it will brace itself by extending its legs. To do the work, it will have both strong arms and delicate arms, as well as bright lights and a TV camera. It will be able to cut apart the entire vacuum vessel, magnet coils and all, using a rotating saw blade, then weld it back together, using a torch. To remove bolted-on parts, it will use wrenches, with the bolts having cone-shaped heads to make it easy to slide the wrench down for a tight grip. For really heavy work, the entire array of surrounding coils can be lifted off with a crane. These coils will have remote disconnects for the cooling water lines and electric power. The crane will carry still another robot manipulator, to work these disconnects. Indeed, from the very beginning in 1975, the TFTR has been designed to allow for all this.

However, because of budget cutbacks, these robots won't be in place till 1988 or 1989. Nevertheless, Meade and his physicists have worked out a very clever plan to postpone the need for them, and to avoid making the machine radioactive, yet still to carry out a true $Q = 1$ experiment. The key to this will be to make only ten full-scale shots in DT. To a fusion physicist, this at first glance would seem entirely absurd, for everyone in that field knows that it takes not ten but ten thousand shots before you can be sure you've learned anything. The cleverness in the plan consists of doing so much preliminary work that the final step of going to full DT will be only a small extension of what will have been done previously. The preliminaries will start in August 1986.

"We will go back to a hydrogen plasma with deuterium in the beams," Meade said, "and in this hydrogen plasma we put in something like a tenth of a percent of tritium. Then we run off several hundred shots. That will tell us how the tritium is moving in the plasma. It will make a lot of DT neutrons, and we can look to see what the activation produced will be. Right now we do very elaborate computer calculations, but they're quite complex, and we could be mistaken. So we will be able to check them. Then, we will up the tritium to maybe one percent, run off fifty or sixty shots. Then up to ten percent, see how that goes. So we will be able to draw a curve through these points."

The tritium will come from the Department of Energy facilities at Savannah River, Georgia, in an initial lot of five grams and in a special shipping cask. This lot will last for two or three weeks of work, and will be carefully accounted for, because of the radioactivity of tritium. Each shot will use no more than a twentieth of a gram. Of this, only a minuscule amount will be burned in fusion reactions to produce the neutrons, but about a fifth of the total lot will stick to the polished metal walls of the vacuum surfaces and be trapped. After each shot, vacuum pumps will draw off the gas-that-was-plasma, and the tritium will be retrieved, though it then will be impure and incapable of being used again. Of the original five grams, about four will be recovered and returned to Savannah River. There the returned lot will be assayed, the DOE scientists will convince themselves that the container really does have four grams, and they will then issue a second lot—of four grams. The most the DOE allows the Princeton lab to have at a time is five grams, total. With this second lot, the Princeton physicists expect to put four grams in and get three and a half out. Tritium will go back to Savannah River three or four times in this way, while the Princeton scientists are reaching for $Q = 1$.

"What will we do next?" Meade continued. "We will convert the beams back to hydrogen, the plasma to pure tritium, so we will see a tritium plasma for the first time at high temperature, heated with a hydrogen beam. So we will learn how to handle that. Now we will gradually introduce deuterium into the beams. We will start off with a very small fraction of deuterium and do this game again, with a tenth of a percent, with one percent, and ten percent. And again I can draw myself a little curve, so I will have two different cases which cover what we need. And now we will have used the first two loads of tritium. It will take a week to have it assayed and get the next lot, so we will sit back during that week and think about what to do next. Then we'll be ready to do the final thing. We'll be there, we'll have all the pieces, we'll know what to do." They will convert the beams to a hundred percent deuterium, and prepare to shoot pure D into pure T, bringing the beams to full power and tuning them up. "It will probably take several days," Meade says. "We hope to do this last part in just two days. I hope it's October 1986. It could drag into the winter, but we're going to try to do it as fast as possible."

Meanwhile, they will have been keeping track of the activation: "Once you get the machine activated, it takes a long time to cool off," Meade has said. The machine by then will be radiating at about fifteen millirem per hour, which isn't much. A man could spend forty working days there next to it before exceeding his allowed radiation dose. Still, there will be no sense in making DT neutrons for nothing, so Meade's physicists will start with some short shots. These will check mainly that the impurities aren't a problem. The diagnostics are so quick that in mere milliseconds they will detect the X rays radiated from impurities, and the computer would then abort the shot, cut it off before it makes very many neutrons. The experimenters will do some deliberate short shots, then project on the basis of their data whether they really will get to $Q = 1$ on the long shot, a full half-second. Once the computer says that this looks possible, probably late at night after a long day for everybody at the consoles, they will go for it. The shot everyone will have been waiting for will be at hand.

Thirty seconds before this shot, the neutral beams will be interlocked and put under control so that they will fire only when directed. The computer will be alerted to take data. At minus four seconds the main magnetic coils will be turned on, the motor-generators being switched to send their currents pulsing through. Then, at minus four-tenths of a second, things will begin to happen fast. Within the central doughnut hole of the tokamak, a powerful electrical coil, a transformer used to heat the plasma will be charged up. Auxiliary magnetic fields will come on. Tritium gas will start to puff into the vacuum vessel. Then, at time zero, the central transformer will be shunted to generate a powerful current within the gas, instantly flashing it to plasma and driving its temperature up steeply. All the while, more tritium will be puffing in. At nearly the same time will come the moment everyone will have been working toward. The neutral beams will come on.

There are 240 holes or ports in the vacuum vessel, most of them for instruments, and at least one will be a window with a color TV camera. A window on the plasma, after all, is quite a good diagnostic; the good old eyeball is far from obsolete. But the camera will need freeze-frame action to follow the rapid events within the machine. The plasma will first form as a bright purple-blue flash, completely filling the chamber. Then, very quickly, it will heat further and change. Most of the glow will come from residual impurities, with their electrons, and the center of the plasma will darken and glow only dimly as it grows hotter. The plasma will appear as a central dark band flanked by two bright stripes, the outer peripheries where impurities still will have some electrons. On the walls opposite the neutral beams will be flashing sparks and scintillations. Not all of the beams will be absorbed in the plasma; some will go straight on through, and will produce these pinpoints of light.

Then in the control room, people will be cheering and passing around champagne. The tension will be broken. The long months and years of intense effort will finally culminate in a computer display, something like "$Q = 1.02319$." Anyone who is there will see how the sudden burst of fusion energy touches off a sudden release of emotional energy: we made it! The overwhelming feeling will be one of success suddenly gained at the end of a very long road.

On the second story, above the control room and off to one side, there is a concourse carpeted in the rust-red of the main lobby. It features large windows set at a slant behind railings; visitors can lean over and watch the people below. An orange wall facing the concourse has a large stainless-steel plaque with a physics diagram, a chart somewhat resembling a map, which indicates the combination of temperature and other plasma conditions needed to produce fusion energy (see Illustration 22). On that night when fusion is to be achieved, the concourse will crowd with TV cameras and journalists, jockeying for position to record the drama. Perhaps Dan Rather or Tom Brokaw will be there, standing near the steel diagram, showing the conditions reached in this most critical of tests. Then, down below, physicists will be shouting and celebrating with their champagne glasses. The word will go out: for the first time, true fusion will have been achieved. For the first time, a tokamak will have generated energy from fusion. The work will have reached fruition, the man-made sun will have shown its first glow.

Quite probably, while Dan Rather has to content himself with watching from the concourse, there will be an honored guest within the control room itself. This will be Lyman Spitzer, Jr., a slim, elderly astronomer, and he will deserve to be there. He started Princeton's entire line of fusion research, back in 1951. What led him to do this was a pronunciamento by Juan Perón, the dictator of Argentina.

Perón was no great judge of scientific talent but was in love with things German, and had fallen in with Ronald Richter, a mediocre scientist from Germany who had come to him with a scheme for fusion power. On March 25, 1951, Perón held a press conference, disclosing that he had backed Richter in setting up a secret laboratory on an island in an Argentine lake. Argentina, Perón proclaimed, had surveyed the efforts that scientists in other nations were making to design uranium fission reactors and had chosen to try an entirely different direction: "On the critical 16th of February there was held with complete success the first tests which, with the use of this new method, produced controlled liberation of atomic energy, as we successfully tested our thermonuclear reactor for the first time on a technical scale. We have a large-scale thermonuclear reactor under construction and it should be possible to have this 'Atomic Furnace' in full scale operation in about ten months or so." Neither Perón nor Richter gave details, and, as *The New York Times* would shortly point out, "only a superficial knowledge of the facts will establish at once that such a claim involves several factors, each known to be impossible, as far as present knowledge goes." Nevertheless, in those days Perón still had his Evita, and he was at the height of his power. His statement made banner headlines in newspapers all over the United States.

At Princeton, Lyman Spitzer, in his mid-thirties, was a professor of astrophysics. He had spent much time studying plasmas in astronomy, the hot dense plasmas in the center of stars as well as the cool, incredibly rarefied ones filling interstellar space. The Korean War was on, the Soviets had recently exploded an atomic bomb, and Spitzer had already agreed to work with colleagues to do research on the hydrogen bomb. Then he got a phone call from his father: "The Argentines seem to have gotten ahead of you. Have you seen *The New York Times?*" Spitzer rushed out to get a paper, and read that scientists in this country were incredulous.

Just at that moment, he and his wife, Doreen, were packing to take the train for a vacation at Aspen, Colorado. Today's hotels and condominiums were still far in the future. The town was little changed then from what it had been during the silver boom of the previous century, except that a number of wooden buildings had burned down or been torn down, leaving vacant lots along the streets. The Spitzers stayed at the Norway Lodge, at the foot of Mt. Aspen, with the ski lift just across the street. It was nice to be able to ski back down the trail to the front door of their hotel at the end of a day, freshen up with a hot bath, and then drift downstairs for dinner in the Norway's restaurant, with its wooden floor and furniture, its hearty Western fare. Equally pleasant were the rides up on the ski lift, and these were what gave Spitzer time to think.

There was only one lift, featuring single chairs, and the view was spectacular. Pine forests ran up and down the mountain, the trees close at hand. From the crest one could look across the Colorado peaks extending away into the distance. Sometimes the ski lift crossed a ravine, falling away a hundred feet below, with skiers in their brightly colored parkas making their way downhill. But after a couple of days, even this scenery began to become familiar. In his single chair Spitzer had no company and few distractions, and the ride up the mountain took a half-hour; Doreen was in her own chair, twenty feet away. He began to think about ways to use a magnetic field to hold in a plasma. On the way back down it was different; he had to concentrate on his skiing or he wouldn't get very far. But on the trips up, he thought about physics.

When he got back to Princeton, he sent in a proposal to the Atomic Energy Commission, pointing out two ways to build a fusion reactor. One was like today's tokamak, with a powerful current being made to flow through the plasma. But Spitzer believed that this current could be made to flow only briefly. He knew that a real reactor would have to run continuously; not for thirty years would anyone figure out how to make a tokamak do that. This approach did not look worthwhile to Spitzer. Instead he opted for a different approach, in which currents would run continuously through coils wound in spirals around the tube containing the plasma. He called this invention the Stellarator, a name reflecting his background in astrophysics, and soon was at work building the first one, Model A. Eventually this would be succeeded by the more powerful Model C. Eighteen years later it would be converted to a tokamak, and would start Princeton down Spitzer's road not taken, a road that in 1986 will stand to lead to that night.

This will not be the end of the night. All too soon everyone will have to go back to work, taking at least an hour to call up their data, look them over, and convince themselves that they really did reach their $Q = 1$. Then it will be time for another shot, and perhaps another, before that most unforgettable night will end. All too soon, like graduating seniors leaving a prom, they will have to go, to leave this place of triumph, where the odd bits of paper and computer printout might as well be party favors and crepe paper left over after the band has left. To walk out into the cool night air, to drive homeward, and onward in their careers.

Then there will be the rest of the DT shots to run off the next day, followed by additional experiments in succeeding years. Much new equipment will be coming in, including the robots and the facilities for remote handling. In a few years they once again will be working in DT, but then it will be for many thousands of shots, each longer and more powerful than the quick bursts just achieved. Then beyond that will be the next big fusion machine after TFTR, and the next one after that, which must run successfully before the first fusion reactors can be built to do their share of the world's work. On that night of success, the TFTR story will already have spanned more than a decade. These future devices will need similar lengths of time in their own turn.

The quest for fusion will span more than the time of a single project, even a TFTR; more than the time of anyone's career; as long, perhaps, as a human lifetime. A Lyman Spitzer may trace in his career the developments from Aspen to this night in 1986. A young graduate student, newly admitted to the company of the control room, may hope to see the day, much later in his own career, when fusion will begin to light our cities. But no single individual will take part in it all, from start to finish.

Still, just in that hour of achievement, it will be enough to see the lights shining in the eyes of those who have shared the long road. To share the emotions, and to savor the champagne.

4

Doing It with Mirrors

WHEN Robert Hirsch took over the fusion program in August 1972, one of the first things he was determined to do was to shut down several experimental projects that he regarded as turkeys. To Hirsch,this was a test of whether he would be able to control the program: "Everybody knew they should have been turned off years ago, but nobody had the guts to do it." These projects were backed by their laboratories' directors, and in the pre-Hirsch days when these directors controlled the fusion program, they were untouchable. Not even Congress had been able to prevail. In 1964 Senator John Pastore, of the Joint Committee on Atomic Energy, had railed against them: "Is this not a very expensive way of getting this basic knowledge? We can build these machines until the cows come home. How long do you have to beat a dead horse over the head to know that he is dead?" At about the same time, the House Appropriations Committee had cut the fusion budget and made the same point in less bucolic language: "The Committee indicated last year that it expected that the number of concepts in this program would be reduced. Nothing was done about it. It is expected that this reduction in funds will achieve the objective." It didn't.

At the top of Hirsch's hit list was a project called Astron, at Lawrence Livermore Laboratory in California. Astron was very much the wild-and-woolly reflection of its creator, the Greek-American physicist Nicholas Constantine Christofilos. A technical reviewer had once described Astron as having the flavor of Rube Goldbergopolis. Christofilos had no degree in physics, and in that field was entirely self-taught, although that did not keep him from being a very good physicist. His Astron featured a remarkable high-energy electron accelerator of his own invention, which in its own right would have been quite enough to make his reputation. The machine was supposed to work by having the accelerator emit burps of electrons, which would then stack or combine to aid in confining the plasma. Some of Christofilos' colleagues predicted from mathematical theories that the Astron wouldn't work this way, that each burp would eject the previous one without stacking. That was exactly what was happening. Christofilos kept insisting there was a way around this problem, but by 1972 he hadn't found it.

In going after Christofilos' project, Hirsch in a sense was taking on an easy target, for Christofilos had always been something of an outsider in the physics community. As one top Berkeley physicist said, "Well, my contacts have been with other members of the scientific fraternity, and Christofilos really isn't a member." In a world of senior scientists who usually tried to appear to be dignified and reserved gentlemen, Christofilos was voluble, passionate, and argumentative. He loved to play the piano, loudly. ("For Nick," said a friend, "all pieces are written fortissimo.") His colleague Richard F. Post was one of the founders of fusion research at Livermore. One day Christofilos gave Post a ride to the airport, and Post found it a hair-raising experience. Christofilos' speed matched closely his emotional mood. Whenever he got excited, which was often, he would push the pedal to the metal. In fact, during his driving career he got so many speeding tickets that a judge ordered him to install a mechanical governor to limit his car's speed.

Even before Hirsch took over the fusion program, he had been enthusiastically supporting his predecessor, Roy Gould, who was already greasing the skids for Christofilos. Gould had commissioned a technical review; it found strong fault with Astron. Christofilos then was told to set forth precise goals and timetables, and to submit monthly reports on their fulfillment. These reports made interesting reading to Hirsch, for, as he expected, they showed that Christofilos was falling behind his goals. Hirsch's Washington associates then set another technical review for the spring of 1973, with a decision to be made then on terminating the program; but by early September 1972, Hirsch's mind was already made up. Meanwhile Christofilos, a compulsive worker by nature, was working on Astron around the clock. The effort was too much even for him. On September 24, at the age of fifty-five, he died suddenly of a heart attack.

Hirsch had not yet consolidated his own leadership, but he went ahead and issued orders to turn Astron off. He also ordered the shutoff of another Livermore project, Levitron. He knew full well these moves would meet opposition. As he told it, "Either the machines were going to get turned off, and I would have gone a long way toward establishing leadership and control in Washington—or I was going to lose my job. But if I couldn't turn those machines off and could not have control of the program, then as much as I wanted that job—I didn't want that job. It was a matter of confrontation, very early on. It was a question of who would back me up, but you know what happened. It bothered the living hell out of a bunch of people [the lab directors], because for the first time they had somebody who was standing up and saying, 'We're not going to do it that way; we're going to do it another way.' It really called their cards. It made a lot of people very uncomfortable and there was a lot of bitching, of running to other people in the Atomic Energy Commission. But the AEC backed me up, and it got done, and I had established a significant amount of control over the program."

Having bagged those two scalps, Hirsch then turned to examine the rest of Livermore's fusion program. What was left was based on a series of fusion devices called magnetic mirrors. In its simplest form a magnetic mirror holds the plasma within a vacuum chamber that has magnetic fields that become stronger toward the outside. A magnetic mirror might resemble an egg; the "yolk" would be a region of weaker fields, the "white" a region of stronger fields. Plasma particles, seeking to escape, encounter the increasing fields and are turned back or reflected; this magnetic reflection accounts for the name. One of the Livermore devices was Baseball II. It had a large current-carrying magnetic coil shaped like the seam of a baseball, enclosing plasma on the inside. Another was 2XII, a rebuilt version of the earlier 2X, which had been "two times" the size of *its* predecessor. Neither one was working very well. The mirror arrangements just were not very effective in trapping plasmas, and they were very leaky. There are degrees of badness in fusion experiments, and Baseball and 2XII at least were not at the lowest rung on the ladder. That distinction had belonged to still earlier mirror devices, which were so unstable that they had flung their plasmas into the chamber walls in mere microseconds. During the 1960s, the Livermore physicists had at least learned to prevent that instability and had raised the confinement time to milliseconds. But this was still far too short. If the plasmas were at least being bodily contained, they nevertheless had very rapid rates of leakage, far too rapid to be capable of producing fusion.

Hirsch wanted the mirrors to succeed, but he knew that the way to get results would be to hold people's feet to the fire. His management technique was to sit down with project leaders, push them to set forth the

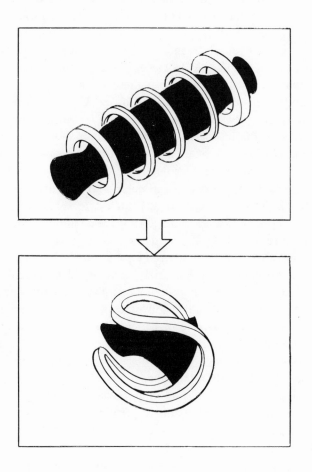

FIGURE *4.* *Early magnetic-mirror ideas. Top, the simple mirror: a tube lined with ring-shaped magnets, plugged at the ends by more powerful magnet rings. This arrangement worked very poorly; the plasma (dark shading) escaped within microseconds. Bottom, the "baseball" mirror, with a magnet in the shape of the seam on a baseball. It could hold its plasma for up to a millisecond, but this still was far too short to be useful.*

most difficult goals they felt they could reasonably meet, have them draw
up schedules and milestones—then hold them to these. That type of
maneuver was what had given him the technical basis on which he could
shut down Astron. Now he decided to apply this technique to Livermore's least-bad mirror, 2XII, and to tell the managers of this project that
it was time to fish or cut bait. In December 1972 he sent one of his top
aides, Stephen Dean, to meet with these managers, Fred Coensgen
(pronounced "Ken-see-en") and Ken Fowler, Coensgen's boss. When
they arrived, they saw that Dean had written some numbers on the
blackboard, defining his view of what their goals should be.

In plasma physics, one of the most important measures of success is
known as the confinement parameter, $n\tau$. It is the product of plasma
density, n, in particles per cubic centimeter, and the confinement time τ,
in seconds. The TFTR is to achieve an $n\tau$ of 3×10^{13}, in getting to Q = 1.
At 3×10^{14}, a plasma would ignite. That is the condition under which, if
the temperature is right, a plasma will produce so much energy that it will
keep burning or reacting with no need for neutral beams or other outside
heating. The intermediate value, 10^{14}, could be called the Holy Grail of
fusiondom. It is a nice round number—100 trillion, actually—and corresponds to a good, substantial fusion reaction. In the tokamak breakthrough of 1969, the Soviet T–3 had reached an $n\tau$ of 5×10^{11}. By
contrast, in 1972 the best that Coensgen had gotten in his leaky 2XII was
8×10^9. Dean had written on the blackboard, "$n\tau = 10^{12}$." He was
demanding a hundredfold improvement.

Coensgen protested that this was impossible. Another of Dean's goals
was for them to use neutral beams to get a plasma temperature of 120
million degrees Celsius. But at that temperature, in 2XII, such an $n\tau$ was
simply impossible according to physical theory. Dean backed off a little.
He demanded not 10^{12} but 10^{11}—still a very difficult goal to achieve. He
also said that Fowler and Coensgen should get their results within three
years, by the end of 1975, but offered no additional funding to help
achieve them. At that point, it was Fowler's turn to protest. He argued
that providing no more money amounted to a decision in advance to shut
down 2XII, and with it virtually the whole of the Livermore program in
magnetic fusion, without giving the Livermore scientists a fair opportunity. Dean relented once again, promising that the money saved by
canceling Astron and Levitron would be plowed back to give more
support to 2XII. This doubled its budget and gave them a fighting chance.

Fowler was a soft-spoken man from Georgia, of medium height, slim,
and with a distinguished-looking head of white hair; he might have been

John Wilkes of the plantation Twelve Oaks in *Gone With the Wind.* He was in charge of the whole of Livermore's magnetic fusion program. Coensgen was in charge of 2XII. Originally from Montana, he had been at Livermore ever since getting his Berkeley Ph.D. in 1953. He was short, stocky, and nearly bald; but what people noticed about him was the determination in his eyes. He and Fowler were under the gun, but they didn't feel that very much was different. They had been working under pressure for a long time. To Coensgen, "having challenges, having extreme pressure has always been a way of life around here. I may look like a relaxed fellow but that's not really true. If you don't believe it, just ask the people who work for me. I used to work sixteen hours a day, six, seven days a week. I don't do that now. Other people do. We've tried to get away from that; it broke up too many marriages in our group, caused health problems. We had had earlier crises that threatened to kill the program, times when we had to save it from extinction. So this was just one more crisis in a long series."

They decided to try to get the 2XII plasma into as good a condition as could be had, then hit the upgraded plasma with neutral beams. As 1973 progressed, however, results in improving the plasma were too slow in coming. The Livermore leaders began to feel themselves under even more pressure than usual. So they decided to forget about improving the plasma, and instead to hit it with the unprecedented total of twelve neutral beams, which were being developed at nearby Berkeley. As Fowler described it, "Our thinking was that if you went at it with a sledgehammer, which in this case was lots and lots of neutral beams, that you had a chance you'd find out something that would surprise you in a positive way. That was the motivation. We had no theory that said it would work, but we had experimental data that said it was worth a try." The new arrangement was christened 2XIIB, where B meant "beams."

This was a gamble, for neutral beams were still in their infancy. Fowler's scientists had built a ten-ampere beam, and were reasonably sure that they could get up to fifty amps. But then they were proposing to mount twelve of them on 2XIIB, for a thousand times more beam power than had been used on Baseball II. This was a little like building an airplane to be powered by twelve piston engines, with all of them being required to work together smoothly. Still, this was the chance they took. Sure enough, a major problem soon cropped up. The beams "cross-talked"; if one fired prematurely, it would make others fire as well, spoiling the test shot. It was necessary to shield them so they would not cross-talk.

In November 1974, as the completed system was coming together, Fowler had a visitor from the Soviet Union. He brought with him a message from the top man in the Soviet mirror program, M.S. Ioffe. It was a photo showing three white lines, oscilloscope traces displaying the escape of plasma in Ioffe's mirror device; there was not even a cover letter. But Ioffe had marked each curve with a note to indicate the amount of residual gas in the vacuum, so Fowler could follow it. One curve plunged down quickly, the second not so quickly, and the third more slowly yet. This looked interesting; the plasma was staying around for a long time. The curious thing was that the more residual gas there was, the better Ioffe's mirror worked. This was quite contrary to the usual expectations. Indeed, Fowler's people had been spending a lot of time working with equipment that would sputter thin films of metal all over the insides of their machines, to soak up gas and get better vacuums. But here was Ioffe saying that a bad vacuum was giving much better performance than a good one.

To Fowler's prepared mind, however, Ioffe's data fitted in with a long-standing mystery in the mirror program: the existence of a missing mode of plasma leakage. Plasmas have many ways of leaking, which are called by different names and studied using different theories. This missing mode was called the DCLC.* Richard Post, Christofilos' erstwhile driving companion, had predicted its existence in 1964, when he had gone down to San Diego to spend a few weeks in the summer working on problems of plasma theory. Post also predicted from theory that there was a cure: inject a stream of cooler plasma into the main body of hot plasma. But no one had ever identified the DCLC mode in data on plasma leakage.

To Fowler, listening to theorists was like being visited each day by life-insurance salesmen: it was hard to sort out the competing claims; there were too many possibilities. He knew that only an experiment could sort things out. Ioffe's data looked as if it might be it. The background gas in Ioffe's vacuum could be a source for Post's "warm plasma," cooler than the main plasma, which by Post's theory then would reduce leakage by the DCLC mode. That might explain why the more residual gas was left in Ioffe's machine, the less his plasmas were leaking.

Fowler knew that Ioffe was interested in finding the DCLC mode, and even had claimed that he had detected it in his data. Fowler had been skeptical, but he could recognize that Ioffe's oscilloscope traces

*Drift-cyclotron loss-cone.

amounted to saying, "There, I told you so." The traces on Ioffe's photo, after all, fitted in nicely with Post's theory, and thus were evidence that this theory was correct. With this, Fowler said to Coensgen, "Look, we'd better develop a way to put this warm plasma in there." Coensgen, however, was ahead of Fowler in this. He had been aware for a long time that a warm-plasma stream might have a very good effect, and was quite prepared to try it.

The rebuilt 2XIIB was finished just about then, but before anyone could start doing experiments with it, there was a time for debugging. The common experience at Livermore has been that it takes about nine months to get all the bugs out of a new machine, and 2XIIB was no exception. Coensgen said this was like commissioning a ship in the navy; "take it out to sea and see what falls apart." Thus, when they turned on the vacuum pumps, they found for a while that, try as they might, they just couldn't get a good vacuum. It turned out that during an earlier instance when the vacuum chamber had been taken apart, someone had used masking tape to mark parts of its lining while they were being reassembled—and then had forgotten to remove the tape. The mucilage was evaporating in the pumped-down chamber, spoiling the vacuum "as if someone had left a mouse in there," in Fowler's words. Still, by June 1975 they were ready to use 2XIIB for experiments. The beams were ready; the vacuum pumps were ready; the equipment was in place to spray clean metal atoms all over the inside to keep the impurities down. It was the cleanest they'd ever had. When they pumped it down, the plasma confinement got worse and worse the cleaner the plasma got and the better the vacuum they could achieve.

This was very discouraging. Contrary to their hopes, the beams were doing nothing at all to improve the plasma confinement; it was leaking away as fast as ever. After all this work, they apparently had nothing to show for it, and if they had been working without guidance from theory, they might have been sunk. Actually, 2XIIB was behaving just as in Ioffe's data. As Fowler would later say, "This was the only thing I ever had Nixon to thank for." He had gotten that photo through a cultural exchange program Nixon had negotiated only a year or two earlier.

At the start of each shot in 2XIIB, an array of plasma guns fired bursts of low-energy plasma, to provide an initial source of plasma for the experiment. Coensgen ordered that the firing of one of the guns be delayed until the middle of the shot. The gun would then inject a burst of warm plasma, which was what they wanted. Fowler called this a "drool" of warm plasma, but Coensgen said that the name was undignified, and they soon changed to calling it a "stream." Fowler recalls what happened next: "He tried it, and probably you would have said there was no effect. But he looked at the data and said, by God, there was. So he modified one of the guns to lengthen the duration of its plasma stream, and just got a bigger effect. Then there wasn't any doubt." Sure enough, the DCLC mode was the culprit behind the rapid leakage, but now this was no longer a missing mode. They had found it, and were taming it. And suddenly sunlight was breaking through the clouds and there was the chance that everything would be all right.

That summer, one of the young experimental physicists working with 2XIIB was Grant Logan, who would rise to play a key role in the years that followed. He had grown up out in the desert near the Colorado River, in Blythe, California. Even in high school, he had wanted to work on fusion. He built a science-fair project that purported to generate a plasma, but he never could get a good vacuum pump down in Blythe; all he got was a fluorescent-tube discharge in a poor vacuum. Still, reinventing the fluorescent light while still in high school was sufficient to win him a science-fair prize. ("It beat dissecting frogs," he said.) He got his Ph.D. at Berkeley, then joined 2XIIB for his first permanent job.

"The karma was good," he commented. "The mirror program had just been creeping along for twenty years; progress was slow. Now a young physicist in his first job wants to make a name for himself, advance his career. But you gotta get into some exciting research where you're starting to make some breakthroughs. If you have the misfortune of picking a problem that has no solution, then even though you might be potentially brilliant and able to make discoveries, Mother Nature can really stymie you. Some good physicists never blossom because of that." Now, however, 2XIIB was about to take off. "I felt like I was a member of a club, because it was really run by consensus; there wasn't a dictator telling us to do this and that. There were about ten of us. We could run the experiments at night, take out pizza while we were running. Those were wonderful times; I still miss them. It was such a hectic pace, but I liked that; it wasn't dull. Really, it was too good to be true. Almost every time we'd button it up for a run, we'd learn something new. And it was so exciting, because every time we ran it, it worked better and better.

"We helped each other out a lot, kidded each other a lot. We'd always be pulling each other's leg, making fun of each other's foibles, but in a good-natured way. For instance, I was kidded for a long time about a makeshift wiring job I had to do one night, when I had to have a power supply for my diagnostic and I jury-rigged one with cables and high-voltage batteries. I was in a car pool with some of these guys; they kept asking me if I could give them a jump start with my power supply, things like that. Anyway, when we were working, we had an x-y plotter that slowly plotted out the data after each shot, four or five curves for us to puzzle over. We'd stand around watching it, even argue over whether a little wiggly curve would show a dip or a rise, while it was plotting. Some of our diagnostics were limited so we had to do a lot of detective work together. We'd sit around a table trying to understand what a few bits and pieces of the puzzle meant. We often did the analysis by the seat of our pants, late at night, with a few pieces of data. Then we'd set up and run another shot. If you wondered if something would have an effect, you'd take another shot with this new thing. You'd look for what you wanted, and it would pop out at you. That gave you an immediate high."

It was July 1975, and Robert Hirsch was convening a meeting of his lab directors and consultants in Monterey, less than a three-hour drive from Livermore. This was the year that by the original arrangements made in late 1972, Fowler would have to prove to Hirsch that he had met his goals. At the end of that meeting, the attendees, who were Hirsch's principal advisors, were to come up to Livermore to see how things were going. It was raining quite heavily when Fowler drove down, leaving Coensgen behind to hold the fort. As the meeting went on during the next couple of days, the rain kept falling, which was unusual for California in July, and Fowler kept phoning Coensgen to ask how the data were coming along. The night before he was to drive back to Livermore, he got hold of Coensgen once again. Coensgen was as dead tired as Fowler had ever heard him on the phone, from working almost continually. He said, simply, "We got it." He had calculated an $n\tau$ of 7×10^{10}, which was close enough to what Hirsch wanted, and the plasma was good and hot, besides. This still was over a thousand times less than would be needed in a mirror reactor. But everyone knew that by going to higher plasma temperatures yet, with more powerful neutral beams, and by making adjustments to the magnet arrangements, then in a much larger mirror they could get at least 10^{13}. This would put them within hailing distance of what a reactor would need. The remaining improvement in $n\tau$ would have to come from further progress in suppressing plasma leakage, but now this was no longer out of reach.

Everyone came up to Livermore the next day. It was raining there, too. Coensgen got a marking pen and drew by hand a briefing chart to show the new data, then carried it to the meeting from a building across the way. There was a big rain streak across his viewgraph as he showed

Hirsch's advisors that he had achieved what they wanted. What was more, Fowler had also been on the phone to the men he called his "dynamic trio" of theorists—Don Pearlstein, Dave Baldwin, and Herb Berk—and they were able to tell him why the DCLC mode had not previously been seen. This meant that they not only had gotten good results, they also knew why their results were good. This combination was enough to turn the mirror program around and vault it to prominence.

Things looked so good that Fowler decided to hold a press conference. Hirsch agreed to this, and got Robert Seamans, head of the Energy Research and Development Administration, to come out as well. Together they made a big splash in the San Francisco papers. Then Fowler took Hirsch on a tour of the lab, and while they were walking through the computer center Hirsch said, "Gee, you just don't realize what a great thing has happened here." He meant that now that they had achieved success with 2XIIB, he would support them in going ahead with a big new project as the next step. Fowler and his mirror designers already knew what they wanted. They had been designing a much larger version of the same kind of magnet used in 2XIIB.

2XIIB featured a main magnet known as a yin-yang, which Richard Post had helped invent back in the early sixties. It was actually two magnets in one, each in the shape of a large and very thick horseshoe, and with these two horseshoes facing each other with a space in between. This space was for the plasma. That would be the basic arrangement to be used in this new project; but whereas the 2XIIB magnet was the size of an automobile, the new one would be as big as a two-story house. It would be superconducting, too: that is, it would be chilled in liquid helium to the point of being able to produce a powerful magnetic field without needing to be fed with electric current. Soon this project would be christened MFTF, Mirror Fusion Test Facility.

Fowler and Coensgen certainly deserved to have it. Hirsch's way of running the fusion program was indeed one of presenting his project leaders with very high hills to climb. But if anyone then succeeded in surmounting such a hill, he would find a very smooth downhill run on the other side. Hirsch badly wanted a second major approach to fusion. He was leery of putting all his eggs in the tokamak basket; moreover, he wanted Livermore to compete strongly with Princeton. By his own rules of the game, therefore, his approval for MFTF had to be virtually a foregone conclusion. There was a problem, though, and it was symbolized by the last letter in the project's name, which was F, not R. MFTF would not be the mirror program's counterpart of the Tokamak Fusion

Test Reactor, because even a very large yin-yang magnet would make a very poor fusion reactor. MFTF might be a wonderful facility with which to study plasma physics, but Hirsch still was the man who wanted a fusion reactor.

When Fowler and Coensgen took their MFTF proposal to Washington, Hirsch asked Coensgen if he really believed they could develop a big yin-yang into a reactor. Coensgen replied that he thought they could make a reactor out of it, but it might not be a reactor anyone would want to buy. Hirsch approved going ahead with MFTF, because he knew such a large mirror experiment was critical to their progress. But he also insisted that they work to find a better reactor design. There was a theoretical limit to how well a big yin-yang could contain a plasma, and it was barely adequate to permit a reactor at all. A yin-yang, like every other mirror design ever studied, had what amounted to an unpluggable hole in its magnetic fields. Coensgen's warm-plasma stream had succeeded in plugging this hole to a considerable degree, but, though smaller, the hole was still there. It couldn't be got rid of by building the yin-yang larger; the hole would still be a hole. In fact, the best anyone could hope for was that a reactor based on a yin-yang would need to use up to four-fifths of its power just to keep itself going, with this power being fed to the neutral beams and the warm-plasma stream.

In our everyday life, if an auto tire has a hole that can't be plugged, we toss out the tire and get a new one. The Livermore people couldn't toss out their mirrors, however, because getting a new magnetic bottle might not be possible. There were only two basic ways to confine a plasma within a magnetic field: with toroidal magnet arrangements such as the tokamak, or with mirrors. Princeton had a hammerlock on tokamak research; Livermore had a similar prominence in mirror research, and had been working on mirrors for twenty years. Having reduced the leakage, partially sealing the hole in the magnetic fields, their job was now to work to seal it still more.

But while their job was to improve the performance of mirrors, there was no reason why they had to stick with the yin-yang. In fact, they had an excellent incentive for looking for alternatives. The yin-yang might be the best they had so far, but the laws of physics put sharp limits on the degree to which its leakiness could be reduced, even in principle. These limits, in turn, put strong restrictions on the performance of a fusion reactor based on the yin-yang. These restrictions were particularly dramatic when Fowler's designers calculated the Q of their reactor concepts.

For mirror reactors as for tokamaks, Q would be the ratio of power out to power in. The power out would be fusion power from the plasma; the power in would be shot into the plasma with neutral beams. Because yin-yangs were far less effective than tokamaks at confining plasmas, their Q would be far less. Thus, TFTR was planned from the start to go for $Q = 1$. Eventual tokamak reactors might reach ignition, where the beams would be shut off altogether; then Q would reach 100 or more.* By contrast, MFTF could at best produce $Q = 0.03$. The best yin-yang reactors anyone could conceive would reach only about $Q = 1.2$, barely produc-ing net power—and even that would require that plasma leakage be reduced to the lowest level permitted by physical theory. Clearly, some-thing better would be needed.

Hirsch and the mirror scientists called this the problem of Q enhance-ment. By this they meant that Livermore had to come up with a new kind of mirror that would be able to achieve larger Q. The demand for Q enhancement meant that Fowler and his people could not rest on their laurels. Their 2XIIB was a success, but it didn't point the way to a reactor. They had invented the best magnetic mirror yet, but now they would have to go back and reinvent it.

By April 1976, President Ford had promoted Hirsch to take over much of his energy research, and Hirsch had left his fusion job to his deputy, Ed Kintner. Kintner made the demand for Q enhancement explicit, saying to Fowler, "Ken, look, I'm only going to do MFTF on one basis, and that is that we honest-to-God find ways to fix Q enhancement." Thus, Fowler now had two tasks ahead of him. He had to continue working to learn how to reduce plasma leaks; that was the purpose of the $94.2 million MFTF, for which he now had official authorization. In addition, he had to look ahead to the day when these leaks would be well controlled, and to invent a new reactor concept, a new form of magnetic mirror, one that could be guaranteed to work well once this control was achieved. The laws of physics ensured that any kind of magnetic mirror would always have some irreducible leakage, because of the hole in its magnetic field. Q en-hancement meant that candidates for this new invention would be judged by whether this leakage was sufficiently low to make a good reactor, and a good reactor would need a Q much higher than 1.2.

Why were the yin-yangs so leaky, even after all Fowler and Coensgen had done? The reason lay in the way the plasma particles leaked through that magnetic hole. At first there was a rapid leakage of electrons, which

---

*Even the best possible tokamak reactor would still need to feed a modest flow of power into the plasma, to keep a strong electric current circulating in that plasma. That power would not be zero, but would be low enough for $Q = 100$ or more.

were much lighter and more mobile than the ions. This quickly produced an excess number of ions in the plasma, and the ions carried electric charge. Hence this charge soon built up throughout the plasma. This "space charge" then pushed on the ions, repelling them strongly. With this force pushing on them, the ions soon were escaping as rapidly as the electrons. The key to it all, then, was this space charge. Somehow, Fowler would have to get rid of it, or find a way to turn it to his advantage.

Certainly there was a need for new ideas. Bill Ellis, who was managing the mirror program in Washington, suggested in March 1976 that they hold a workshop on Q enhancement. This would be a meeting of the world's mirror researchers, where they could talk about various ideas for inventing new kinds of mirrors, and try to pick the best approach. They scheduled the workshop for September. Still, Fowler's group was the world's most advanced, and was in the best position to make this invention on their own. As Fowler described their state of mind, "We had had the big machine approved, the MFTF, and we knew we were on a different track. Also we had guidance from theory for the problems of plasma leakage. All this sort of freed our minds to think in a different way. We had had our stumbling block, which was the DCLC problem, but once we got over it our minds were really clear to think of the alternatives."

As early as 1967, George Kelley of Oak Ridge had thought of an approach to the problem of this space charge. He argued that there should be not one but three magnetic mirrors, linked together in tandem. The space charge that developed in the central mirror would then be balanced by the space charge in the end cells, the two mirrors that flanked the center. But no one had ever followed up on this idea. It was difficult and tricky enough to make a single mirror work, let alone three. To Tom Simonen, one of Fowler's top experimental leaders, "it was hard to imagine running more than one mirror."

Fowler was talking quite frequently with Richard Post about the need for this new invention, looking for an idea. They lived only about half a mile apart, in the town of Walnut Creek, and they liked to carpool together in Post's old Volkswagen Bug. One afternoon in early July, they were talking about electric effects and Kelley's idea. Then, as Post put it, "a light bulb turned on in Ken's mind: Why not go the Kelley mode one step further?" Fowler was not put off by the complexities of running several mirrors in tandem. He was thinking that it might be possible to make the space charge help them, not hurt them. The way to do that would be to create even more space charge in the end-cells, the two outboard mirrors, than in the central mirror. It would be like turning a tall hill into a valley by raising up taller hills on either side of the central hill.

"Ken had the idea; I was ten steps behind him," Post continued. He raised a variety of technical objections. But then it began to dawn on him how this new idea could be made to work. For Fowler, the idea really came together a day or two later, on July 4, the day of the bicentennial: "I was lying on my belly on my living-room floor, and my kids wanted to go see the fireworks. I was calculating away, and the numbers came together that night, before we went off to the fireworks. I could see there might just be a way to do that." He wrote up some notes, talked some more about it with Post, and then went off on a vacation to his home in Georgia.

Meanwhile, Grant Logan was also thinking about Q enhancement, about the need for a new invention. He had left 2XIIB and was working on new mirror designs, a job that suited him well: "Ideas pop into my head all the time," he said. His friends, still teasing him about how it had been on 2XIIB, now kidded him that his promotion had come because he had made so many goofs in the lab that he was too dangerous to be let back in. Logan was trying to learn more about plasma leakage. He made some calculations to study a fairly standard problem in plasma theory, the escape of plasma through a nozzle as if it were rocket exhaust. His calculations showed that the escaping plasma would set up a space charge that dropped off in a way that sucked the plasma out, making the flow go faster. "That was bad. And since it was bad because it was a drop, I said, 'Hey, it'll be good if it's a rise. Now how do I make a rise?' " He thought of plugging the nozzle with a mirror like 2XIIB, which he already knew about. It produced plenty of space charge, which by his calculations was just what he needed. Soon he was merrily pursuing the idea of a tandem mirror.

He walked on down the hall to talk about it with Post, and when Fowler got back from vacation he found that Logan, working quite independently, had come up with nearly the same approach as his own. At first Logan was a little disappointed that his ideas weren't unique, but at the time they still thought it was entirely their own invention. Soon afterward, Logan received a thick document in the mail.

The International Atomic Energy Agency would soon be hosting its biennial fusion conference, just as it had sponsored the 1968 Novosibirsk conference, at which Lev Artsimovich had announced his tokamak breakthrough. This 1976 meeting would be in Berchtesgaden, Germany, once the location of Adolf Hitler's alpine retreat but now a resort. The IAEA organizers had assembled a volume of abstracts, short summaries of the papers to be presented, and had mailed copies to all the attendees. One abstract was by G. I. Dimov of the Soviet Union. He was proposing a tandem mirror, too. Reading this, Logan knew his idea had to be right. The tandem mirror had been invented entirely independently, by three different physicists in two different labs, at nearly the same time.

They went ahead and held their workship on Q enhancement. Scientists came not only from within the United States but also from Japan and the Soviet Union. Dimov was not there; the Soviets had followed their familiar practice of sending to the United States a man less knowledgeable but more politically reliable. Fowler had been planning to present several ideas, the key one being Coensgen's concept for what he called a field-reversed mirror. This would bend the magnetic field back on itself, sealing up the leaky ends of the mirror entirely. But tandem mirrors stole the show. Then Fowler told Logan to nurture the idea, bring it to the point of being a design for a new device they could actually build. Coensgen began to work closely with Logan on the design of what soon would be called the Tandem Mirror Experiment, or TMX.

In Kintner's Washington office, Stephen Dean, whom Kintner had inherited, was following matters closely. He flew out to the West Coast and met with Fowler and Coensgen. Coensgen urged them to shut down Baseball II, which was still going forward as a program, and to transfer its people and facilities to TMX. As always, one of the issues was that if the program was terminated, would they lose the money that supported it? Being a good manager, before agreeing to shut down Baseball II, Fowler got Dean to agree that its funding indeed would go to TMX.

Coensgen immediately started putting together a staff of engineers and physicists to prepare a formal proposal for TMX. By the time they got to Berchtesgaden in October, they had determined most of its essential features. At that meeting, Kintner himself said that it was quite probable that TMX would be built. This was rather a bold thing to say, because TMX represented an entirely new departure for the mirror program. Yet it had not even been completed as a paper proposal, let alone been reviewed and critiqued. Still, as Fowler said, "The Russians had been calculating, we had been calculating, and we all got the same answers. So we had to be on to something."

They zipped the proposal together in near-record time; it was ready by January 1977. It laid everything out: the design, the computer simulations of plasma behavior, the performance goals. Coensgen took it to Washington and then sat on the hot seat as people from Kintner's staff questioned him closely. Like every other holder of the Ph.D. degree,

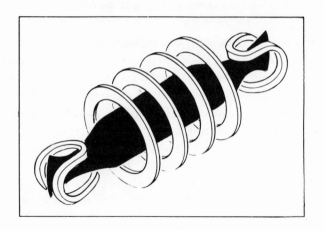

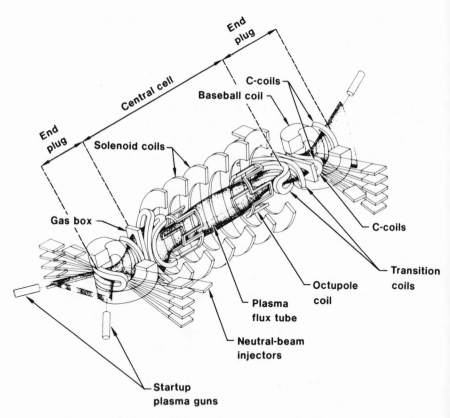

FIGURE 5. *The tandem mirror, which overcame the problem of space charge. Top, its general approach: a tube formed from ring-shaped magnets, closed at the ends with baseball-type magnetic mirrors. Dark shading is plasma. Bottom, a simplified diagram of the magnets and neutral beams used in the Tandem Mirror Experiment, or TMX.*

Coensgen had once written a dissertation and then defended it in an oral examination by his professors. However, he told me, "this was much more difficult. When I defended my thesis, I thought it was the toughest exam I'd ever have. It turned out to be one of the easiest." The proposal was formally reviewed in February and approved for funding in March. Within days, workmen were dismantling Baseball II and its equipment, then digging a deep pit in the floor of its building to prepare a construction site. In Fowler's words, "We went from a basic idea to a hole in the ground in nine months flat."

Building it took eighteen months, which was also a record for so large a system. Leading the charge was Charles Damm, who had formerly been the manager of Baseball II. His project had been plowed under to make room for TMX, but he was an enthusiastic supporter of the new project. He had reason to be. TMX would use not one but two of his baseball magnets, one in each end-cell.

By October 1978 TMX was complete. Then came the usual nine months of Coensgen's "commissioning," marked by their share of glitches. For instance, the scientists had a total of 24 neutral beams to ride herd on. As on 2XIIB, which had had 12, there was the problem of keeping one faulty beam from perturbing the operation of the others. That was solved by arranging to control the beams, not with electric cabling, but with a sophisticated optical system in which control signals would be carried as pulses of light traveling within light-pipes. Then some idiot drove a cart loaded with heavy batteries over the light-pipes, crushing them. However, they were soon replaced. Despite the rapidity of its design and construction, TMX quickly began turning in outstanding performance. As Logan described it, "We had a group of people with a good can-do attitude, and we made things happen, and fast. And it worked."

They had set out predictions in the proposal, stating in advance what its performance would be, as measured by certain parameters such as plasma density and plasma confinement time. They made these predictions from scratch, for an idea that had never been tested and that in a way was radically different, because for the first time they were calling for three mirrors to work together. Within a year after turning TMX on, they actually got close to achieving their predictions—in Logan's words, "if you don't insist on factors of two. I mean that in our business, to call the shots to within a factor of ten is more than some people can expect. To predict energies of two hundred electron volts and get a hundred fifty routinely, two hundred on occasional shots: I think that comes pretty close, given the uncertainties and that you just had to literally do the calculations by the seat of your pants." This meant that they had very good understanding of TMX. Rather than present them with unpleasant surprises, it was confirming their theories.

For the TMX experimenters, it all came together in about a week, during July 1979. At first they ran shots only for a millisecond, then gradually lengthened them to 25 milliseconds, which was as long as they could achieve with their equipment. Tom Simonen, who was leading the tests, appreciated that the way to see if tandem mirrors really worked would be to shut off one of the end-cells and check whether the leakage out that end would increase markedly. Late one afternoon, he said, "we took the shot, and turned one end-cell off early, during the 25 milliseconds. Immediately after the shot the digital recorder displayed the information, and we saw, hey, the leakage went up. Everybody got excited, but we were a little restrained because there could have been lots of other things going on; maybe we were being fooled. We weren't really sure yet. So we tried a shot where we did the opposite, turned the other end off and left the opposite end on. Sure enough, now the leakage increased out the other end. Now this was another exciting moment, and now our confidence was higher, that what we thought was happening really was happening." In time, Simonen and his people would make improvements to the point where the end-cells confined the plasma nine times better than if they had not been there.

Already the Washington political climate was turning out to be very favorable to mirrors. President Carter had upgraded energy programs to the cabinet-level Department of Energy, and his Energy Secretary, James Schlesinger, was determined to steer fusion onto a new course. His eventual policy in 1978 called for delaying a commitment to any major new tokamak beyond TFTR, while letting the mirrors catch up. This meant that as with 2XIIB, the success of TMX had put Fowler in a very strong position to go ahead with still another mirror project, bigger than any to date. This project would answer the pesky question, which in 1978 was still unanswered, of what to do with the big yin-yang of MFTF.

MFTF had been designed and approved before anyone had ever thought of inventing the tandem. It was based on the idea of building the largest superconducting yin-yang they could reasonably come up with, without worrying overly much about whether it could be made into a reactor. But when TMX came along, it became natural to think of a much larger tandem mirror, whose end-cells would use yin-yang magnets the size of MFTF. The original MFTF would plug one end, and a second one would plug the other. Fowler and his people called this follow-on project MFTF-B, but it was as much a trap as an opportunity.

If TMX was the size of a fifty-foot railway tank car, a tandem MFTF-B would be as large as the fuselage of a Boeing 747. It would cost $350 million, including operating expenses and the cost of the original MFTF. It would fulfill the promise the original MFTF had left unfulfilled, of being the mirror program's counterpart of TFTR. It would say to the world that mirrors had advanced to become heavy hitters like the tokamaks, and were able to play in the same league. But to make such a statement, the tandem arrangement they chose would have to be good. And when people studied reactor designs based on TMX, these designs fell short.

Such reactors had a projected Q of about 5, a vast improvement over the 1.2 of reactor concepts based on the simple yin-yang of MFTF, but still not of tokamak class. And even to get $Q = 5$ would call for much better suppression of plasma leakage. A reactor based on the TMX would need exceptionally powerful neutral beams, which would use up a lot of the power this reactor would generate. Such a reactor also would need very strong magnets, whose magnetic fields in fact would be too strong to be provided by superconductors. These magnets, in turn, would demand additional power.

Nevertheless, the responsibility for facing these issues lay entirely with Fowler. Kintner was not about to push him by driving a hard bargain. He was not about to tell Fowler that his price for approving MFTF-B would be that Fowler had to go through still another round of Q enhancement, try once more to reinvent the magnetic mirror. On the contrary, in early 1979 Fowler knew that Kintner wanted very much to go ahead. The tandem-mirror version of MFTF-B was already designed, its approval ready to be triggered by the anticipated success of TMX. Still Fowler was dissatisfied: "We had ambitious goals, but we didn't have a design that would meet the goals. So we kept working on it." He knew that tokamaks were still way ahead, and if mirrors were Number Two, he regarded them as at best a very distant Number Two. His mirror people simply had to try harder: "We were not even like Avis, we were more like Budget Rent-A-Car or National." Finally, in January 1979, he called a meeting of his associates, reviewed their work on tandem reactor design concepts, then said bluntly, "It's time we conclude this is crap. We gotta do better."

The man who started the next round of invention was John Clauser, known to all as a maverick. One senior manager who knew him well described him as "a very bright guy who causes a lot of trouble. He gets quoted in *Scientific American,* he is very productive, but when you get him in a meeting, he tends to think that he is the only one who knows how to do anything right. He proceeds to attack everything in sight. He gets asked to these meetings because he has a lot to offer, but he can't point out what you're doing wrong. As far as he's concerned, if you're not doing it his way, it's the wrong way. So tempers heat up, the meetings get very volatile. He has a tendency to do everything himself, he wants to manage the whole nine yards, he wants to do it all and nobody can do it right. So he tends to be very hard on people. He works for Coensgen, and if I were in charge I wouldn't dream of getting rid of him—but he doesn't work for me." Clauser was in his mid-thirties. In lots of other organizations he might long since have been fired for being a troublemaker who didn't fit in, who wasn't showing proper attention to relationships with people. But at Livermore he did just fine.

Some time earlier Clauser had been working on 2XIIB, which was still operating. He was making measurements of the temperature of electrons. Now there was a result in plasma physics that Lyman Spitzer had shown many years earlier, and that was known to anyone who had studied the subject. This result stated that the electron temperature would always be the same along any line of force in a magnetic field, because the electrons could freely shuttle back and forth along that line.

Clauser wasn't out to challenge Spitzer's work; he just wanted to see what was really happening. To everyone's surprise, he got a much lower electron temperature on the outside of the hot plasma than on the inside. He couldn't be talked out of it. This result was striking; it stuck in people's minds. Then, while TMX was going forward, some of the theorists had gone back to look at his data and find an explanation. It turned out that what was making the difference was the warm-plasma stream that Coensgen had installed. This stream was tweaking the plasma within 2XIIB in such a way as to allow two regions to exist with electrons at different temperatures. Spitzer had not considered such tweaks in his theory.

Clauser's findings, and the theoretical explanation, had for some time been sitting in the back of Grant Logan's mind. Looking for new ideas, he thought of these results. Then he realized that Clauser's two electron regions could be the key to an entirely new type of tandem mirror, one that would offer much better performance and a much higher Q. This invention would avoid the problems of the existing type of tandem, which would call for power-hungry neutral beams and magnets; this invention instead would demand much less power to run it.

The key to it all would be to produce two regions in which the electrons would be at different temperatures, and to keep these regions separate. Logan thought he knew how to do it. He went off to talk about it with Fowler, who looked over his work and said, "That looks like you're trying to set up a thermal barrier." Then Logan talked about it with Dave Baldwin, one of Fowler's "dynamic duo" of theorists. (The original "dynamic trio" had been reduced since Herb Berk had gone to the University of Texas.) Baldwin added another essential idea, showing how to arrange a set of neutral beams to get a much better thermal barrier. And suddenly Fowler had his better tandem-mirror design. Eventually, thermal barriers would give projected values of 20 and more for the Q of a reactor.

Thus, during the winter of 1979, Fowler had several issues to face:

Plasma leakage. His experimentalists were still working with 2XIIB as well as with the new TMX tandem mirror, trying to learn more about these leaks, trying to reduce them to the minimum level permitted by the laws of physics.

Q enhancement. Provided this plasma leakage could be reduced to this minimum level, Logan's ideas were offering the hope that it would be possible to build a fusion reactor based on magnetic mirrors, with a high enough Q to stand as a real competitor to the tokamaks.

Thermal barriers. Logan's new idea looked very promising, but it was no more than a theory, and it had to be tested in an experiment.

MFTF-B. If thermal barriers were to work, and to prove sufficiently promising, then this enormous new project would stand in a new light. Certainly it would be a very bold extension of the original MFTF, with its magnet the size of a two-story house. But if it could be built with thermal barriers, then it would offer greatly improved performance in confining a plasma. It thus would be much more attractive to Kintner, more significant for the future of fusion.

Logan and Baldwin had come up with the thermal-barrier idea only a few weeks after Fowler had said that ordinary tandem mirrors wouldn't be good enough. Obviously, the next thing to do would be to test the idea in experiments with TMX. Tom Simonen, its experiment director, tried for months to find a way to do these tests without making major modifications to TMX, but finally decided he couldn't. This meant that TMX would have to be torn apart and rebuilt completely, with entirely new arrangements of magnets and neutral beams, in order to prove out the idea of thermal barriers. The rebuilt TMX would be called TMX–Upgrade, or TMX–U. It would cost $14 million.

That summer, the original TMX gained its success, as Simonen turned the end-cells off and on to show that they really did reduce the leakage. With that, Fowler and Kintner were prepared to push onto an especially bold course. They would go ahead with the $226-million MFTF-B, and they would base it on the unproven idea of thermal barriers. If Kintner had been following traditional conservative approaches, he would have demanded that this idea first be thoroughly tested on the smaller TMX-U before allowing it to go into the much costlier MFTF-B. But by now nobody wanted to hold back.

Why were they able to go ahead so boldly? Because they had been testing the thermal-barrier ideas using computers, and the tests were coming out right. The Livermore people had developed powerful theories and excellent computer programs to predict the behavior of plasmas and of the magnetic mirrors that would contain them. These theories and programs had shown their worth on both the old 2XIIB and the newer TMX. Thus, armed with their mathematics, Fowler was well prepared to raise the stakes of the game. He could not only propose to build TMX-U to test the thermal barriers; he could also propose to use them as the basis for MFTF-B, anticipating that the TMX-U experiments would be successful. He would go ahead with both machines, at the same time. But so bold an approach meant that he had little more than the last half of 1979 to put together complete proposals for both the $14-million TMX-U and the $226-million MFTF-B.

Twice before, in times of trouble, Fowler had been fortunate enough to have people from the Soviet Union come to his rescue. In 1974, Ioffe's photo of his oscilloscope traces had given him the clue to warm-plasma streaming. In 1976, Dimov's abstract had convinced Logan that tandem mirrors indeed were the key to Q enhancement. Now, late in 1979, Fowler arranged for still another Russian to come visit. He was Boris Kanaev, a member of Ioffe's group in Moscow.

"We were really struggling," said Fowler, "because we really wanted to build MFTF-B." Now, however, thermal barriers had raised the stakes. To get the most out of them, it would be essential to achieve better control over plasma leaks then they had ever gained before. Kanaev came with data from his Moscow experiments, which Fowler's people could compare with their own. This new data gave them what they wanted. In the course of making the thermal barriers, they would also be able to control the leaks. These two issues, plasma leaks and thermal barriers, were separate and distinct. Nevertheless, Kanaev's data helped them see that the methods to be used to address these issues fitted together so well that they would be able to deal with them both.

That was in the fall of 1979. As Fowler recalls, "We had to get our proposal out, and it was really sweat and tears around here. It was pouring rain, and people were working nights and weekends. We turned to, and turned that concept into a proposal in less than a month. The document was really thick. Lots of technical appendices and everything, written by a great fleet of people. We were proposing a project costing hundreds of millions, the MFTF-B, and you didn't do that on the back of an envelope." The critical element in it all, the suppression of plasma leaks, really came together within two months of the date when Fowler and his scientists had to stand up and say, "This is it."

Where they had to stand up was the technical review. The reviewers were to meet on Monday, January 28, 1980. On the previous Thursday, Fowler's people had a dry run for the review. The theorist Don Pearlstein put up a viewgraph, the drawing of a tandem-mirror design that would reduce plasma leaks to the minimum allowed by the laws of physics. Then he said, "This is the first time that we have ever had a completely stable mirror design." To Fowler, "it was so momentous, such an earth-shaking moment, that right then we had an earthquake. Magnitude five-point-five on the Richter scale."

A quarter-mile away, the world's largest laser was knocked off its supports. The quake spalled concrete off the large blocks lining the wall in the bay where the big MFTF magnet was being built. A welder was on top of its vessel with a welding machine, fifty feet above the floor, when his whole world started to shake. He was in a harness, and he thought he had been paralyzed, because he couldn't move out of the harness. Four days later, the review group came. There wasn't even heat in the building, but they met anyway, went over the design for MFTF-B, and approved it.

With this, it was time to get on with building its big magnets. These would be made by winding cabling around a central form, to produce what everyone called a magnetic coil. Back in 1978, Fowler had set up a coil-winding shop to build the two-story-tall yin-yang magnet for MFTF. It featured a coil-winding machine that amounted to a combination of a rotisserie and a lazy susan, but big enough to hold an elephant. This was actually the support for a radio telescope, able to swing a heavy load up and down while swiveling in any direction. Then a crew of four workmen proceeded to wind the magnets with superconducting cables or conductors. The conductor had a silvery core of niobium-titanium alloy, surrounded by a wrapping of copper, with channels in the copper to let in the liquid helium that would cool it all. Short lengths of this conductor looked like candy bars a half-inch across, with square cross-sections, but were much heavier. Longer lengths resembled steel reinforcing rods for concrete. It was quite stiff; you could not bend it by hand. The conductor came in mile-long batches wound on big wooden reels, fifteen feet across, like the ones that hold coils of telephone cable.

To wind the coils of a big yin-yang, they started by hoisting a heavy twenty-foot-wide horseshoe-shaped forging of stainless steel onto the coil winder. Then for the next several months, those four people would be winding away. The conductor was paid out from one direction, and pulled taut under tension. One man operated the big coil-winder, swinging the big forging up and down or rotating it amid the loud whine of the winder's motor, pulling out the conductor the way a fishing reel pulls fishline. Two other men helped lay the windings in place, working on a platform that rotated with the coil-winder, hitting the conductor with mallets to make it lie correctly; the fourth man was a quality-control inspector. "We wanted to offer it as a ride to the Great America amusement park," said one manager of that coil-winder, "only it was too slow." The conductor cost about $18 a foot. Each yin-yang, with its two huge horseshoe magnets, had over thirty miles of conductor, all wound in this way by hand.

Then Chicago Bridge and Iron took over. Their workmen took a year and a half to build a thick stainless-steel casing around each horseshoe magnet. The casings had to be up to five inches thick, because when running with current, each horseshoe would produce a magnetic force of 11,000 tons, acting to spread the horseshoe wide. Then these bridge-builders brought in four cranes, each capable of lifting ninety tons. Each completed horseshoe magnet weighed over 150 tons. They lowered one of these big horseshoes onto a sand pile, and using all four cranes, hoisted the other one and carefully placed it in position atop the first, using the cranes to maneuver it as well as to lift it. Then the complete yin-yang could be welded together. Next, it had to be moved to its final location, about 300 yards away. They dug sand away from one side of the pile, causing the yin-yang to tip over and roll. Then they pulled on it with a crane, and maneuvered it onto an iron sledge as wide as a two-lane highway. Its load was the biggest magnet ever assembled, over 25 feet across and the size of a house.

They picked a weekend to do the moving, so the sledge wouldn't block traffic on the lab's main street. To roll this most advanced development of modern technology, they relied on a technique the Pharaohs had used to move blocks for their pyramids—wooden rollers. These were logs, rolling beneath the sledge on plywood set on the roadway to make the rolling smoother and keep these logs from being pressed into the asphalt. One of the ninety-ton cranes pulled the sledge slowly along. This move was made in May 1981, taking the magnet to the entrance of Building 431, the enormous concrete hall that would someday house the entire MFTF-B. There other people worked on it further for several months, enclosing it in panels that would hold liquid nitrogen. Liquid nitrogen would not be nearly so cold as the liquid helium that would permeate the inner windings, but it would be much cheaper, and having it in an outer shell would make it much easier to conserve the costly liquid helium. The completed yin-yang, still resting on its sledge, then weighed nearly 400 tons.

At the beginning of September, the sledge and its load were at one end of Building 431. At the other end, mounted on steel supports, was the 36-foot-wide cylindrical steel vessel that was to house it. Between the two was a deep pit fifty feet across, part of the excavations for the MFTF-B project. The dirt from the pit had been piled high half a block up the street. Someone later put a sign at the top: MT. KARPENKO, for the project leader. The immediate problem facing the magnet-movers was how to get their load across Karpenko Chasm. They built a bridge across it, using steel I-beams three feet thick. They pulled the sledge across the bridge and into the vessel at the other end, then welded the yin-yang into place with support rods. Finally they knocked the sledge apart underneath it, disassembling it. They had moved and installed their 375-ton load without once being required to lift it with cranes.

"This was shortly after Princeton dropped their load," said Ted Kozman, who was in charge of these magnets. "So there was a lot of consternation as the magnet made the move, a lot of people worrying about the magnet falling from its bridge into the pit and destroying the project. This was the only one of its kind; it had taken us four years to build it. The pit was forty feet deep, and the speculation was that little damage would be done to it if it fell—but then how would you get it out of there? Plus, there'd have been safety investigations like there were at Princeton. We'd have gone on for years and years, analyzing what went wrong, could we avoid doing it again. We'd have been analyzing rather than building." But when they made the move, everything went smoothly.

Five months later came the main test of the magnet—its power supply, its liquid helium pumps and valves, the vacuum equipment, the computer control system. Kozman and his engineers took two weeks to cool it down, slowly bringing all 375 tons to liquid helium temperature, −451.6 °F. Amid this frigid cold, the niobium-titanium could carry thousands of amperes of current while requiring almost no power, and the field was as strong as anyone wished. Kozman climbed on top of the magnet vessel and walked around carrying a steel flashlight. The field there was only about one-fortieth as strong as in the center of the yin-yang, where the plasma would be. Still he had to strain his wrist to flick the light from side to side. The flashlight was acting like a compass needle, aligned with the magnetic field.

Then it was time to warm the magnet up. They couldn't simply blow warm air at it; the air itself would have frozen solid in a thick layer around it. Indeed, anything other than helium gas would have frozen solid. So they had to proceed gingerly, circulating cold helium through the magnet's cooling channels, then gradually warming the helium. Earlier, while cooling it down, someone had made a mistake and turned the wrong valves. Helium gas had flowed into a big blimplike rubber bag on the roof of Building 431. Then more helium had flowed in, till early in the evening, it popped with a loud bang. In a moment, $10,000 worth of helium gas was blown away in the winds. Fortunately, nothing like this happened while they were warming the magnet, although the warming took longer than they had planned. After two weeks, they finally got the magnet up to about the temperature of an iceberg. That was when they could open its steel vessel and let it warm the rest of the way in the air. But when people looked in the bottom of the vessel, they saw puddles of water. Moisture in the air had condensed on the ice-warm metal surfaces, and trickled down to make these puddles.

One of Kozman's friends drew a cartoon about all this, showing the big yin-yang encased in an iceberg, floating in the water. A yellow puddle on the iceberg was inspired by a rumor, entirely untrue, that someone had tried to speed up the warming by urinating on the magnet. A fish looks on with love in his heart, waiting for the magnet to thaw; a penguin sitting on a floating ice slab wonders what is going on; a shark fin protrudes above the water nearby. Atop a seawall, "Captain Koz," dressed as a South American dictator and carrying a swagger stick, says, "Okay, troops, let's warm the mother up!" Nearby, a blimp marked BADYEAR blows a hole in its gasbag and collapses limply. "Dr. L. Ponderous Valium" (Larry Valbe, a project engineer), dressed in thick fur coat and earmuffs, his nose red from the cold, replies, "That didn't help much, Ted."

Construction of MFTF-B will be going forward during the next several years, and the main components should be complete by September 1985. The rest will be finished within a few months. Then, in Fred Coensgen's words, "We'll bring it all up together." Magnets, neutral beams, microwave heaters for the plasma—all should be finished as part of the whole system. Along the way, other people will be running mirror experiments to prove out many of the key ideas. Already, in July 1982, the completed TMX-U gave encouraging evidence that the ideas of thermal barriers were correct. These tests were continuing during 1983, with good success. Then, in 1984, at MIT, a small tandem mirror called TARA is to test the arrangement of end-cell magnets known as an "axicell," which is being built on a much larger scale for MFTF-B. When the Livermore people start working with MFTF-B, they will not be flying blind.

They will be running shots on MFTF-B as often as once every five minutes, with each shot going for as long as thirty seconds, in comparison with the one- and two-second-long shots on Princeton's TFTR. On MFTF-B the confinement parameter, $n\tau$, is planned to be a tokamak-level 5×10^{13}, which if achieved will match that of TFTR. But unlike Princeton, at Livermore they don't have to fool around with motor-generators. They have something better: the Pacific Intertie.

The country near Livermore is flat, but a mile or two to the east it rises in a range of rounded brown hills, undulating toward the Altamont Pass. Along the ridge, amid this stark rangeland, run steel towers carrying powerlines. This is the intertie that links together the great power projects of the West—dams like Hoover, Bonneville, Grand Coulee; nuclear plants like San Onofre and Rancho Seco; coal-fired ones like Four Corners. Linked through this intertie, the major utilities can sell power to distant regions if they are producing a surplus, or import needed power if some generators are out of service. Behind Building 431 is a substation or electrical switchyard, from which powerlines run on steel towers, marching across the nearby low hills, to connect to this intertie. When operating, MFTF-B will draw up to 250 megawatts of power. But the Pacific Intertie carries 2,000 megawatts, sometimes more. When running their shots, the MFTF-B operators will not be dimming the lights of San Francisco, or Spokane.

These operators, of which there will be only four or five, will be spending their days (and nights) sitting in comfortable swivel chairs, each one in front of an array of color TV screens. Their computer displays have a most remarkable resemblance to the ones in the cockpit of the Pan Am spaceliner in the movie *2001: A Space Odyssey*. When *2001* was filmed in

1968, computer technology was not adequate to produce those displays; they were drawn on plastic sheets by animation artists. But the MFTF-B computer control arrangements will bring *2001* to life. Six large screens, directly in front of each operator, will present color diagrams of parts of MFTF-B such as a magnet assembly, or display other diagrams indicating what controls have been turned off or on. To work with any of the systems, you begin by identifying yourself, inserting a coded badge into a badge reader. Then on one of two smaller screens, you call up a diagram of the system you will work with. To start a pump, for instance, you touch the symbol representing that pump on the diagram. The screen responds to your touch, and the pump symbol takes on a double outline, indicating it has been selected. Also squares marked "Start" and "Stop" on the screen, which up to now had a dotted outline, take on a solid outline, indicating they can now be pushed. All other pushbuttons on the screen change to a dotted outline, indicating they are no longer active. The computer keeps a log of who is operating it, what buttons they push, and what happens as a result. Thus if anything goes wrong, it is easy to find out who did it, and how.

For their part the physicists, operating the diagnostics and taking data, will have their own computer terminals. Only a half-dozen or so people will suffice for that, each controlling several diagnostics, and the camaraderie stands to be just as strong as on 2XIIB. As always, they will be running sequences where the "shot leader," the experiment director, will be changing dial settings from shot to shot. Or they may take the same shot, over and over, to see how repeatable it is. After each shot they will be calling up some of the data, arguing back and forth over what it means. They may split up and go off to dinner, particularly if there is some dead time while something is being fixed. If the evening is busy, though, the most anyone will do will be to dash off for a bite, or send someone to get sandwiches. But in other respects it will be quite different from 2XIIB.

"Things have changed," said Grant Logan. "It used to be ten guys sitting around a plotter. Now there's this giant computer over there that sucks in eighty data channels at once, and you never see it at all. The next day people call up their data, but then it's more individualized, more of people individually analyzing their data. They'll still be arguing back and forth, but not so much. It's a different way of doing it, when each guy has his own computer terminal. Often they'll run through a series of shots without even looking at the data. They'll just go through, twist the knobs and punch the shot button. The planning will be different. The skull sessions will take place the next day when they get together, not right while they'll be taking data. So there won't be the emotions, the immediacy of making changes just as you go along. This business of going through a shot plan and going over the data later—it's different. Oh, there'll be a few extra scopes down there, and anybody who's interested will be able to look at what's happening at the moment, if they take the trouble. But we can't go back to the plotter, do things the way we did them in the old days. It would take forever to get all this information and process it, the eighty data channels."

And what will the fusion program be gaining, if MFTF-B succeeds? Why go to all the trouble in the first place, since tokamaks have been working so well? The answer is that if the mirrors succeed, they may offer reactors superior to tokamaks in several key respects.

Any reactor will have to have a great deal of equipment hung around it. It will be necessary to get at this equipment, and to get inside the vacuum vessel, which is surrounded by magnets, to replace parts and do maintenance. The doughnut-shaped tokamak will need to squeeze a lot of items into its central hole, where they will be crowded together and hard to get at. By contrast, a mirror like MFTF-B will have the basic shape of a cylinder—the simplest possible, and the most accessible, so far as allowing plenty of room for all its equipment is concerned.

A tokamak requires a powerful current driven through the plasma. The current is driven by an electrical transformer, and the transformer works by driving a steadily increasing voltage through a coil. But the voltage can go only so high. Then the tokamak must be shut down, the transformer set back to its starting point, and the whole arrangement restarted. This is inconvenient; repeatedly starting and stopping a large, complex system will produce all sorts of opportunities for something to go wrong. There are ways around this problem, in which the central plasma current will be driven by powerful microwaves. Thus, future tokamaks may well operate continuously. But the most straightforward way for a tokamak to operate is on-and-off; continuous operation calls for more complexity. By contrast, a mirror is inherently continuous, because it needs no central plasma current.

This lack of a central current helps in another important respect. When people speak of a steady-state or continuous tokamak, what they really mean is steady-state except for disruptions. In the disruptions, which are poorly understood, it is the central current that causes the plasma to writhe violently and hurl itself against the walls. Such disruptions will

surely have to be understood and controlled before there will be tokamaks anyone will want to buy for commercial use. But in mirrors, the
lack of a central current renders them inherently free of disruptions.
What is more, control of plasma impurities is an important issue in
tokamaks, since they can quench the plasma, keep it from getting hot
enough. But in mirrors, impurities can be made to escape through the
ends, through the holes in the magnetic fields. Mirrors thus again achieve
naturally what tokamaks will achieve only with difficulty.

The last advantage involves beta, the ratio of plasma pressure to
magnetic pressure, the restraining pressure of the magnetic fields. Beta
should be as high as possible, for it is a measure of how efficiently a fusion
system uses its magnetic fields, and magnets are expensive. If the plasma
pressure becomes too great for the allowed beta, the plasma will go
unstable by expanding outward, like air bursting a balloon. In tokamaks,
the twisting of the magnetic fields, which are bent into a doughnut shape,
amounts to straining these fields so they can contain only a modest
internal plasma pressure. Tokamak researchers would love to see a beta
of 10 percent, but no large tokamak has yet even reached 5 percent. By
contrast, there is no such strain on the fields in a mirror. The TMX got up
to beta values of 40 percent, and the same is expected of MFTF-B.

It is worthwhile keeping in mind the basic differences between mirrors
and tokamaks. A mirror like MFTF-B features a long tube lined with
magnets in the shape of hoops, and sealed at its ends with additional
arrays of magnets. This difference in shape leads to the mirror's advantages in ease of access, beta, and impurities. A mirror confines its plasma
with its magnets alone, but a tokamak must drive a powerful electric
current through its plasma and requires a great deal of special equipment
to drive this current for more than a short time. This, in turn, leads to the
mirror's advantages in steady-state operation, as well as in disruptions—
which are driven by that central current.

How could the mirrors lose, with all that? The reason tokamaks still are
in first place is that they continue to have the best advantage of all: they
hold their plasmas better. Confinement in magnetic bottles is still the
name of the game, and the standard of comparison is known as classical
confinement. Here the leakage comes only by a process of plasma
diffusion, which is not speeded or made more intense by waves within the
plasma. Thirty years ago, Lyman Spitzer was inventing the stellarator and
Richard Post was working on the earliest mirrors. They used the equations governing this diffusion process to predict the minimum rates of

leakage attainable, if the plasma were completely free of such waves and disturbances. This predicted confinement came to be called classical confinement, and was good enough to make controlled fusion appear to be worth pursuing. In a tokamak, however, classical confinement means that the plasmas slowly diffuse to the sides of the machines. This involves a far lower rate of leakage than does classical confinement in mirrors, where the leakage is out the ends. As Post has said, "We *gotta* get close to classical. Tokamaks don't." Tokamak builders can live quite happily with confinement hundreds of times worse than classical.

So the goals of MFTF-B will be more subtle than those of TFTR. TFTR is to produce a sizable amount of actual fusion power, using tritium. By contrast, MFTF-B is to study issues of plasma physics, with the goal of getting close to classical confinement. The MFTF-B physicists will have to prove out its thermal barriers, which separate a plasma into two distinct regions, each with its electrons at a different temperature. This is a key to demonstrating the promise of the mirrors, and to showing that they can serve to build a fusion reactor with a high value of Q, one which will not use up much of the power it produces merely to keep itself running.

It is quite reasonable to think of mirrors and tokamaks as running a race. The prize in this race is the next major project after TFTR and MFTF-B. That project will cost up to several billion dollars, and will be built either as a tokamak or as a mirror; but no one today knows what it will be. That is why Princeton and Livermore are running hard, battling to make their major projects succeed and meet their goals. We could say that Coensgen and Fowler are coming up fast on the outside track, while the Princeton managers, Dale Meade and Harold Furth, though still in the lead, are nevertheless beginning to hear footsteps. It will be some time yet before Fowler and Coensgen can take the lead, let alone ease up as the other runners falter. But the mirror program's goals are already about as advanced as they can be; there will be no latter-day Robert Hirsch to urge its leaders on to still more advanced goals. The pressure will continue to be on them. As Coensgen has said, "There is a number on the blackboard, nevertheless."

And beyond the questions posed by the thermal barriers, mirrors today may face an entirely new issue: plasma leakage out the sides. In a series of tests with TMX-Upgrade, during mid-1983, the thermal barriers worked fine; they sealed up the ends just as everyone had hoped. Fowler's experimenters had anticipated some side leakage, on the basis of predictions from theory. But to their dismay, it was up to ten times more rapid than they had expected. Thus, this today stands as a new problem on the road to success with MFTF-B. When I asked Fowler about it, early in November, he kept his cool: "We're not sure that it's serious. It could be gone by the end of next month." Still, even if it should prove serious this would be only the latest in a long series of problems, each of which at one time loomed as similarly intractable. And if MFTF-B gets into trouble over this issue, presumably Fowler will have yet another Russian to come and help him out.

5

Lasers and Microspheres

T HERE is a restaurant in Ann Arbor, Michigan, called the Rubaiyat. Before it moved to its present location, there was a suite of offices over it, and in the early and middle 1960s, one of the tenants was Keeve M. Siegel, known to his friends as Kip. He was president of a firm called Conductron, headquartered in those offices. "When we'd work late, food would be sent up to us, coffeepots," recalled his secretary. "They used to have big buffet lunches. The aroma of the food used to drift up into the offices. It smelled so good: soups, onions, things like that."

Siegel had pioneered in the development of side-looking radar, which is used to produce high-quality maps by means of large radar antennas mounted on the sides of aircraft. He had founded Conductron to build these radars for the Air Force, and during the sixties he built the firm up to a level of over $50 million a year in sales. In 1965 Conductron moved to a new suite of offices in the north of town; Siegel, a lover of good food, was the last to move. Then about 1967 he sold his firm to McDonnell Aircraft, the aerospace firm. He wanted James McDonnell, the chairman, to help Conductron expand, but McDonnell was completely tied up in the arrangements he was making to acquire Douglas Aircraft and to form the larger firm of McDonnell Douglas. So Siegel resigned as president of what had become McDonnell's Conductron Division. Having sold his company, he took his money and proceeded to buy several other companies. Soon he was heading his own mini-conglomerate, KMS Industries, named for his initials. He set up a research center in El Segundo, California, and brought in Keith A. Brueckner to lead it.

Brueckner was a physics professor from the University of California in San Diego, and had been a frequent consultant and advisor on classified research programs for the military. He had also done similar work for the fusion program.

In 1969, Brueckner came up with what to him was an entirely new approach to fusion. He decided that it should be possible to start with a small pellet of fusion fuel, zap it with an intense burst of light from a laser, and thus compress and heat it so strongly that it would explode, producing fusion energy. KMS filed patent applications, and Brueckner convinced Siegel that these ideas were so important that KMS should put money into their development. Siegel's enterprises were producing sales of $50 million a year or more, so some money was there. He set up a new outfit in Ann Arbor, KMS Fusion, to pursue Brueckner's ideas. Brueckner's calculations promised early success. His goal was breakeven, a situation in which a pellet would explode like a microscopic hydrogen bomb, producing the same amount of fusion energy as was in the laser pulse that zapped it. Indeed, Siegel was convinced he could get breakeven in 1975. Their first goal was to produce appropriate pellets, and by the end of 1972 they had that problem licked.

Brueckner had the idea that he could use microscopic glass spheres for his pellets, microballoons a fraction of a millimeter across. These are made by commercial manufacturers such as the 3M Company, a billion or more to the pound, for use in reflective paints or as fillers in polyester resins. His idea was to find a way to cut a sphere open, fill it with a mixture of deuterium-tritium fusion fuels at high pressure, then seal it shut. Working for him was a physicist, David Solomon, whose job it was to find that way. However, early in June 1972, Solomon conceived a very different way to solve the problem of filling a microsphere.

He was sitting with his feet up on his desk, thinking about how there really was no way to weld a microsphere shut and still have it possess high enough strength or be sufficiently smooth on its surface. But in earlier work at Bendix, Solomon had had a continuing problem with hydrogen gas permeating deeply into glass, so deeply he could never get all the hydrogen out. It occurred to him that perhaps he could fill a microballoon by getting hydrogen atoms to permeate or diffuse their way through the thin glass wall. The main problem was that he wanted to hold hydrogen at hundreds of atmospheres pressure. The glass would then have to be nearly as strong as steel.

He turned loose his team of engineers, and sent them out to buy hundred-pound bags of these spheres. He knew the spheres would be far from identical. They were made by a process that guaranteed large differences from one sphere to the next, in the strength of a sphere and in the quality of its glass. Thus, he could set out to find a few good ones out of hundreds of billions. "I felt that somewhere in those billions of microballoons there must be good spheres, and perhaps a few nearly perfect ones," he said. "We devised a way to crush the bad ones, and wound up with only some thousands of survivors." His engineers then put some of these fittest ones into a chemist's pressure vessel, a small thick-walled tank resembling a pressure cooker. They filled the vessel with high-pressure hydrogen, and heated it to about three hundred degrees. As Solomon recalls, "We left it for a couple of weeks. When we took the lid off, it sounded like popcorn popping." The pressure was bursting many of the spheres. "We sprinkled them onto a dish filled with water and found that many of them sank. Those were the ones that had fractured. But there were a few survivors."

Those that had stayed intact might contain hydrogen, but the question was, how much? The microspheres were tiny enough to fit into a hypodermic needle. A chemist took a syringe with a plunger that ran down the center, and sealed the end of the needle. He loaded a single sphere, stuck it under water, then pushed the plunger forward to break its glass shell. Solomon watched the bubble of gas come out of the syringe. The bubble was much larger than the original hollow glass microsphere, so a simple calculation told how much gas was in it. Tests then confirmed that the gas was indeed hydrogen. By November 1972, Solomon was able to report these results to Siegel, who appreciated their significance. Siegel expected that by means of appropriate industrial processes, he could manufacture strong, perfect spheres in very large numbers. Now, Solomon's work meant he could fill these spheres with fusion fuel, too.

Siegel wanted to put most of his efforts into developing such pellets, because they would be a commercial product he could manufacture and sell, once laser fusion became a going concern. Still, he needed a powerful laser to test his pellets, so he arranged to buy one from a French affiliate of General Electric. His pellets would be small enough to rattle around on a pinhead like a marble on a coffee cup's saucer. His laser would emit a pulse with as much energy as in a one-pound weight hitting a wall at 150 miles per hour. All that energy was to be focused onto the microsphere, in less than a billionth of a second. Brueckner had calculations showing that would suffice for breakeven.

Early in May 1974, they got their first neutrons from a laser-zapped pellet. These neutrons were proof that they were getting fusion reactions within the pellet. There were only a few thousand such neutrons, which wasn't very many. A few years earlier, a Los Alamos scientist had remarked that he could produce more neutrons by rubbing a stick containing deuterium against the seat of his pants. Still, KMS was the first to produce such neutrons from a pellet microexplosion. By 1974 laser fusion was looking very interesting to the Atomic Energy Commission, and large programs were under way, in both the United States and the Soviet Union. But KMS' success swept that firm into the improbable position of world leader. Siegel's Washington lobbyist went around town leaving behind samples of microspheres so small that ten million would fit into a medicine vial.

However, all this research was costing money. Between 1970 and 1975, Siegel put some $20 to $25 million of KMS money into laser fusion research. Yet none of this was coming back; their laser fusion work was producing no salable product. He had started during a bull market on Wall Street, and early on he had made money in the stock market. He got a loan of several million dollars from Detroit's Bank of the Commonwealth, and another loan from New York, but the Detroit bank called the loan in early, demanding repayment. Siegel's only recourse was to sell off operating divisions of his conglomerate. This was obviously not a course that would carry them far. Moreover, it brought the likelihood of delays in getting needed money, when a sale would get hung up for a while on points that had not been resolved. Siegel began to have trouble meeting his payroll. By 1974 he had sold off the most profitable parts of his empire. Creditors were beginning to get anxious as bills went unpaid, and it became clear that somehow Siegel had to find a new souce of money.

The obvious place to look was the U.S. government. As early as January 1973, just after they had proven the basic methods for making fusion microspheres, they had had a visit from the Atomic Energy Commission. Robert Hirsch himself had headed that group, along with Sol Buchsbaum of Bell Labs, one of his advisors. They had come away tremendously impressed with KMS' achievements. Still, that did not mean KMS had its financial support assured. For one thing, Siegel was still expecting he'd have breakeven in 1975. He wanted the sole glory and profit of inventing laser fusion as a source of power, with no need to share anything with the federal government; all he wanted to owe them was his taxes. However, in addition there was the very real question of whether he could get government money at all. There were a lot of people high up in Washington who felt that Brueckner had no clear right to his laser-

1. Robert Hirsch

2. Ken Fowler

3. Fred Coensgen

4. Grant Logan

18. LEFT: Lowering a TFTR neutr
beam box into its work area

19. BELOW: Beginning the final asse
bly of the TFTR. The large steel ring
a toroidal-field magnet. Left to rigl
Harold Furth, Don Grove, Paul Rea
don, and an official from the Depa
ment of Energy whom Reardon w
about to ask for more money

fusion patent applications, which had got KMS Fusion started in the first place.

Quite simply, these people felt that Brueckner's work properly should belong to the government. Brueckner had been a consultant to the fusion program, an advisor to the Department of Defense and the Atomic Energy Commission, as well as a frequent visitor to Livermore and Los Alamos, the two labs most strongly involved in pursuing laser fusion. In particular, people at Livermore had been developing ideas of laser fusion as far back as the early 1960s. That work had been highly classified, but Brueckner had been aware of at least part of it. Indeed, as early as 1962, the director of Livermore had told Brueckner of his interest in powerful lasers, and had written to him, "I would appreciate your bringing to the attention of the appropriate people our interest in the development of a high-power laser." Brueckner's Washington critics took the attitude that he was entirely smart enough to figure out why Livermore wanted such a laser.

Their attitude was that Siegel and Brueckner were stealing the crown jewels from the Atomic Energy Commission. Senior officials of the AEC wanted to know how many of Brueckner's alleged innovations could be traced to the privileged information he had received from their laboratories. Within Congress, Chet Holifield, chairman of the powerful Joint Committee on Atomic Energy, wrote, "I and the other members of the Joint Committee have supported and obtained the authorization of hundreds of millions of dollars for controlled fusion research over the years. It is, at the very least, distressing to contemplate the entire discipline being put in a position of economic disadvantage relative to an individual or group whose main source of information has been from research funded by the Government."

Had Brueckner in fact stolen their crown jewels? In fact, while he had indeed benefited from his extensive contacts with the government labs, his most important innovation was his own invention. That was a way to make a pellet that in his judgment stood a good chance of reaching breakeven with the kind of laser that could be built in 1969. Livermore's earlier pellet concepts had called for lasers of much higher power, which would not be available for years to come. Brueckner's pellet design thus seemed to be a real advance.

He didn't know there was a key error in the computer programs that he was using to study his pellets' performance. Moreover, his competitors at Livermore were using their own computer programs to criticize these designs and to undercut Brueckner's position. In fact, the KMS managers had agreed that this be done. They had signed an agreement with the AEC whereby AEC scientists would review the KMS pellet designs. It just happened that the scientists who were reviewing these pellet concepts were all at Livermore.

These designs were studied with the help of bomb codes, powerful computer programs used in the analysis of hydrogen-bomb designs. By 1973 Livermore had concluded that with Siegel's laser, Brueckner's pellets would fall short of breakeven by a factor of a thousand or more. These calculations were well known to the AEC officials from whom Siegel was seeking funding. Still, the KMS computer disagreed with the one from Livermore, and there was no way to be certain who was right. Then in 1974, the KMS scientists produced fusion neutrons by zapping their pellets. They were the first to obtain such neutrons, which was quite a coup; but the number of neutrons thus produced was far below what they had expected. That was in line with the Livermore predictions, and it put quite a different light on things. Siegel had hoped to do these experiments, win a big breakthrough, and have the AEC come to him on his terms. Instead, he had stirred up the AEC to the point where they were putting a lot of money into Livermore, which would soon be in a position to run far ahead of KMS. Thus, he had no choice but to try to deal with the AEC on its terms.

By late 1974, Siegel had nevertheless concluded that his only hope was to get a contract from the government. There simply was no other way to get money. In early 1975 he was spending some three days a week in Washington, working up to sixteen hours a day, meeting with anyone he thought could help. At the same time he was staving off his creditors while trying to meet his payroll. None of this was very good for his health. He had hypertension, and he was grossly overweight. He stood only about five feet eleven, yet weighed over three hundred pounds. He loved to visit good restaurants; in fact, he was a compulsive eater. Not for nothing had he located Conductron's offices on the floor above the Rubaiyat. He liked to start the day with a big breakfast, featuring sausage and eggs with breakfast rolls. At lunch he would munch away on deli sandwiches, usually while dictating letters to secretaries in his office. Then for dinner he liked a full spread, starting with an appetizer like shrimp and continuing with an entree like a double cut of prime rib with baked potato, topped off with cheesecake. Meanwhile, between meals he was pushing, pushing, pushing, desperately working to save his company and to keep alive the hope of succeeding with laser fusion.

It all came down to an afternoon in Washington in March, where he was to testify at hearings before the Joint Committee. He was to make a final appeal, to say that KMS must have a government contract or they would go under. One of his close associates was with him as the afternoon dragged on, his frustrations mounting as other witnesses, also appearing to testify, went on and on with their statements. Finally, much later in the day than he had planned, his turn came. He sat down behind a long table covered with green baize, with a microphone and a water pitcher in front of him, facing Holifield and a few other committee members. They were sitting solemnly on a dais behind an elegant wooden partition bearing their names—MR. HOLIFIELD, MR. PASTORE, and so forth—like so many judges at the bench. Siegel began to speak from his prepared statement. Suddenly he stopped. As his aide looked on with alarm, he half-rose out of his chair, clapped his hand to his forehead, and painfully forced out the word "s-t-r-o-k-e." Then he fell over the green table and collapsed. Someone called for an ambulance, but it was too late for him. He had indeed suffered a stroke, and died in the hospital a few hours later. He was fifty-three years old.

Yet his death saved KMS. He had taken out "key-man" insurance, which would pay large benefits to a corporation if its leader should suddenly die, and the insurance benefits came to more than a million dollars. As David Solomon recalled, "His final act, his death, assured our survival. We didn't know how we were going to make our payroll in March, let alone April. The insurance kept us alive for those months. Without it, we'd have gone bankrupt. Then on May 19 we got the letter of intent from the government, for our contract. We signed the contract almost at midnight on June 30, the last day of the fiscal year when the government could direct funds to us. Without it we would not have survived, and those of us who lived through that time have developed a pretty steel will." The contract was for from $8 to $10 million a year, and ran through 1981, when it was renewed. It ensured that KMS Fusion would indeed maintain itself as a leading center for research on pellets for laser fusion.

What was laser fusion that an experienced businessman like Kip Siegel would drive his company to the edge of bankruptcy for it? To begin with, it is useful to try to appreciate something of the really remarkable physics that goes into it. Much of this physics involves the lasers.

A laser will bear comparing to a charge of explosive. In this explosive, energy is stored within its molecules. With a shock from the detonator, the explosive suddenly releases its energy, which comes off as blast, smoke, noise, and a flash of fire. In a laser, energy also is stored within its molecules. The trigger which releases this energy is a flash of light of the appropriate wavelength, which is made to pass through the material of the laser. But when the laser releases its energy, it does not do so in blast and smoke. Rather, the energy comes out noiselessly, as an extremely bright, intense, and sudden beam of light.

How powerful is this beam? We are accustomed to thinking of energy in kilowatt hours, as on our electric bill. One could just as well speak of a kilowatt second, 1/3600 of a kilowatt-hour. A kilowatt second is the basic unit in which we measure the energy of a high-power laser. It corresponds approximately to the energy in a baseball as it flies off the bat of a slugger who has just hit a home run. Physicists refer to a kilowatt second by a different name: a kilojoule, after the nineteenth-century British physicist James Prescott Joule. Siegel's laser was designed to have an energy in the beam of a kilojoule, although in fact its performance fell far short of this goal. Modern lasers are rather larger, and we speak of lasers of 10 kilojoules, 100 kilojoules, and so forth.

How long does the laser flash? Here the basic unit of measurement is not so easy to visualize. It is the nanosecond, a billionth of a second. This is the time light takes to travel a foot. The speed of light is 186,000 miles per second; this is very nearly a foot per nanosecond. In our ordinary experience, a second is about the shortest length of time we can directly experience, and then only when we are in a footrace or trying to dodge an auto accident. A nanosecond is to a second as a second is to the time since Eisenhower was first elected President, over thirty years ago.

If a laser puts out a kilojoule of energy in a nanosecond of time, then during that nanosecond it is putting out very high power. Physicists have a word for it, from *watt* and the Greek for "monster": this level of power is called a *terawatt*. It is just about twice the total electricity-generating capacity of the United States, and equals a trillion watts. Thus, we could say that a kilojoule is a kilowatt for a second; but in laser land a second is not something people talk about. It's much too long; it is about as appropriate as measuring track and field records in centuries. We would rather say that a kilojoule is a terawatt for a nanosecond. Again, though, a terawatt is the small beer of the laser world. Its hope is for a power of a hundred terawatts—for a vanishingly brief fraction of a second, naturally. That is the power that would let the laser physicists do some really interesting experiments. The record to date is 26 terawatts, and in a few years they expect to go higher.

It is not hard to say a lot about these lasers; much information about them has been openly published, and the people who work with them are willing to talk more or less freely. The same is not true of the pellets. A pellet is a carefully designed charge of fusion fuel, and there are many ways to prepare it. Its purpose is to amplify energy. It receives a certain number of kilojoules of energy from the laser and then, if all goes well, it produces a much larger number of kilojoules from fusion reactions. The criteria for a successful laser fusion microexplosion (explosion of this pellet) are similar to the criteria for success in magnetic fusion. The fuel in the pellet must be heated to a hundred million Celsius degrees; the laser thus acts as an enormous match, which heats the fuel to its kindling temperature. Also, the pellet fuel must have the right confinement parameter, $n\tau$, the product of fuel density n and confinement time τ. In magnetic fusion, the confinement parameter must have the value 10^{14} or more. In laser fusion $n\tau$ should be 10^{15}, to make up for inefficiencies in the laser and pellet.

But there is a very basic difference between laser fusion and magnetic fusion. In magnetic fusion, we usually use plasmas with densities around 10^{14} particles per cubic centimeter. Hence, the required confinement time is close to one second. In laser fusion, there is tremendously violent motion within the tiny confines of a pellet, and the pellet begins to blow apart in ten trillionths of a second, or 10^{-11} second. Hence the fuel must be compressed to 10^{26} particles per cubic centimeter, a hundred times the density of water, or a thousand times the density of liquid deuterium and tritium.

How can laser light compress fusion fuel to such densities? The intense light pulse violently heats the outer layers of the pellet. These layers explode outward, blowing off at very high speeds. As they do this, they squeeze and compress the pellet interior. If you surround a sphere of some material with a thick shell of dynamite and set it off, the dynamite will compress the material quite strongly. That is what happens to the interior of a laser-fusion pellet. In fact, however, the compression achievable with lasers far exceeds what can be done with dynamite. This is easy to understand; if good compression could be had using explosives, no one would have ever gone to the trouble of trying to build large lasers.

The details of specific pellet designs, however, have been kept under strict secrecy. The reason is that they bear close relationships to the details of specific designs for the hydrogen bomb. For example, when Brueckner filed his patent applications, one of the things that most enraged the AEC officials was that Brueckner had made use of such basic concepts as the Teller-Ulam principle, the key to building a successful H-

Burn

Thermonuclear burn spreads rapidly through the compressed fuel, yielding many times the driver input energy

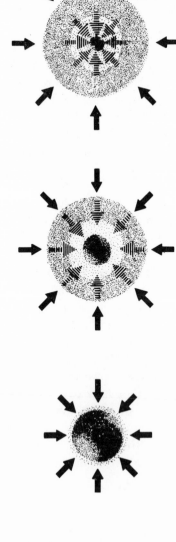

Inward transported thermal energy

Atmosphere Formation

Laser or particle beams rapidly heat the surface of the fusion target forming a surrounding plasma envelope.

Compression

Fuel is compressed by rocket-like blowoff of the surface material.

Ignition

With the final driver pulse, the fuel core reaches 1000–10,000 times liquid density and ignites at 100,000,000° C.

FIGURE 6.   How laser fusion works.

bomb. This was the AEC's secret of secrets. Even today, an understanding of pellet design gives valuable clues to H-bomb design, and only limited descriptions of pellets have been declassified.

Laser fusion has actually been an attempt to make practical use of the physics of the hydrogen bomb. Since the early 1950s it has been known that an atomic bomb serves as the fuse that sets off the much more powerful H-bomb explosion. However, a description of the actual method by which this is done was published only in 1979, in Howard Morland's celebrated article in *The Progressive*. When the A-bomb fuse explodes, it first gives off an extremely powerful pulse of heat, which appears as an intense burst of soft X rays, with wavelengths a thousand times shorter than those of ordinary light. Traveling at the speed of light, these X rays outrace the slower debris expanding from the A-bomb explosion, filling the H-bomb casing. That casing will very soon be blown apart, but it holds together for the fraction of a microsecond that it takes to set off the H-bomb. Within that casing is a carrot-shaped rod of fusion fuel, a mixture of deuterium and tritium chemically combined with lithium to form a solid. The X rays heat and compress the rod to the point where it ignites and begins to explode; the explosion then propagates into a larger mass of lithium deuteride. Neutrons from the fusion reactions transmute the lithium to tritium, and the explosion proceeds apace. As advocates of laser fusion like to point out, this is the only way that fusion has yet been made to produce energy this side of the sun and stars.

Laser fusion research started several years after the invention of the H-bomb, but before the invention of the laser in late 1960. As early as 1957, at Livermore, the physicist John Nuckolls was wondering just how small an H-bomb explosion could be, if it didn't use an A-bomb as a fuse. Edward Teller, co-inventor of the basic method for setting off an H-bomb, had founded Livermore in 1952 as a center for work on nuclear weapons, so Nuckolls' work fitted right in. By early 1960, he had come up with designs that would shrink the H-bomb down to the size of a child's marble.

He envisioned that someone would invent a means for producing a short, intense energy burst to be directed onto a pellet. The pellet would be a microsphere of fusion fuel, contained within a shell, and with a gap between shell and microsphere. This gap meant that the microsphere would be inside a cavity. When the energy burst entered the cavity, it would be trapped there; it would go in but it couldn't go out. Being trapped, the energy would turn to heat. The principle here was the same as that when sunlight streams through your window and enters your

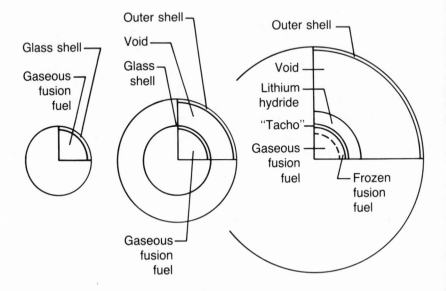

Glass shell

Gaseous fusion fuel

Outer shell

Void

Glass shell

Gaseous fusion fuel

Outer shell

Void

Lithium hydride

"Tacho"

Gaseous fusion fuel

Frozen fusion fuel

FIGURE 7. *Pellets for laser fusion. Left, a simple pellet consisting of a glass microsphere filled with fusion fuel. Center, a cavity-type pellet. Within the cavity, or void, the inner part is suspended by a membrane of thin plastic film. Right, a design such as might be used in advanced tests. The outer shell vaporizes in a preliminary laser pulse; the resulting vapor cloud absorbs the main pulse in the manner of a cavity. The thick shell of lithium hydride produces a much stronger and more prolonged blowoff than a simple glass shell, and therefore gives a much more powerful compression. "Tacho" is a plastic seeded with the metal tantalum; it guards the fuel against being heated too fast. Most of the fuel is frozen on the inner shell interior.*

room, it turns to heat. Because of the high power involved, the heat within Nuckolls' cavity, shot in by his powerful energy source, would take a special form: soft X rays, like those from an atomic fireball. The X rays would fill the cavity, compressing and heating the fusion fuel at the cavity center, very much as in the H-bomb. If everything went right, the pellet of fusion fuel would explode.

Nuckolls' approach was magnificent in that it left completely open the choice of the driver, the means for producing the energy pulse. Any form of pulsed energy, entering the cavity, would turn to heat and produce soft X rays. Hence any invention that could produce intense energy pulses would be a likely choice for the driver. The only requirement would be that it conform to Nuckolls' calculations as to how powerful it would have to be, how rapidly it would have to emit its pulse, and how its pulse would have to grow in intensity as the trillionths of a second went by. As he later said, "I worked out the power, the energy, the focusing that had to go with it, and I had all these parameters worked out at the time the laser was invented. It was as though I had fashioned a wagon and somebody marched a horse along."

Nuckolls had a colleague at Livermore, Ray Kidder, who was also interested in this research. A friend of a friend of Kidder's was Ted Maiman, the inventor of the laser. His invention came out late in 1960, and a few months later his associates showed how to produce powerful light pulses of short duration. Kidder made calculations and decided that the laser could be used to set off a fusion microexplosion. His calculations showed that the laser would need an energy of 100 kilojoules, to be shot into the pellet in ten nanoseconds. These requirements were far beyond the state of the laser-building art in 1961. The question in Kidder's mind was, could lasers grow to meet such a challenge? Early in the spring of 1962, he went down to Los Angeles and met with Maiman himself. Maiman told him there were no basic reasons why such powerful lasers couldn't eventually be built. With this assurance, Kidder was able to launch Livermore's high-energy laser program.

The work went forward at a desultory pace during the 1960s. There was too large a gap between the low-power lasers then available and the much more intense lasers that would be needed for fusion work. Moreover, all this work was under strict wraps of secrecy, because of its connection with the H-bomb. But in 1969 Nuckolls came up with a way to make effective use of the simplest possible pellet: a droplet of liquefied fusion fuel, such as might be produced by an eyedropper. The key was "pulse-shaping":

the careful control of the power or intensity of the laser pulse from one nanosecond to the next, with this power rising sharply to a peak.

He had been looking at liquid-droplet pellets and pulse-shaping as early as 1961, but hadn't gone very far. Not till the late 1960s did he have adequate computer programs. Moreover, he wanted to avoid the need for enclosing his droplet within a cavity. His 1969 solution called for surrounding the pellet with a thin shell, to be vaporized by a preliminary laser pulse. This would be followed by the main pulse, which would hit that surrounding cloud of vapor. The cloud then would help greatly in absorbing the main pulse, as well as in smoothing out its imperfections. What was better still, he was able to show how to achieve the best possible pulse shape for such a pellet, which would make it work with the best possible laser energy.

Nuckolls had a collaborator, Lowell Wood, whose calculations showed that this pellet might perform much better than earlier designs. Indeed, it might reach breakeven with a laser energy as low as one kilojoule, far less than the 100 kilojoules of Kidder's calculations. The difference was Nuckolls' particularly careful use of pulse-shaping; Kidder had not taken advantage of the opportunities it presented. Significantly, one kilojoule represented a laser energy that might be available even then, in 1969. This liquid-drop pellet thus could stand as the basis for a much larger program in laser fusion, which would move forward to exploit this new promise.

The liquid droplet offered good performance at low laser energy. At nearly the same time, Brueckner at KMS invented a different pellet, which also offered good performance but was more complex, featuring spherical shells nested inside one another. He too was highly impressed by its prospects, so much so that he went off to lead Kip Siegel into the laser-fusion business. Of course, no one could talk about the details; they were all classified. But Nuckolls' and Wood's simple liquid droplet didn't look like an H-bomb, and Nuckolls knew that the officials in charge of classified nuclear information might be willing to let it be talked about publicly. As Nuckolls recalls, "I realized around 1970 that I had the keys to getting something declassified." Declassification would move laser fusion out of the confined restrictions of a weapons lab, and into the broader world of energy programs, fusion research, and physics in general.

These developments were known in detail among only a relative handful of people, the "Q-cleared community," who held the highest security clearance. These were the physicists knowledgeable about the details of nuclear weapons and related matters. There were no more than a few hundred such people. Still, within this community, the pellet ideas of Nuckolls and Brueckner were stirring a lot of interest, and other physicists began trying to get involved. This led to the first public disclosure of laser fusion. It also led to two classic goofs, as laser builders, equipped with good enthusiasm but poor lasers and measuring instruments, set out to score an early coup.

At the University of Rochester in New York, the physicist Moshe Lubin wanted to develop a major lab for laser-fusion research, and he hoped for funding from the Atomic Energy Commission. However, because Lubin was a newcomer to this field, the AEC would not readily support his work. He felt that he could win AEC attention by building a very impressive laser and then letting people know how impressive it was. He measured a full kilojoule in his laser pulse, an astonishing performance, and proceeded to send telegrams throughout the Q-cleared community, announcing his achievement. He also wrote an article, "Fusion by Laser," which appeared in *Scientific American* in 1971, complete with photos and diagrams. That was the first chance anyone outside the Q-cleared community had to learn about these ideas.

All this activity certainly got Lubin the attention he wanted. There was just one problem: his laser couldn't be used in fusion experiments. To be useful, the laser beam would have had to be focused down to a spot smaller than a pinhole. Lubin's laser energy came out in a broad beam, like that of a flashlight, which meant it couldn't be focused. He later claimed that he had not been deliberately overselling the work of his laboratory, that the claims were the fault of the public-relations office at the University of Rochester. But his protestations didn't carry much weight, because he himself had sent out those telegrams.

Another group of enthusiasts was working at Sandia Laboratories, in Albuquerque. They had instruments to measure the characteristics of their laser pulse, and the resulting measurements looked very exciting. Apparently their laser was putting out a lot of energy in a very short pulse, which was just what they wanted. John Nuckolls became interested, had a batch of special pellets made up, and sent them down to Sandia to be tested. Nothing happened in the experiments. It turned out that their laser was indeed putting out a very short pulse—but that that pulse wasn't carrying much energy. The laser was putting out lots of energy, all right, but not in the short pulse. Most of it was in a long, lower-power pulse superimposed on the short pulse. The long pulse was several nanoseconds in duration, hopelessly long for laser-fusion work, yet still so short as to be difficult to detect with their instruments. Like Lubin, these Sandia people had been fooled, misled by their own high hopes.

Similarly, large lasers built by Kidder at Livermore and Brueckner at KMS Fusion also fell far short of their intended performance. Obviously, something more was needed than to build a big laser and start zapping pellets. It would be necessary to build vastly better and more powerful lasers, as well as to develop ultrarapid instruments to make the needed measurements. In addition, there was the problem of developing computer programs that could predict what would happen when a specific pellet was hit with a pulse from a specific laser. Such programs were necessary for studies of pellet design, by simulating the detailed physics of a laser pulse striking a pellet and causing it to compress, heat, and undergo fusion. During 1969, Nuckolls' work in this area received a powerful boost.

He had been joined by Lowell Wood, who had worked with him on the development of the simple liquid-droplet pellet. Wood was a protégé of Edward Teller, and would in time become one of the nation's leading laser experts. In 1982 Teller would go to the White House and give an hour-long talk on lasers to an audience that included President Reagan, Vice-President Bush, White House Chief of Staff James Baker, Presidential Counselor Ed Meese, the President's science advisor, and his national security advisor. That briefing would lead Reagan, a few months later, to give his "Star Wars" speech proposing laser weapons for defense against missiles. Many of the ideas in Teller's briefing came from Lowell Wood and his associates.

Wood had a keen eye for talent; one of the people he brought to Livermore was George Zimmermann. As an interviewer for a fellowship program, Wood went to Harvey Mudd College, near Pasadena, California. Zimmermann had not applied, but some of his professors had arranged for him to meet the delegation from Livermore, which included not only Wood but Teller. Wood arranged for Zimmermann to spend the summer of 1969 working at Livermore, on the strength of his bachelor's degree. Then Zimmermann went back to start his graduate work at Caltech. But there was a problem: the Vietnam War. During that fall, Zimmermann realized he was in line to be drafted. Wood saw an opportunity. He said, "If you have to serve your country, you should do it here, and not slogging through the rice paddies in Southeast Asia. Would you be interested in coming back here if we could arrange for an essential-occupation deferment?" Zimmermann indeed preferred Livermore to the alternative, and quickly joined Nuckolls' group. He got his draft deferment, too.

His job was to write the needed computer programs to simulate the physics of laser fusion. "I was carrying with one hand, in a not very inspired fashion, the development of codes or programs to do the laser fusion modeling," said Wood. "I was delighted because I knew how extremely capable George had been." Zimmermann arrived in time to help produce calculations for a very important review of laser fusion, right after the New Year, in Robert Hirsch's Washington office. "We were very eager to have the best calculations we could. George worked very effectively with me on developing the codes. Then, over New Year's, John and I used them to do the calculations that we flew away with to Washington, to show the people there why it should be believed that lasers could be particularly effective." Over the next few years, Zimmermann went on to write much of LASNEX, the main computer program used for laser fusion studies. It was an outgrowth of bomb codes, programs used to simulate and study the detonation of nuclear bombs, but it went far beyond them in incorporating much new detail.

"LASNEX is probably the most powerful physical simulation tool that's ever been developed, in terms of the breadth and accuracy of the physical phenomena it models," Wood said. "It was unprecedented in my experience that one person could develop a huge physical simulation code as rapidly and as perfectly as Zimmermann did. LASNEX convinced people that there was very substantial reality behind these concepts, that they did conform to what can be done in the real world. He did a stunningly effective job. Frankly, the program was founded at least as much on George's calculations as it was on whatever concepts and early calculations John Nuckolls and I had been responsible for."

Zimmermann wasn't the only young genius Wood recruited. Another was Rod Hyde, who came to Livermore for a very specific purpose. He was to design a rocket with a thousand times better performance than any built to date.

"I grew up in Oregon, playing chess and reading sci-fi," Hyde said. "Mostly Robert Heinlein, Gordon Dickson, and Keith Laumer. That was when I started getting into the idea of designing rockets." He graduated from high school there. Then, he "faced the standard choice of MIT or Caltech. Anyone on the West Coast who's any good faces that choice." To Hyde, "any good" meant a string of 800s on the College Board exams, the highest scores possible. He had also taken college courses on the side, at Oregon State, in math and physics. Those courses meant that the choice was a bit sticky. He'd always intended to go to Caltech—"it was an ego thing"—but if he went there, his first two years of courses would be prescribed for him. He didn't want that: "I knew that crap, I didn't want to just waste my time doing it again." MIT, by contrast, was "the sort of place where if you live or die, no one cares. So if you're good, you can get away with a lot more."

During his first six months there, he made a mistake. He took the advice of his guidance counselor, which was not to work too hard till he was sure he could adjust to life at MIT. This meant that by his way of doing things, he wasn't getting anywhere. So after six months, he said, "I turned on the burners and got out in a year and a half." He got all A's, except for one B in an economics course. Then, as graduation approached in the spring of 1972, he was nominated for a fellowship. The interviewer for that fellowship was Lowell Wood. Wood looked over Hyde's transcript, his string of A's, his massive quantities of courses, including most of what he'd need for a master's degree. Then he asked Hyde, "What's your hurry?" In his laconic way, Hyde replied, "Who's hurrying?" Wood went on, "Well, you've got five years of courses packed into two years here; it looks like you're hurrying." Hyde replied in a sort of slow drawl, "Oh, there's no hurry to it at all. They charge you the same amount per semester around here, whether you take n units or $3n$ units of courses. I discovered that early on. I'm paying a lot of money to get an education in this place, so I take all the courses I can. But it's not a hurry. I'm just economizing on what it costs per unit."

They fell to talking about rockets and space, and at the end of the interview, Wood said, "By the way, you want to come out and work at Livermore this summer? We've got a rocket idea we'd like you to work on." Hyde had spent previous summers working in a beet cannery in Oregon, so Livermore would be a definite step up. The rocket concept was one that Wood, together with John Nuckolls, had been fond of for some time. Some years earlier, Nuckolls had thought about using laser fusion microexplosions to drive a rocket. The rocket would be equipped with a blast plate, what amounted to a big manhole cover mounted to its rear with shock absorbers. The laser pellets would explode just behind the plate to produce thrust, their explosions pushing on the plate. Then Wood had decided that the way to go was to wrap a magnetic field around the microexplosion, rather than use the pusher plate. The magnetic field then would act as a nozzle, so that the expanding plasma fireball from the microexplosion would blow out the back as a rocket exhaust. This basic idea would allow such a rocket to achieve performance a thousand times better than anything yet flown.

Any rocket with such advanced performance will come equipped with large wings. They will not be for flight, however; they will be radiator panels glowing red, to get rid of excess heat. Every rocket produces plenty of heat, but ordinary rockets, as in the Space Shuttle, have the nice property that they can be cooled by flows of hydrogen fuel. In the Shuttle, though, these cooling flows amount to over 440 pounds of hydrogen per second. Fusion engines will also produce plenty of heat, but unlike the Shuttle they can't dump this heat into the rocket exhaust; they must shunt it to the radiators and let their red glow carry it away. Hyde put a lot of effort into thinking about these radiators and their heat-carrying pipes. He also worked on the laser and the magnetic thrust chamber. Once he knew what to do about these things, designing the rest of the rocket wasn't so hard. As he put it, "Everything else was just standard rocket junk."

Still, things did not go entirely smoothly that summer. Hyde didn't get his security clearance in time, so he couldn't work on the real problems, which included highly classified questions of pellets and laser fusion. His bosses coped with that by having Wood design the pellet. Wood then described the microexplosion to Hyde in very general terms, which at least gave him enough to get started in his calculations. But Hyde had a terrible handicap. He was a chess fanatic, and 1972 was the year of the world championship competition between Boris Spassky and Bobby Fischer. This meant that during that August, the one place in the world where Hyde absolutely had to be was Reykjavik, Iceland, where the match was being held. Going there was nearly as important to him as breathing, but by late July he really was just getting well under way in his work.

So once again it was time to turn on the burners. As his deadline came closer, he went into a typically Hydean way of working, just as he'd done at MIT. He closeted himself in his office and worked for four straight days and nights to write his paper, without any sleep. He just sat there and wrote and wrote, doing a lot of calculations along the way to tie up loose ends that kept cropping up. His eventual paper was only about as long as one of the chapters in this book, but it had well over a hundred math equations as well as nearly two dozen tables and diagrams. He really put an amazing amount of detail into his calculations, his approach to the overall systems of the rocket. Then he emerged, papers in hand. As Wood recalls it, "There was this staggering performance of a hundred hours straight, whereupon I drove this zombie to the airport and stuck him on the plane to Reykjavik."

The next thing was to present this work to the world of aerospace engineering. They picked for this the Propulsion Specialist Conference in New Orleans, late in 1972. There was no lab budget to send Hyde to give the paper, so Wood bought him an airline ticket out of his own pocket, and they shared a hotel room there. Hyde didn't do much wandering around; he didn't go off to the Vieux Carré or visit the fleshpots. His main memory was that "New Orleans is a damn flat place." Their paper, however, was anything but flat. Hyde presented it at a session titled "Advanced Propulsion Concepts." To most of the people there, advanced propulsion meant the engines then being designed for the Space Shuttle. Rocket performance is determined by its specific impulse, which is the length of time one pound of fuel would last, while being burned to produce one pound of thrust continually. For the Shuttle's engines, this duration is 455 seconds. The highest-performing rockets anyone had then proposed had values of a few thousand seconds. Hyde was talking about an engine with performance of at least 300,000 seconds, which perhaps could be extended to be close to a million.

Hyde stood up there at the podium and laid it all out: the physics, the design concepts, the calculations, the performance equations. Almost nobody in the audience had the background to follow him. Still, the paper was a blockbuster, for it had about it the aura of classified technology. People didn't know much in those days about laser fusion, though it was considered quite a hot topic, but nobody could say why. Also, the word got around that Wood had had a hard time getting the paper cleared through classification so that it could be released. The government had just barely, and reluctantly, let it out. Then the requests for copies of the paper started pouring in. Wood had ordered two hundred reprints, which looked to him like a wildly optimistic guess as to how many he'd need. They were all gone in two months, and he was reduced to sending out copy-machine copies. A few of the requests even came from the White House, from the office of the National Security Council under Henry Kissinger. At this time Hyde was nineteen years old.

By then he had received his clearance, and he could do more work on his rocket. But he soon realized he was up against a problem: "You can't discuss the rocket design without getting into the pellet design, and the moment you do that, it's classified. There's really no easy way to partition it. Either the whole thing is classified or you write a fairy tale." The New Orleans paper had not been a fairy tale, but it was so short that he had been able to sidestep the real issues. In a longer paper there would be no way around them, and he would have to face up to them. They were how big the laser would have to be, or, more to the point, how much thrust and performance could you get with how small a laser. The only way to say anything about that was to say quite a lot about the pellets and how they work, which simply couldn't be done, at least not openly.

Hyde went back to Livermore and spent the following January working on a lengthy, massive report, perhaps ten times longer than his New Orleans paper. He redid his designs with much more detail, greatly refined them, and improved them in many ways. This report, UCID–16556, has been updated since with new material. Quite likely it is the world's most authoritative reference on such rockets, which might be suitable for man's first missions to the stars. It has never been published.

The people in charge of classification had let out the New Orleans paper, but then they were very unhappy that they had. Afterward, they claimed they had been misled and hadn't understood, and for a while they even refused to let Wood and Hyde distribute copies. This was obviously absurd; the New Orleans paper had been openly published, and anyone could order a copy simply by sending two dollars to an address in New York. So Security let up on that. But they also said that they would never let Wood or Hyde publish anything in that area again. They could not add to the paper; they could not delete from it; that was all there would ever be. In the ensuing years, Hyde did a huge amount of follow-on work, detailing exactly how the engine would work and what it would do, detailing the laser, the radiation shields, the power equipment. The government just sat on his reports and refused to release them. Then about 1980, the security people said that if Hyde resubmitted this updated report, they'd probably let it through with small changes. Hyde was all for this, but first he wanted to tidy up some features with which he wasn't quite satisfied. By the time he had done that, classification policies had changed again, and once more the curtain of secrecy had rolled down on his report. It remains unreleased to this day.

Still, the reason why Hyde had been able to give his New Orleans talk at all was that by 1972 security restrictions had eased a bit, just enough so that it was possible to write and talk publicly about some of laser fusion's basic ideas. This meant that for Nuckolls, Wood, and their associates, the thing to do was to write about their work in the physics journals, so far as they would be allowed to. The problem was that if they wrote a scientific paper and submitted it to the *Physical Review,* which was the natural choice, the editor would send it out for review. That was the procedure followed by nearly all scholarly journals: to have such a paper reviewed by referees, knowledgeable technical experts, who then would challenge any weak arguments in the paper.

Nuckolls and Wood, however, didn't want this to happen. They wanted their paper to be published without this review. They knew that laser fusion was new, little known, and controversial; a paper on it would offer an open invitation to criticism. The referees would almost surely find points about which to object. Yet the classified status of laser fusion meant that Nuckolls and Wood would not be allowed to supply details to answer referees' objections. The way around this would be to publish their paper in an unrefereed journal. Few such journals existed; fewer had any quality or prestige. However, there was one that not only would be eminently suitable; to publish in it would be a considerable coup. The journal was *Nature,* and its editors accepted their paper. With this, laser fusion reached rarefied heights indeed.

Often compared to *The New York Times* for authoritativeness and completeness, "the good grey *Nature*" is something of a house organ for the scientists at Cambridge University. Cambridge has been home to Isaac Newton, Charles Darwin, the Huxleys. John Maynard Keynes did his work there, as did Bertrand Russell. The electron was discovered there, as was the structure of the DNA molecule; when Watson and Crick made that discovery in 1953, they published the announcement in *Nature.* Within Cambridge, Trinity College has won more Nobel Prizes than all of France. Within *Nature,* on pages gray with the ghosts of Newton and Darwin, on September 15, 1972 these Livermore scientists set forth their manifesto:

> Hydrogen may be compressed to more than 10,000 times liquid density by an implosion system energized by a high energy laser. This scheme makes possible efficient thermonuclear burn of small pellets of heavy hydrogen isotopes, and makes feasible fusion power reactors using practical lasers. ... One kilojoule of laser energy may be sufficient to generate an equal thermonuclear energy.

The article then went on to describe LASNEX calculations involving the liquid-droplet pellet of Nuckolls and Wood. These calculations stated that if this pellet were hit with 60 kilojoules of laser energy, it would release 1,800 kilojoules of fusion or thermonuclear energy, the energy in a stick of dynamite. As the article stated, "Since a 60 kilojoule input of laser light was used, net electrical energy production would be possible."

Sixty kilojoules would be a lot of laser energy, but if it offered the prospect of using fusion to generate electricity, it might be well worth going after.

How could such powerful lasers be built? The approach everyone favored was to begin with a master laser, of low power. It would produce a precisely tailored version of the laser's eventual pulse, having all the characteristics needed, with one exception: it would have very low energy. Then this master pulse would have its energy boosted, with laser amplifiers. The master pulse would be split into several parts, or sub-pulses, somewhat in the way that a garden sprinkler splits a flow of water into a spray of subflows. Each laser subpulse then would feed into its own set of amplifiers, the amplifiers being long cylinders in a line, one after the other. The amplified subpulses, their energy boosted as much as a thousandfold, would all emerge from their amplifiers at the same time. Then an array of lenses would focus their energy into the pellet.

What made the whole effort so promising was that by 1972, the scientists at Livermore had already built master lasers that could produce the kind of pulses they wanted. These master pulses had astonishingly short durations, 0.01 to 0.025 nanoseconds, and if only their power could be boosted sufficiently, success would be at hand. Lowell Wood recalls what the program was like in those days: "It was a very hectic period, very exciting and full of promise. Rapidly growing programs always are. When your budget is increasing by fifty to a hundred percent a year, things are necessarily hectic. And when it is increasing in those terms, it's because the people who control budgets share your belief that there's an enormous promise to be realized quickly. People were saying, gee, fusion has been very hard, we've been working on it for twenty years, now perhaps here's the straight, clean, easy approach, so we could have micro-hydrogen bombs working in power plants.

"We had a large laser-building effort under way, aimed at what we felt would be breakeven lasers, a few kilojoules output in 0.01 to 0.025 nanoseconds. And it would be possible to build these lasers up in something like a year to a year and a half, though the team that was building them hadn't existed six months before. That's why it was hectic. The master lasers were straightforward, they worked, that's why there was this enormous promise. We had the pulses we needed from the master lasers. All we had to do was beef them up in total energy by that last factor of a thousand, and we were there. All that stood between us and breakeven was designing and building these laser amplifiers. So it was, hey, let's get these amplifiers built, fellows."

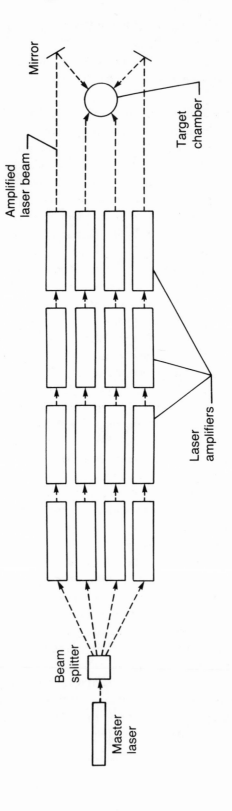

FIGURE 8. *The basic layout of a large laser, this one having four arms, or chains of laser amplifiers. Not shown are the lenses used to focus the beams down to pinpoint size.*

Their most powerful laser eventually would be called Shiva. In the mythology of India, Shiva is a god who grows as many arms as he needs for any task. In the laser, these "arms" were the lines of laser amplifiers, twenty in all. "The real miracle of the god Shiva," said Wood, "was not that he had so many arms, but that all of them worked." One of the project leaders, Hal Ahlstrom, had an Indian wife, Radha, and a father-in-law who was president of an Indian university. Near that university was a studio casting bronze copies of a ninth-century sculpture of Shiva, in his incarnation as Nata-Raja, Lord of the Dance, gracefully swaying with four arms and showing a most feminine waist. A bronze copy, four feet high, on a granite pedestal, would eventually stand in the lobby of the main Shiva building—next to a display of a laser amplifier.

This was not to say that building Shiva would be easy. As Wood was saying about then, "There are three categories of people who don't tell the truth: liars, damn liars, and laser builders. Laser builders lie an awful lot, not only about what they *can* do if you shower them with gold— they're like alchemists in that respect—but about what they *have* done. So you have to ask, how was their power measured? And can you focus that laser power? High-power lasers tend to be like flashlights: they put out this enormous beam that can't possibly be focused, which means it's worthless for fusion work. They'll say, 'My laser puts out such-and-such a power.' Well, you say, did you focus that through a fifty-micron pinhole and put your energy and power meter on the other side? And they'll whimper and whine, 'That's not fair, that's not relevant, you really don't need to go through a fifty-micron pinhole now, do you?' So you know that you're dealing with the laser-builder class of liar. What they had was a power meter this big"—he would make a large O with his thumbs and forefingers—"and they could barely coax the beam into it. They wouldn't stand a ghost of a chance of focusing it."

A fifty-micron pinhole has a diameter of 1/500 of an inch, and without mentioning names, Wood in his oblique way was alluding to the story of Moshe Lubin at the University of Rochester. He also had a similar allusion to the fiasco at Sandia Labs: "Then you ask, what sort of time interval did you measure? Did it ramp up for ten nanoseconds [a very long time in the world of laser fusion]? Or is it actually short enough that it might be useful?

"And, can you reproduce it? Or was that the one shot that destroyed your laser system?" Here he was referring to an incident in the Soviet Union. For two or three years, the record for high laser power was held by Nikolai Basov, in Moscow, who had shared the 1964 Nobel Prize in physics for contributing to the invention of the laser. At that time, about 1973, Basov claimed a pulse of eight to ten kilojoules. This greatly exceeded anything claimed in the West, where nobody had got more than a fraction of a kilojoule. It was the sort of news that might have had congressmen fretting worriedly over this latest Soviet breakthrough. The true story came out only later.

As with all big lasers, Basov's featured a master laser, to produce an initial pulse that would then be sent down the train of amplifiers. This pulse was several tens of nanoseconds long, much too long to be useful. Therefore, in its path there was to be a Pockels cell, a kind of fast-acting shutter. It would chop out a two- or three-nanosecond portion of this pulse, containing perhaps one-tenth of its energy, throw away the remaining nine-tenths, and send the subpulse on to be amplified. However, a technician installed the Pockels cell backward. When the pulse came through, it chopped out that central portion, but then threw *it* away, sending the remaining nine-tenths on through the amplifiers. These promptly shattered and were completely destroyed. The power meter registered a total energy of about ten kilojoules before it too gave up the ghost. It was the only really high-power laser in the entire Soviet Union, and rebuilding it took the better part of two years. What was more, that accident knocked the Russians out of world-class competition. When Basov was asked about the fate of the man who had made the mistake, his face clouded over, the face of a Soviet Nobel laureate asked to contemplate the fate of a lowly technician. He said simply, "He does not work for us anymore."

As Livermore's big-laser program began to hit its stride, everyone was aware of these mishaps, as well as of others. Clearly, there would be lots of things that could go wrong. The obvious way to proceed, then, was to bring in the best man they could to ride herd on the design of these lasers. This was John Emmett, who had made a name for himself managing a high-power laser program at the Naval Research Laboratory. Emmett was a maverick. He drove a black Corvette, at very high speeds, and liked to think of himself as the dashing young technologist-about-planet. He liked to concoct rare and exotic liqueur cocktails to try out on his friends. A licensed demolitions expert, he had long been fascinated with high explosives. Occasionally he took friends on boulder-blasting expeditions, going out into the desert and blowing large rocks apart with dynamite. He had graduated among the top ten percent of his class at Caltech, a group known as a breed apart for their fascination with science and engineering, their very high intelligence, their enormously deep and broad education.

He had studied for his Ph.D. at Stanford under Arthur Schawlow, who had shared in the invention of the laser and who later won the Nobel Prize for his work with lasers. Then Emmett had gone on to build up his world-class laser project at NRL.

The first thing he did, soon after getting to Livermore, was to complain that his predecessors had asked for much less money than would be needed. He argued that they had asked for only $20 million but that Shiva would cost at least $50 million. When he proceeded to ask for more, he was told, "Sorry, that's all the money there is; you'll have to do the best you can with what you have." One thing he could certainly do was to demand lots of overtime from his technical staff. However, these physicists and designers had to go home for dinner. The lab cafeteria was open for breakfast and lunch, but after their long afternoons of work, people had a way of getting hungry. By six or seven in the evening, they were either wilting or leaving. He arranged for the lab cafeteria to stay open for dinner as well, and, what was more, to let families in. Wives could bring the kids and sit down with their harried husbands as they bolted their food before getting back to the lasers.

One of Emmett's main problems was the tendency of a high-power laser beam to spread out like a flashlight. This problem had plagued the early laser-builders. Within a laser amplifier, at high intensities, the beam would not be focusable into a sharp, clear spot. Instead this spot would be fuzzy, and would be surrounded by a number of bright streaks and smears, which represented energy lost from the main beam. This smearing or fuzzing of the beam was due to optical effects arising within the optical components themselves. It had to be overcome, if Emmett was to be able to get anything useful out of Shiva.

The solution lay in the fact that the beam would not become fuzzy immediately but would accumulate these distortions only gradually. Therefore, Emmett could use laser amplifiers to boost the beam power in the usual way, up to the point where these distortions began to be serious. Then the beam would pass through lenses and be focused through a pinhole. The central part of the beam would pass through, but the surrounding smears of light, the accumulated distortion in the beam, would hit the sides of the pinhole and be stripped off. Emerging from this pinhole, the beam would again be clean and distortion-free. It could then be amplified further, until new distortions began to build up. Then another set of lenses would focus the beam through another pinhole, and clean it anew.

This solution was so simple that it took only a few years to make it work. The laser beams were so intense that when they were passed through a pinhole, the surrounding fuzziness was bright enough to vaporize metal from the plate in which the pinhole had been drilled. This metal vapor then acted like a very fast camera shutter and blocked the main laser beam. Thus it was necessary to use special materials that would resist vaporizing, and to limit the beam power to a degree. By 1975, however, Emmett's laser-builders had their pinholes working well. This meant that Shiva would succeed and, what was more, that it would be possible to build still larger lasers in the future.

Emmett's group built several preliminary lasers before going ahead with the construction of Shiva. Shiva was to be an extremely ambitious venture, with twenty arms or laser-amplifier trains, able to produce over ten kilojoules. Thus, Emmett started with a laser named Janus, a two-arm laser with relatively small amplifiers. Then came Cyclops, an early version of one of the Shiva arms, which showed that there were significant problems with then-current ideas about how to design Shiva. The third laser, Argus, then was built to test solutions, including the use of those pinholes, though the Shiva design was frozen before Argus could actually be tested. Argus amounted to one-tenth of Shiva, with two of its arms, and featured the same optical components that would go into the eventual Shiva. In addition to studying laser-design problems, these lasers were also used to conduct experiments, shooting pellets of various designs. In 1976, Argus pioneered in shooting cavity-type pellets, which featured a microsphere of fusion fuel enclosed within a somewhat larger shell. These were prototypes of the pellets Nuckolls had been working on since the late 1950s; but no one had ever shot them in tests. The earlier tests had featured complex versions of the pellets studied at KMS Fusion, which included a tiny glass sphere filled with fuel. These cavity-type pellets offered much greater flexibility in design as well as improved performance. They also were very similar to hydrogen bombs in design, which meant the results of these tests would be highly classified. This brought about a bit of a flap, in 1976.

A Soviet scientist named Leonid I. Rudakov was pursuing similar fusion ideas, and he wrote a paper about it. He proposed that cavity-type pellets could be driven to produce fusion using an intense beam of electrons, as a substitute for the intense light of a laser. Somehow he got his paper published. Somehow he got permission to visit the United States and talk about his ideas. Nobody at Livermore was permitted to talk about their work, so Rudakov stirred a great deal of excitement. *Aviation Week,* the aerospace industry's trade magazine, which was ever alert for rumors of a Soviet breakthrough, worried in print over how far ahead the Soviets were. Rudakov showed up at Livermore and gave a talk. The story—which Nuckolls flatly denies—got around that when Rudakov had finished with his talk, security officials came and carried away the blackboard on which he had written. Yet months before, Nuckolls had invented a special version of the cavity pellet, suitable for testing on Emmett's 0.1-kilojoule Cyclops. These Livermore scientists had successfully tested these pellets in actual experiments, about the time of Rudakov's visit. They just weren't permitted to talk about it; that was all.

Still, by 1978, Nuckolls and his associates knew they were in trouble. Using Argus, they had launched the "100 X campaign," a series of experiments aimed at compressing the core of a pellet to a hundred times the density of liquid deuterium, in the course of a microexplosion. This would be a long step on the way to the thousandfold compression needed to achieve fusion. They weren't able to do it, and for a while they didn't know why. Nuckolls tried to get around the problem with new cavity-type pellet designs, but his pellets were so complex they could not be fabricated until 1980, and it took until then to achieve their 100 X. And even then, it took the power of Shiva itself to reach this goal. As Nuckolls described the earlier attempt, using Argus: "It was like the Battle of the Bulge. We had gotten too far ahead of ourselves, we didn't have the reserves in hand, the enemy had encircled us, and we just had to stand out there and fight." They had been over confident, and in going for 100 X, they fell short.

By then, several years of intensive work in laser fusion had shown that there were some very severe problems, which had not been apparent during the early optimism of 1972. These problems meant that the lasers would not perform as well as the scientists wanted. Also, even if they could somehow fix their lasers, the pellets would not perform well enough. With the performance of both lasers and pellets falling short, the whole concept of laser fusion would be much more difficult to make work than had been thought only a few years earlier.

The original hope was to produce laser pulses with durations as short as 0.01 nanoseconds. However, there were abstruse and little-understood optical effects that made it impossible to turn a kilojoule pulse on and off in such a short time. There was no problem doing that with the low-energy master pulses, one-thousandth as powerful. The physicists just couldn't get that last factor of a thousand, which Wood had been counting on and which was needed for breakeven with the liquid-drop pellet.

The second problem lay in the use of pinholes, with their associated lenses, to scrape off the laser-beam distortions and prevent the beams from spreading out like flashlights. Shiva had not been originally proposed with these additional optical components in mind, and they cost money. A high-quality lens cost up to $20,000; there were two to each pinhole, and Shiva mounted 102 pinholes. Shiva was built under a fixed budget, and the cost of these components came directly out of the funds Emmett had hoped to use to buy more laser amplifiers. In addition, Emmett was unable to get full performance out of the amplifiers he could afford. The beam power and energy had to be reduced to keep from vaporizing the pinhole plates, and then part of the amplified laser-beam energy got scraped off by the pinholes, so Shiva could not reach the 100 terawatts that represented its original goal. It could achieve less than 30 terawatts, and this power loss greatly reduced the amount of fusion energy that could then be produced using Nuckolls' pellets.

The third problem lay in the pellet. Throughout the early and middle 1970s, Nuckolls and other pellet designers were continually changing and refining their ideas, as new calculations and theoretical studies came to the fore. Still, only data from experiments could show where the true difficulties lay. The first really good data began to come in with the 1976 tests using Argus, and, sure enough, there was a serious problem. The pellets were heating up in the wrong manner.

It was very important that the pellets heat from the outside inward, like a hamburger cooking on a barbecue. The center of the pellet was to stay cool till the last possible moment. If it heated prematurely, it would expand and blow apart before the pellet could be compressed properly. Nuckolls had all along been counting on the pellets' heating in this barbecue manner, and his LASNEX calculations had said that was what would happen. The Argus data showed that these calculations were wrong. The pellets were heating as in a microwave oven; their centers were becoming hot prematurely. This was what kept Nuckolls from success in the 100 X campaign, with Argus. It meant that he was left without a pellet he could count on for reaching breakeven with Shiva.

Still, Nuckolls was not ready to give up. He was one of the world's leading experts on pellet design, and he felt there would be a way to create pellets that would overcome the problems they had discovered. He continued to argue that with these new pellet designs, Shiva might yet come close to breakeven. But by 1978, the odds were against him. His pellets could neither be fabricated nor be made to work effectively using the wavelength of light available with Shiva.

Shiva had originally been conceived as the laser that would at least approach breakeven, and possibly demonstrate its achievement. In the political world, where the advocates of laser fusion were competing for money and attention with those who favored magnetic fusion, such a demonstration would have been very important. It would have shown that laser fusion was the road to follow, and that it was capable of producing fusion energy years in advance of its magnetic competitors. By 1978, though, it was obvious that Shiva would fall far short of breakeven. It would simply deliver the best experimental results anyone knew how to get at that time.

6

Arms and the Man

SHIVA was completed in the fall of 1977, and in the laser world it immediately drew rave reviews. It represented highly precise optical technology carried out on a huge scale. Nothing remotely like it had ever been built before. The laser amplifiers were sealed in blue cylindrical pipes, eight inches across. These pipes, together with the lenses, filters, and other optical components, resembled oversize versions of long telephoto camera lenses and conveyed a similar impression of cost and exactness. All these components were stacked in six rows on a steel lattice, a frame big enough to support a grandstand for a thousand people or more. That frame, three stories tall and painted a dazzling white, had been assembled by riggers fresh from a job on the Alaska pipeline. They were so glad to be in from the cold that they finished the job in one-fourth less time than had been planned, and with twice the precision.

The amplifiers and other laser components were made from neodymium glass, a heavy dark-purplish glass that could be formed as large slabs and cylinders. It was, and is, the best material for producing highly intense and very brief laser pulses. It was made with top-quality optical glass, better than in camera lenses and refined to high purity. Originally, the hope had been that it would produce pulses with power of 100 terawatts, but this had proved impossible. "Laser media do not cheerfully handle hundred-terawatt power loads," said Lowell Wood. "Neodymium glass begins to commit suicide" at such power levels. Still, Shiva could easily achieve ten terawatts, and at times it did better. That was impressive enough.

Shiva was housed in a large white bay with a white linoleum floor. The bay was very brightly lighted and kept very clean. Visitors undeniably felt that they had stepped into the future, that this was the shape of things to come. Walking through a door into the target room, one would see the target chamber at the focus of the laser beams, a stainless-steel sphere looking like an old-style naval mine from World War II, full of protrusions. Those near the top and bottom served to let in the beams; the rest were ports for diagnostics, to observe the microexplosion.

Shiva's tests were run by a "shot director," who sat in its control center. Five minutes before a shot, he would press a button to initiate a computer-controlled firing sequence. During this sequence, electricity would flow into the laser power supply. At the proper moment, the director would push the "fire" button. Instantly the electricity from the power supply would spark like lightning through some 2,000 photographic flash lamps, which were built into the amplifiers. These primed the amplifiers with the energy that would boost the power of the laser pulse. A thousandth of a second later, the master laser would fire, producing its low-power pulse. That pulse would be split by mirrors into twenty beams, which would separate and pass through Shiva's twenty arms, its trains of blue amplifier pipes and other components. Emerging from the amplifiers, twenty powerful bursts of laser light would pass through lenses and focus into points. A few billionths of a second later, these focused light-beams would hit the fuel pellet from twenty different directions and with a momentary power dozens of times greater than that of all the electricity generated in the United States. The pellet would explode in a brilliant burst of green light, as high-speed instruments recorded the data.

Shiva did its work very quietly. The power supply produced nothing more than a slight pop. The laser beams, hitting the lenses, blasted dust particles with an audible snap. The pellet explosion took place within a vacuum and made no sound. When I visited it as a guest of Lowell Wood in 1979, its target room was brightly lit but displayed no flashing lights or buzzers, no countdown clock. As we stood there, I asked how we would know when a shot took place. He replied, "How should I know? So far as I can tell, that pellet may have already gone to glory."

Shiva featured a computer-controlled system to keep all its components aligned, continually tweaking them to keep them from getting out of position. This system, too, was something of a leap into the dark. No one had ever before built one like it or made it work on such a large scale. At first no one knew if it would work, and Emmett and his associates were betting the success of Shiva on its achievement. The alternative was that technicians would have had to work continuously to try to keep everything aligned. In fact, it would have taken them twenty-four hours a day to do this, so that there would have been no time at all for experiments, and Shiva would never have been aligned to the point where it could be used. The computer system automated the work of those technicians. Then in January 1980 came the earthquake, just as the directors of the magnetic-mirror program were having their review of the MFTF-B design.

A quarter-mile away, in Livermore's laser fusion area, John Emmett looked up from his desk and saw a big oak bookcase, full of books, swaying over and about to fall on top of him. He ducked under the desk and probably saved his life; the bookcase crashed right where he had been working. Other people weren't so lucky. Several dozen people in the lab were hit by falling bookcases; they were cut or bruised, and had to be sent to a hospital to be checked for broken bones. To Wood, "the building looked like a bomb blast had gone off inside. The fluorescent lights in the ceiling fell in; all the partitions in the offices were turned over sideways. Emmett and I went in there, and the place looked worse than a building trashed by vandals. The ceilings were strewn all over." The quakes rolled and swayed for over a minute, and with this long-drawn-out shaking, the 200-ton Shiva frame sheared some of the bolts that held it to the floor and rocked out of position. But within a week, riggers brought in jacks and got the frame back into place, putting in double-strength bolts for good measure. The computer alignment did the rest. Within two weeks after the quake, Shiva was back in operation.

In fact, Shiva was so remarkable that it was featured in a Hollywood movie. A year after the earthquake, down at the Walt Disney studios in Burbank, the producer Steven Lisberger was working on *Tron*. One of his employees called the public-affairs office at Livermore, explaining that he had heard there was a big laser there that might be a good high-tech environment for their movie. Mike Ross, who handles visits to the lab, was dubious. He had never hosted a film crew before and was concerned that it would be quite an imposition on the laser researchers. So Lisberger's man sent copies of the script to Ross. He read it and liked it, as did the laser people, and they agreed to give it a try. Soon Lisberger and his art director came up, and Ross was quite surprised to meet him: "He was a kid! Thirty years old, but he looked twenty-three, with jeans and some sort of New Wave–type slinky jacket." When Lisberger took a tour of Shiva, he said, "This thing only cost twenty-five million dollars and it *works*? I couldn't build this for twenty-five million out of plastic!"

Lisberger's crew filmed at Livermore for four days, over Memorial Day weekend of 1981. They filmed not only Shiva, but also the Livermore computer center and the word's largest hinged door, which was eight feet thick, weighed fifty tons, and was used to shield a powerful radiation source. There were eighty-five people in the Hollywood crew. Many of the laser people were very concerned at first; they had millions of dollars' worth of optical equipment that was essentially to be used as a prop. But all went well. As Ross described it, "In laser tradition, they finished on time and under budget."

What did Shiva accomplish? It set laser records by delivering pulses with up to 15 kilojoules of energy. When operating with very short pulses of 0.1 nanosecond duration, peak power was 26 terawatts. This was a long step toward the level of 100 terawatts, at which pellets might actually undergo fusion. The experimenters got pellet compressions of a hundred times liquid density, and in different experiments they heated the centers of pellets to 100 million Celsius degrees. They would have needed more than a thousandfold compression at this central temperature, all in the same pellet, to reach breakeven. As Wood said, "We just ran out of power. The power at which Shiva effectively worked, that you could use it again and again, was not the record-breaking twenty-six terawatts. It was more like ten terawatts. The promised land begins to become accessible at one hundred terawatts of effectively applied power, and a factor of ten short makes an enormous difference." Still, they were able to use LASNEX to predict the performance of pellets having specific designs, then test these predictions by shooting the pellets with Shiva. When it was shut down in December 1981, Shiva was three times more powerful than any other laser anywhere in the world.

For all this, something was missing. In laser parlance, Shiva operated "in the red." This had nothing to do with budgets or cost overruns; it had to do with the wavelength of its light. Shiva operated in the infrared, at long wavelengths. All laser media have characteristic wavelengths at which they emit their light; for the neodymium glass used in Shiva, the wavelength was infrared and no one could do anything to change this fact. The problem was that pellets did not work well with infrared light; they worked much better with light at short wavelengths. Still, it was possible to get such shorter wavelengths. The way to do so was with "frequency-multiplying crystals," which looked like big slabs of window glass set in frames. The idea was to shine the laser light through a window whose panes were made from these crystals; the infrared would go in and green and blue light would come out. These colored beams, at one-half and one-third the wavelength of the original infrared, would be just what was needed.

Experiments were made on the Argus laser, in 1979 and 1980, which showed that the use of such colored or short-wavelength light was the key to overcoming the problem of the pellet interiors' heating prematurely. This preheating had kept the pellets from being compressed satisfactorily. Unfortunately, these experiments came along too late to influence the final shape of Shiva. John Emmett had needed to save money, and he thought he had pellets that would work well with Shiva's basic infrared light. Hence he had built Shiva "without color"—without the crystals. Originally, Shiva had been intended to have color, to be able to produce laser beams at several wavelengths. The decision to delete the crystals cost Shiva a great deal of flexibility.

The final Shiva, then, was both less powerful and less flexible than had been originally planned, and its pellets would perform more poorly than originally expected. Those facts did not keep the Livermore managers from trying to put the best possible face on things. Following their first successful tests of Shiva, Livermore's *Laser Program Annual Report— 1977,* released in mid-1978, had this to say:

> The High-Energy Laser Facility (Shiva) Project was conceived in 1972. . . . When the project was initiated in 1973, its goal was to provide a laser output of 10 kilojoules in a subnanosecond pulse. On November 18, 1977, at 9:35 p.m. . . . all 20 arms were fired simultaneously and obtained an output of 10.2 kilojoules in 0.95 nanosecond. This performance *exceeded* [their emphasis] the design objective of 10 kilojoules in less than 1 nanosecond. . . . In addition, Shiva is much more versatile than originally conceived.

George Orwell would have been interested. The 1973 annual report had described Shiva as a 10-kilojoule, 0.1-nanosecond facility. Thus, in five years a hard, honest objective of 0.1 nanosecond was weasel-worded into merely "less than a nanosecond." The statement about Shiva's being "much more versatile" than the original design referred to the addition of the optical alignment system and of a computerized control system. Still, that same 1973 report had explicitly stated that Shiva would be built with frequency-multiplying crystals, to cover a broad range of laser wavelengths. By 1977 these were long gone.

"To some extent, people were misled by our optimistic projections," said John Nuckolls. "But none of our publications ever said we'd get breakeven with Shiva." That may be true so far as explicit statements are concerned, but they came awfully close to making such a promise. On page 2 of that 1973 *Laser Fusion Program Semiannual Report* there was a discussion of Shiva and a table describing laser requirements to meet various fusion milestones. These items together made it clear that Shiva was expected to achieve at least one-tenth of breakeven: that is, to produce one-tenth as much fusion energy from a pellet microexplosion as was being shot in with the laser beams. It might even achieve full breakeven, according to that report, if all went well. As late as 1977, in an address to the American Physical Society, Nuckolls stated in writing that "breakeven will be approached and possibly equalled" with Shiva.

So Shiva started out promising 10 percent of breakeven, possibly more. By 1979, following all its design compromises and after more than a year of operating experience, its official goals had been reduced substantially. Now they were merely to compress pellets to a thousand times liquid density, and to achieve 0.15 percent, one part in 600, of breakeven. When Shiva shut down in 1981, it had reached one hundred times liquid density, and the program managers had stopped publishing goals of a percentage of breakeven. But within their published reports were data showing that their best energy release was less than one ten-thousandth of breakeven.

These issues were brought into sharp focus early in May 1981, in an article in *Science* written by the energy consultant William D. Metz:

Last month, for the first time, the Administration decreased research funds for one of the most exotic energy technologies on the books: laser fusion. . . . Laser fusion is now headed on a course that reemphasizes basic research, reduces its importance as an energy option, and may return it to the veiled world of classified research from which it originated. . . . In years past the congressional budget hearings for laser fusion were open, well-attended events, but this year the principal hearing was a closed-door, classified briefing with no part open to the public.

In 1972 . . . John Nuckolls and Lowell Wood projected that breakeven-level experiments would occur during 1973 and that the next step, net energy production, would occur "sometime around 1975." No breakeven had been achieved by 1975, and the program managers officially projected it for sometime between 1979 and 1981. In the next 3 months, that schedule slipped 2 years. By the time . . . Shiva reached full power in May 1978 the laboratory reported that breakeven was a milestone for future systems to achieve "in the mid-1980s." The latest program plans . . . have dropped all reference to energy milestones, showing instead pellet physics experiments through the mid-1980s. . . . In his 1972 paper, Nuckolls predicted that breakeven could be achieved with a 1 kilojoule laser. These estimates were revised upward in the mid-1970s, but not nearly enough. Now Nuckolls has raised his estimate to 300 kilojoules. He and many others in the field would not be surprised now if it took 1 megajoule [1,000 kilojoules].

Metz's article stirred considerable controversy. Nuckolls and Wood had hedged their early predictions with numerous caveats and qualifications, none of which Metz had mentioned. Nuckolls thus was quite upset at being quoted out of context. Back in Ann Arbor, Alexander Glass, president of KMS Fusion, decided to confront Metz head-on. He wrote out a point-by-point rebuttal and submitted it to *Science*. That journal has a very liberal policy on publishing lengthy comments on controversial articles, but its editors asked Glass to shorten his statement to 200 words. The reason was that Metz believed he had done his homework and was loaded for bear, quite prepared to defend his statements. Glass thereupon contented himself with circulating his rebuttal privately, among his colleagues. He did not try further to publish it in *Science*.

Still, Kip Siegel did not die slumped over a table at congressional hearings in the hope of achieving breakeven at one megajoule. So it is worth looking at what Emmett's group will be doing during the next few years, what the country can hope to get from their work, and what it will take to restore laser fusion's bright promise.

During the 1980s, their main efforts will involve a much more powerful version of Shiva, called Nova. "Nova got started because Shiva was insufficient for its purpose," said Wood. "The laser that follows Nova, if and when it comes into existence, will do so because Nova was insufficient for its purpose, which is to demonstrate breakeven. We knew what Shiva was likely to be able to do. There just was no way we could come up with any pellet design—and there is an enormous richness of pellet designs— that would do the job. We tried for several years, but Shiva just didn't have the oomph. It was like telling a guy, 'I'm sorry, but due to budget limitations we're not going to be able to supply you today with the two pounds of oxygen that you need to stay alive. But if you can stay alive with two-tenths of a pound for a while, maybe later we can help you.' That was how it was for us, falling a factor of ten short with Shiva. We needed a hundred terawatts and could get only ten. People asked, can't you somehow get within a factor of ten of break-even? Well, it's like breathing oxygen. If you fall a factor of ten short, you don't just slow down your metabolism to ten percent; you die in a few minutes. That's what hit us on

Shiva, when we fell off by a factor of ten in peak power. So then we said, we've got to have a laser that will deliver the few hundred terawatts that look necessary, and we've got to have color, green and blue light."

Nova got its start in 1975, partly as a response to technical developments. Emmett's laser-builders had succeeded in inventing new types of neodymium glass, which could handle higher power levels within a laser. They had raised the efficiency of the laser amplifiers, too, and had developed means for fabricating particularly large components. In addition, they had proved that by the use of pinholes and lenses, even very powerful laser beams could be kept clean and well-focusable. All this meant it would be possible to build a laser much larger and more powerful than Shiva.

What was more, at about that time Nuckolls had made another invention in pellet design, which offered better performance in the pellets that would be used if laser fusion was to be the basis for a power plant that would produce electricity. This new pellet offered ten times breakeven at 200 to 300 kilojoules—a long step up from Nuckolls' one-kilojoule calculations of 1969, but still a feasible goal for a future laser. With Shiva now well under way, the time was opportune to begin designing this new laser. After all, Emmett had a large staff of laser designers who had to be kept busy and for whom Shiva now offered diminishing new work.

There were political considerations, too. By 1975, the magnetic fusion program was going ahead with Princeton's TFTR, which was to produce a good deal of fusion energy. Emmett knew the laser-fusion program would also need to be able to produce fusion energy by the time TFTR was to be operating. Otherwise, people in Washington would say, "You had us thinking you had an alternate approach to fusion, but now Princeton is making fusion energy and you aren't." To protect his budget, to keep from falling behind, Emmett had to be able to say that laser fusion would be at least equal with the tokamaks when it came to producing energy.

On the basis of Nuckolls' pellet work, Emmett's people set the goals for Nova. Like Shiva, it would have twenty arms. But it would be twice as big, and would put out more than ten times the energy, a total of 250 kilojoules in a nanosecond, for a peak power of 250 terawatts. Emmett and his associates had advanced the laser-building art to such a point that two of Nova's arms, its trains of laser amplifiers, would together have more power than all of Shiva. That was the measure of the improvements that had been gained in building large laser amplifiers. Nova would cost $200 million, eight times more than Shiva. Emmett wanted to go ahead with this new project, and he won the backing of the man who directed the laser-fusion program within the Atomic Energy Commission, an army general named Alfred T. Starbird.

Then Nuckolls went back to work some more on his pellet, made new calculations using data from Shiva experiments, and decided this pellet wasn't quite as good as he had thought. Some physicists asked him to assent informally to the assertion that the full twenty-arm Nova would reach ten times breakeven. This was somewhat off the record; he wasn't being asked to sign his name to a formal document. Nevertheless, he refused to do it, stating that Nova would reach breakeven but not ten times breakeven. This refusal undercut Emmett's case for building the full-size Nova and gave considerable ammunition to those who wanted to build only a smaller, half-size version, with just ten arms.

For at that moment, pork-barrel politics was raising its head. With laser fusion falling short of its goals, there would be only so large a pot of money available, and Livermore was far from the only center for laser fusion. Similar work was going on at Los Alamos and Sandia labs, in New Mexico; the University of Rochester in New York; and the Naval Research Laboratory near Washington. All these labs contained scientists who wanted a cut of the $200 million that Nova would cost. To Wood, "there were a lot of people saying, 'For only twenty million dollars I can do thus-and-such. Wouldn't it be a lot better if you gave Emmett a hundred eighty million and gave me the twenty million?' " Various congressmen were arguing over whether New York or New Mexico was getting a fair share. Of course, Nova had to have enough to do something significant; otherwise, in Wood's words, "you would just fritter the money away. Like giving a dollar to each citizen and telling him to go support his favorite fusion program—unless he'd prefer to spend it on a six-pack."

This was the sort of issue that called out for a high-level technical commission to come out and take an unbiased look. The program managers in Washington set up such a commission, which did just that. It featured two former directors of Livermore, as well as Burton Richter of Stanford University, who had won a Nobel Prize in physics. The chairman was John Foster, who had been director of Livermore from 1961 to 1965, and who in those days had been an early and enthusiastic leader in pushing for research on laser fusion.

This Foster Committee was sharply divided. Each member had a different view of the proper course for laser fusion. Some of them wanted to use part of the Nova money for a different laser project, which would be less powerful than Nova but would give particularly short wavelengths. One thing they could agree on was that Nuckolls' new pellet results, combined with other experiments, indeed had undercut the case for building the full-size Nova and that there was no compelling reason to put in its full twenty arms. After much debate, several members favored upgrading Shiva and improving its performance, rather than building Nova at all. To Emmett, such a course would have meant disaster.

At that point, Foster made the key argument: breakeven would not be good enough. It could be achieved by producing a momentary flash of fusion energy in the center of a pellet. But to go beyond breakeven, to produce fusion energy in useful quantities, this flash would have to propagate into the surrounding fusion fuel, as a flash from a blasting cap propagates into a charge of dynamite. Such "propagating-burn" experiments would be essential in demonstrating the feasibility of making high-performance pellets, and in defining the size of the laser needed for this performance. But even the full-size Nova would not be capable of such propagating-burn tests.

Short of this, however, Nova could advance beyond Shiva and do a very valuable set of experiments. These would compress the pellet to a thousand times the density of liquid deuterium, while producing the momentary central flash. But it would not be necessary for this flash to give breakeven, for Nova could demonstrate the pertinent physics results at lower levels of fusion energy. These experiments, in turn, did not require all twenty arms, but only ten.

This argument amounted to advocating a radical redirection of the laser-fusion program. Always before, the goal had been breakeven, which was a measure of energy production. Breakeven was a vivid, tangible goal, easily grasped by congressmen and by the press. Now, instead of seeking breakeven, Nova was to seek to study the momentary central flash. The next laser beyond Nova, in turn, would go after the propagating burn. But these were physics milestones, not energy milestones. They were nowhere near so vivid, so easily understood. Adopting this view of laser fusion thus was tantamount to declaring that the purpose of this program would not be to produce fusion energy, but instead to do work in experimental physics. It amounted to an admission that laser fusion was not so advanced or so mature as some of its proponents had hoped.

In June 1979, Foster called a gathering of his committee at his home amid the hills south of Los Angeles. It was a bright blue day, and everyone soon sat around his swimming pool, which faced the Pacific.

Naturally, the talk was of laser fusion, and as the discussions proceeded, a compromise bubbled up, a solution to the Nova problem. The committee would recommend building only half of Nova, with just ten arms. After a few years, if the physics results looked good enough, Emmett might get the go-ahead to put in the other ten arms.

Emmett was not happy with this compromise. He wanted all twenty arms, and he was angry at the entire Foster Committee. He also was mad at Nuckolls, who had been a consultant to that committee and who had endorsed the compromise. So far as Emmett knew, next year Nuckolls might come up with another good idea in pellet design, which would once again allow him to exceed breakeven with the full twenty-arm Nova. Nuckolls was painfully aware of the budget limits: "It was like people out in a lifeboat: there's not enough food; somebody's gotta go overboard." What went overboard were ten of the arms. But in their place, Nova was given the frequency-multiplying crystals. Unlike Shiva, Nova would have color.

Compared to Shiva, said Wood, "Nova is a completely different era. It's ten years later. Nova is built out of different materials, different glasses used in the lasers. Nova is built with far more powerful computer programs used in its design. Much more experienced people; everything's different. And it ain't working yet. It'll be years before it's working. Nineteen eighty-five—that's the schedule." Meanwhile, as usual, Emmett has been using ingenious means to push things along and save money. Visitors to Livermore will notice a large multicolored canvas structure bearing a most remarkable resemblance to a circus tent. That is exactly what it is. While building Nova, Emmett realized he'd need storage space for laser components. The trouble was that Livermore is a federal lab, and any new construction, even of a warehouse, involves an enormous amount of red tape. Congress literally has to consent to every building, the General Services Administration has to certify that there are no other federal buildings in the area that can be used, and it can take years to go through the competitive bidding, the authorizations and appropriations, and everything else. So Emmett simply leased one of these tents, to cover dry areas that would serve for storage space. Similarly, when he needed extra office space, he rented some trailers. They cost one-fourth as much as a new building, and were available in weeks, not years.

Nova will not get to breakeven, but it should come close. It is expected to compress the centers of pellets to a thousand times liquid density and a central temperature of 30 million Celsius degrees, producing what is called a hot spot. "We will watch the hot spot flare up and go out," said Nuckolls. "We can diagnose its shape, its burn efficiency—we can get all the data you need to make a reliable prediction of what would happen if you did that same experiment with a five-megajoule laser." (This, in a backhanded manner, was his way of agreeing that the ten-arm Nova would be big enough to do its experiments; the extra ten beams wouldn't demonstrate enough new physics results to make them worth paying for.)

"We'll be producing enough neutrons, from fusion reactions in the pellet, that you can actually photograph the hot spot with a neutron camera. Or we can image it with the X rays it will give off. We'll not only image the hot spot, watch it flare up and go back down, we can also image the fuel around the hot spot, see if it's imploded to the right density. All within hundredths to tenths of a nanosecond. It's our professional opinion that that data will be sufficient, when combined with other data, to predict with confidence what would happen if you did that same experiment with a five-megajoule laser." The "other data" will include results from hydrogen-bomb tests. The combination of H-bomb data and Nova results, together with data from other experiments performed at Los Alamos, should give them the information they will need to make tomorrow's laser-fusion systems work.

In particular, Nova's pellets will be precise models, to one-tenth scale, of the full-size pellets for the laser-fusion reactors of the future. Such reactor pellets will be the size of a pea, containing about five milligrams of fusion fuel. With their smaller sizes, the Nova pellets will leak heat from their interiors when zapped with laser light, and will not achieve temperatures high enough for efficient burn. But they will suffice to prove that their reactor-size cousins will ignite with a propagating burn, when zapped with a large enough laser.

Nova will not be ready for a while, however, and in the meantime the scientists will be working with a smaller laser called Novette. Novette features two arms of the design that will go into Nova's eventual ten. It thus represents only one-fifth of Nova, yet has more power and energy than Shiva. But while Shiva filled a steel scaffolding three stories high, Novette is almost low enough to see over. Like Shiva, Novette features trains of cylindrical blue pipes housing optical components, including laser amplifiers eighteen inches across as compared with Shiva's eight, all assembled within a room of futuristic brightness and whiteness. In addition, Novette has the long-sought frequency-multiplying crystals. It thus will not merely replace Shiva, but improve upon it.

Back while Emmett was preparing to build Shiva, he also built Argus,

which featured two Shiva-type arms. Novette is a similar precursor to
Nova, and will similarly serve to test out its basic designs. However, Shiva
was shut down in December 1981, a full year before Novette was ready,
and during 1982 Livermore was left without any large laser at all.
Moreover, at the time it was shut down, the potential of Shiva was still far
from wrung out. As Wood put it, "Up to the week before they started
tearing down Shiva, we were doing experiments that were extremely
interesting. The day it died, it was the highest-power laser research
facility in the world. It was torn down to save money." It was torn down
for other reasons, too.

By 1981, with Nova under way, the laser fusion program badly needed
data from experiments made at ten kilojoules with green laser light that
had a wavelength much shorter than was available with Shiva. The data
were needed to show that Emmett and his associates could overcome
pellet problems that had been seen when they used Shiva. There were two
ways to get such data: rebuild Shiva for improved performance, or tear
down Shiva and cannibalize some of its parts to build Novette. Either
move would cost money and leave Livermore for some time without a
major laser, but Emmett had to choose. He chose the latter course with
an eye to the future. An upgraded Shiva would lead nowhere in laser
design, but Novette would pave the way to Nova. With only two arms
compared to Shiva's twenty, Novette would also be less costly to operate.

Emmett also had his eye on politics. Any move to upgrade Shiva would
give ammunition to his critics, who would ask whether in fact Nova might
not be worth delaying while the program went forward with improved
versions of Shiva. Moreover, Emmett still wanted all twenty arms for
Nova, and it happened that Shiva was occupying the space in which he
proposed to mount Nova's second group of ten arms. Emmett had not
been able to win support within the Department of Energy for his plans.
However, he had the backing of Mel Greer, a key staff member of the
House Appropriations Committee, and with Greer's assistance he might
have had enough clout within Congress to prevail. Emmett's critics have
charged that when he tore down Shiva, he was really engaging in a
political ploy, to make room for that second group of Nova arms. In any
case, Greer's help proved unavailing.

Still, Novette did go into operation in December 1982. An early goal
will be to compress pellets to two hundred times liquid density, and then
even to four hundred times. This will improve on Shiva's hundredfold
compressions, and set the stage for the thousandfold compressions that
Nova is to achieve. Also, with Novette, it will be possible to repeat many

of the important Shiva experiments, but at shorter wavelengths of the laser light. Novette's frequency-multiplying crystals will make this possible. The Q-cleared community is anxiously awaiting these tests, since they will be the first experiments to show the value of using short wavelengths at high power.

And what is the purpose behind all this work; what does the country stand to gain from it? The answer is implicit in the ambience of the offices where Emmett's people work.

The nearby freeway, Interstate 580, has green signs near the Vasco Road and Greenville Road exits announcing $\boxed{\text{Lawrence Lvmr Lab}}$, but you cannot just drive in as if it were a college campus. There are gates with security guards; you must arrange your visit in advance and be issued a badge. Once you show your badge to the guard, he will look at it very carefully, but then wave you in. The laboratory sprawls over a square mile, and to get from place to place you can drive on the roads, but you may prefer to take advantage of a very civilized custom. The lab has a number of its own bicycles parked all around, and you can just hop on a bike and ride off, your badge flapping in the breeze. You park the bike at your destination and just leave it there for the next person who comes along. Or if it's raining, dial 2-TAXI and ask the lab taxi service to send a van over to pick you up.

The laser offices are in a pair of spread-out two-story buildings faced with tinted glass window panels. One of them has panels of a rosy amber; perhaps the people inside are the optimists, with rose-colored opinions. But if that is so, the building next door must have people with the opposite view, for its panels are tinted black. People call it the Darth Vader building. If you walk around back, you may hear a rhythmic sound that has been compared to a Trinidad steel band. Livermore has not gone reggae, however. The Calypso-like rhythms come from a neutron detector featuring a long pipe hooked to a vacuum pump, which resonates with the pipe. If you enter one of the buildings—say, the rose-tinted one—you will find yourself in a small lobby, at the end of which is a closetlike room resembling a phone booth, or one of the isolation booths used on the quiz shows of the 1950s. You will need to be in the company of an official escort, such as Mike Ross from the public-affairs office. Step inside and insert your badge into a small plastic window with a TV camera on the other side. If everything is OK, a buzzer will sound and you can open the other door, to enter the office area. In the hallway outside the offices you will see warning signs, like those that are set up to mark potholes in a road. They will read CAUTION—UNCLEARED VISITOR UNDER ESCORT and UNCLEARED VISITOR—UNCLASSIFIED DISCUSSIONS ONLY. The uncleared visitor is you. The security precautions reflect laser fusion's close ties to Livermore's longstanding involvement with nuclear weapons, and they reflect more. A laser fusion microexplosion will amount to a scale model of an H-bomb explosion, and thus will be used to produce nuclear fireballs on a laboratory scale. These will be studied with an eye to designing the bombs of the future.

"We are already doing weapons physics," said Hal Ahlstrom, whose father-in-law had obtained that Shiva statue. "We can study the behavior of materials within a nuclear explosion, creating the temperatures, the pressures, and densities that are appropriate to nuclear weapons. There are no other laboratory systems that can achieve the conditions we can. Then there are nuclear weapons effects. To really do significant experiments on weapons effects, we need a laser able to deliver a couple of megajoules to a pellet, and have it give off a couple hundred megajoules of thermonuclear yield." That would be the energy of a hundred pounds of TNT: a fifty-milliton bomb, if you will. "That's what you need to do really significant weapons-effects experiments. We would get lots of neutrons and X rays from the explosion. That would be enough energy that we could put a test chamber next to the reaction chamber, where you could put a whole satellite or missile nose cone. Expose it and determine its response, how its electronics holds up, or other components. See how well it would stand up to reentering the atmosphere in the presence of a nuclear blast.

"We do nuclear tests in Nevada, and we could supplement them with laser fusion. We do four kinds of tests: effects tests, stockpile-confidence tests, weapons physics experiments, and weapons development tests. Laser explosions would supplement the effects and weapons physics tests." They still would be testing new bomb designs, and also picking occasional bombs from the stockpile to set off and make sure they still work, but laser fusion then would certainly add a new dimension to the underground testing in Nevada.

"People think we like to go out in the desert and shoot off bombs, but that isn't true," said one nuclear-effects expert who has spent many years doing just that. "A test can take a year or more from completion to detonation, and they're usually oversubscribed with experiments. It's hot, you get sand in your relays, and things don't always go as planned. Believe me, it's no picnic. Nothing survives in the thermonuclear environment. You do an experiment once and it's done. If the results were not as you expected, it is very difficult to repeat. But with laser fusion in the laboratory, you can just repeat your experiments at will."

Laser fusion could be especially useful to weapons designers if a treaty should ban all nuclear testing, the underground tests as well as the aboveground ones that have been banned since 1963. "We would maintain a cadre of trained thermonuclear designers," said Alexander Glass of KMS. With laser fusion, "you could train people in thermonuclear design, train people to understand thermonuclear physics. If there were no nuclear test program, the nuclear-design laboratories, Los Alamos and Livermore, would have a hard time maintaining a cadre of competent designers, because they would have no data to work with. They could maintain their design capability by working on laser-fusion designs."

A variety of pellet designs have been openly published, but it is known that the real pellet designs are quite different, and the differences are not trivial. "The drawings you see in those articles look very elementary," declared Nuckolls. "But the weapons laboratories have been working on the design of macro-pellets, if you want to call them that, for some thirty years now; they've invested tens of billions of dollars. Then there are these enormous computer facilities down there, five Crays* sitting down there, along with a large number of very complex computer programs used to design these weapons. And so obviously there's something going on there that's much more complicated than what's been drawn in these simpleminded articles. We have LASNEX, of course, which is related to these bomb codes, but there's a great deal of LASNEX that's never been published. This is an art form, and it has an enormous technological investment in it."

It is quite a jump between thinking of laser fusion as a source of clean, inexhaustible power to thinking of it as a means for simulating nuclear explosions. But there is a great deal more to the nation's energy programs than the highly publicized pursuits of solar energy or magnetic fusion, the efforts to develop a synfuels industry, or the regulation of the price of oil and natural gas. Much of the work of the Department of Energy has all along involved the design, testing, and production of nuclear weapons. That was the original purpose of the Atomic Energy Commission, which was the core around which the Department of Energy has grown in recent years. Still, laser fusion has had a continuing appeal to energy idealists. Can its early promise be restored; can the prospects that motivated Kip Siegel be realized?

*Cray-1 computers, the world's most powerful.

When laser fusion is first achieved, it will be done by brute force, with laser pulses totaling about a megajoule of energy. A megajoule laser pulse is a remarkable object. If it is delivered within a nanosecond—and it can hardly be of much greater duration—it will be about a foot in length. The speed of light, which everybody learns as 186,000 miles per second, can equally well be given as very nearly a foot per nanosecond; hence a nanosecond pulse is a foot long. If the pulse is focused by lenses to an inch or so in diameter, it will amount to a stick of dynamite flying through space at the speed of light, and made entirely of photons. Its length, width, size, shape, and energy all will be the same as those of that dynamite stick. Still, before anyone will build a megajoule laser, we have to start with something very basic: the pellet. We have to understand everything there is to know about pellets.

The way to learn about pellets is to calculate their performance using the LASNEX computer code, then to shoot them using the Novette laser or the larger Nova, when it becomes available in a year or so. Then the test of our understanding is whether LASNEX can make reliable predictions. Working with LASNEX is a little like playing Space Invaders or a similar computer game, although it's less lucrative for the people who wrote its program. For instance, George Zimmermann, who wrote LASNEX, doesn't get a royalty for each pellet run.

You sit at your console and watch as a Cray computer chugs away with its calculations of what is happening to the pellet when hit with the laser pulse. The computer slows time down by a trillionfold and more, so a pellet shot that would take a billionth of a second in real life actually extends over an hour. Sometimes it takes longer; Rod Hyde, the rocket designer, has done some of the longest calculations, which have gone to several hundred hours. Your console does not show the vividly colored and rapidly moving scenes of a videogame screen, but there is a slow-scan portrayal: blink ... blink ... blink .... Once every second or so, the diagram on the screen changes as the microexplosion develops before your eyes. It's not the rapid blinkblinkblinkblink at sixteen frames per second, which would give the impression of continuous motion. Even so, after fifteen minutes you can look at the screen and say, "Ah, the pellet is starting to ablate," as its outer layers begin to blast away.

You can call up LASNEX and work with it, but the main code is sacrosanct. You never see it or touch it; you can't fiddle with it. You can't copy it into your file of programs and modify it or see what's in it. This is also like playing Space Invaders on an Apple II, for there too you can call up the game and play it, but you can't get into its detailed programming to change it. The LASNEX code is entirely in the hands of a high priesthood, consisting of Zimmermann and a few others. They alone know its true inner workings and can change it or update it. That way there is only one standard LASNEX, rather than the blizzard of variations that would develop if everyone could get at its innards. LASNEX is much more serious than any arcade game, and is played for higher stakes and on a much more sophisticated level. Still, it has one more feature in common with Space Invaders. It has been used to shoot down other people's rocket ships.

During the mid-1970s, a group of people in the British Interplanetary Society made a study they called Project Daedalus, which was an attempt to design a rocket somewhat similar to Hyde's. Their final report filled nearly two hundred pages in the *Journal of the British Interplanetary Society,* and the London newspapers noted wonderingly that the Daedalus ship would be taller than the dome of St. Paul's Cathedral. Hyde then took a look at it: "I read in detail only the parts on the thrust chamber and the pellet design. Those I did not think too much of, and so I never really read the rest because it's fairly standard rocket-design stuff." Then he ran their pellet on LASNEX. His conclusion: "You would not get thrust. If you got thrust, if you were able to redesign the pellets so that they lit in the first place—you would melt the engine. The problem is, you gotta have a pellet design. Even if it's trash, you gotta have one to build around. They assumed one based on the unclassified literature and they designed a rocket around it. The problem is, the unclassified literature available to them just is not the full literature. And it's not small changes, either."

Nevertheless, LASNEX still has a way to go. It has two major parts, a hydrodynamics code and a laser-plasma interactions code. The hydrodynamics, the behavior of pellet material while being heated and compressed, comes from bomb codes, computer programs for simulating nuclear-weapons explosions. That part is considered reliable. The other part calculates how the laser energy interacts with the hot plasma blown off from the pellet surface. That part is much less reliable and has changed quite a lot as people have learned more. Sometimes it has had to change by a very large amount. In the laser-fusion world, people still recall what happened when the Livermore physicists set out to use the Shiva laser to get the most energy possible from a pellet.

The measure of this energy was how many neutrons the pellet would give off. One of the Livermore pellet designers, who worked for Nuckolls, used LASNEX calculations to predict a yield of 1.3×10^{12} neutrons.

The pellet incorporated a feature that couldn't be calculated using LASNEX, so this designer proceeded to do the calculation using what amounted to an equation from his freshman physics text.

At the University of Quebec in Montreal, Dr. Tudor Johnston held a contest to see who could make the best prediction. He invited each of his colleagues in the laser fusion community to write down a prediction on a dollar bill and send it in, with the whole pot going to the winner. Johnston called this SWAT, Shiva Welfare Altruistic Trust. When the Livermore physicists ran the experiment, they got 2.7×10^{10} neutrons, fifty times less than their prediction. In computer parlance, it was a case of Garbage In, Garbage Out. The SWAT winner was a man from KMS Fusion who had made a lucky guess, 2.75×10^{10} neutrons. The runner-up prediction was 2.997929×10^{10}. This was nothing more than a number familiar to all physicists, the speed of light in centimeters per second. It had absolutely no connection to predicting the number of neutrons. The witticism went around that this was precisely the equivalent of using an ultrasophisticated theory to predict the result of an experiment and then getting a prediction more than four hundred times better simply by writing down the number of your driver's license.

That was in the fall of 1978. Since then, Emmett's associates have shot many more pellets, on Shiva and on other lasers, and have learned a great deal. They will learn even more with Novette and especially with Nova, which should come close to breakeven, producing its momentary hot spot in the pellet center. The combination of data from Nova, from Los Alamos experiments, and from H-bomb tests then will teach us the basic physics of pellets. By 1990 we will probably know enough about them to proceed with confidence to the next step.

For the pellet data, of course, will not correspond to some wonderful laser from never-never land that we don't know how to build. The data will tell us what sort of pellet will work with the lasers we *can* build. The next problem will be to get the needed lasers built. There are several promising candidates, and one of the most promising is known as the krypton-fluoride laser.

It features a tank or chamber filled with a mixture of the gases krypton and fluorine. Krypton is similar to xenon, the gas used in flashlamps, and like a flashlamp this laser would produce short, intense bursts of light, but millions of times more energetic than the flashlamp's. These bursts will be of particularly short wavelength, so they will work well when they hit the pellet; pellets work best with short-wavelength lasers. Also, the laser would be rather efficient as lasers go, and could be fired rapidly, many times a second. The biggest advantage is that such a laser now has actually been built.

Late in June 1983, scientists at Los Alamos turned on the power and produced the first laser beams in what is called, prosaically, the Krypton-Fluoride Laser System. This is a large laser amplifier, rated at twenty kilojoules, which puts it in a class with Shiva and Novette. Significantly, it was developed in the remarkably short time of fifteen months. Following its first test, its project manager was ecstatic: "This is a tremendously exciting alternative to other large laser systems. We have shown that this laser is neither complicated nor expensive."

Building large lasers isn't quite like building aircraft for dusting crops, so it will be a while yet before we can really be sure that this Los Alamos design can be scaled up an extra hundredfold in size, to the point of being useful. Still, laser builders are nothing if not optimistic. They profess few doubts that if they are given enough money—say, half a billion dollars or so—they can go ahead and build a krypton-fluoride laser of two megajoules or thereabouts, suitable for a power plant.

What would a laser-fusion power plant look like? In the world of the electric utilities, there still are quite a few power plants surrounded by acre-size cylindrical tanks, holding the heavy fuel oil that they burn. A laser-fusion plant may well look rather similar, but its big flat tanks will not hold fuel oil. Instead they will hold lasers, with each laser shaped like a low cylinder perhaps two hundred feet across. Such plants will have several interesting features. The lasers, which will be among the expensive and high-tech parts of the plant, will be separate from the fusion reactors; the laser light will be guided to these reactors through long pipes. This means that the laser will not become radioactive and will not call for complex robots. Instead, it can be worked on by ordinary technicians in shirt-sleeves. Eash laser can serve several fusion reactors, with the laser beams being switched rapidly from one fusion chamber to the next.

These fusion chambers, in turn, will be steel vessels some 75 feet tall. Within each one will be what amounts to a very large showerhead, producing a heavy spray filling much of the chamber. The shower will not be of water, however, but of lithium dissolved in molten lead. The lithium will absorb neutrons from the fusion microexplosions and will then be processed to extract the tritium that these neutrons produce, tritium that will then serve as part of the fusion fuel. The lead makes this alloy melt at a lower temperature, and makes it fireproof; pure lithium would pose a very dangerous fire hazard in case of a leak.

The pellets will be injected amid the sprays of molten metal, and the arrangement of showers will be designed to allow laser beams to pass between them. A fusion microexplosion set off within such a metalfall will resemble a firecracker set off inside a heavy flow from a bathroom shower. The liquid-metal flows will effectively absorb the force of the microexplosion. In addition, they will absorb nearly all of the radioactivity. The energy from the rapid microexplosions, in turn, will keep the metal hot and molten. The heat will then serve to boil water, the resulting steam will run turbines, and the turbines will spin electric generators.

Advocates of laser fusion see several advantages to this approach. One of the most significant is that much of the radioactivity will occur as radioactive forms of lead, which quickly cool off as their activation decays and becomes safe. This radioactivity will be particularly easy to handle, and stands as an important example of the point discussed in chapter 1, that designers of fusion reactors will have considerable freedom to avoid choosing materials the activation of which would pose problems like those of nuclear power plants. Also, laser fusion may become available in small-size power plants, featuring only one or two such reactor vessels, which can then grow and expand by adding additional vessels. Utility executives will appreciate this flexibility, the ability of such power plants to be built in a wide range of sizes and power levels.

Of course, all this assumes that laser fusion will succeed and fulfill its promise. In view of its checkered career, this is no mean assumption. Besides, there is the little matter of its budget cuts: $209 million in fiscal 1982, $190 million in 1983, $170 million in 1984. Much of this reduction stems from the falloff in construction costs as Nova approaches completion; still, nobody likes a budget cut. Nevertheless, hope springs eternal, and some of the optimism even involves something better than blind faith. For instance, there is the candid assessment of Stephen E. Bodner, the director of the laser-fusion program at the Naval Research Laboratory:

> Although the laser fusion program has had a disreputable past, in the last few years the program has been generally well run. Most of the research money now goes toward investigations of the relevant physics issues, and we have made substantial progress. I have watched many people swing from one extreme—a belief that laser fusion was an easy extrapolation from the nuclear weapons program that would leapfrog to an early success—to the other extreme—a belief that the failure of the past pellet ideas indicates that the concept has no value for energy production. It is my

own view that laser fusion is still a very simple and beautiful concept, with many potential physics and engineering advantages over magnetic fusion. But there are many uncertainties: in laser-target coupling, in laser technology, and in reactor design, and it is simply premature to evaluate the concept. If the physics breaks right, and enough of it might, then laser fusion could be a winner. That is why I continue to work in the field. But just because the program has gone from extreme oversell and domination by empire builders, to the inevitable disillusionment, please do not dismiss it.

There are still solid grounds for believing in the future of laser fusion. The future of this technology rests on being able to understand the problems of pellets, and being able to build big lasers that can deliver suitable pulses. Today, both these problems appear to be yielding to the efforts of people like Nuckolls and Emmett. Among the most impressive achievements in this program, in fact, have been the successful construction and operation of such large lasers as Shiva and Novette, with the promise of a successful Nova close to realization. Emmett may have lost the battle for his full twenty arms on Nova, but the fact that he was able to build any kind of Nova at all was testimony to the promise of laser fusion. When Emmett began working to build Shiva, many laser experts thought he would never succeed. It had proven hard enough to build a successful laser with one kilojoule in the pulse, let alone the ten kilojoules planned for Shiva. By contrast, no one today in the laser community is astonished when Emmett speaks of a 100-kilojoule Nova. The difference is that Emmett succeeded in turning large lasers from plans and calculations into reality.

The problems encountered along the way, both in pellets and in lasers, have been significant and serious. Moreover, because so much information about laser fusion has been classified, the leaders in its development have rarely been able to show the candor and openness that would win them understanding and support. Indeed, at times they have been quite defensive.

Their reaction to William Metz' critical article in *Science* was a case in point. Metz had not interviewed anyone at Livermore, and his article contained significant technical errors. John Foster, for his part, sometime earlier had submitted to *Science* an article on laser fusion, based on his committee's work. This article stated that there were "no insurmountable roadblocks" to success, and pointed to important potential advantages of laser fusion. *Science* had declined to publish Foster's article, but had published that of Metz. To a number of fusion leaders, this suggested that *Science* was biased against laser fusion. In Nuckolls' words, "Metz did not invent these errors out of thin air. But by failing to interview the key Livermore program scientists, Metz set up himself and *Science* to be the unwitting tools of guerrilla fighters within the laser fusion program, and of critics outside the program."

Yet the fact remains that laser fusion has been a new venture, and any such venture in physics will encounter difficulties and disappointments as it goes along. Laser fusion began with calculations and theories; it grew into a program tempered by hard-won experience. It was inevitable that as studies and simulations gave way to real data from experiments, early enthusiasm would yield to a maturity marked by a true understanding of its problems.

In the words of Emmett's deputy, John Holzrichter, "In the short time of ten years we have seen the laser fusion concept come alive, suffer its early growing pains, break through serious physics obstacles, and give a program that is running close to predictions." Moreover, its close involvement with the nation's programs in nuclear weapons development could actually help laser fusion grow to the point of being useful for producing electric power. Weapons studies and simulations are not the sort of topic one ordinarily likes to talk about. But they are important to some key people in Washington, and these people are willing to spend money on them. This weapons work could well prove a niche for laser fusion, an initial application that will justify putting more money into its development. Eventually, then, it might grow out of that niche and enter the world of commercial power generation.

As laser fusion advances, so too will the prospects for Rod Hyde's rocket, which could become another important use for this technology. The rocket could fly to Mars in as little as nine days, and thus allow us to fly around the solar system about as readily as we travel by ship around the world. No one in Washington is interested today in spending money on such rockets. Fortunately, this rocket work can ride piggyback on existing parts of the laser program, particularly on the development of the krypton-fluoride laser. That laser looks attractive for use in producing electricity by laser fusion. It has short wavelength, good efficiency, and rapid fire, all of which are important for a rocket. In addition, by good luck, it can operate at high temperatures. That isn't important in a powerplant, but it's vital in a rocket. All lasers produce excess heat, which has to be got rid of, and in space the only way to get rid of it is with large radiators resembling wings. Hot lasers mean hot radiators, and the hotter the radiator, the smaller it can be. The best possible radiator is one that can be built to be smallest.

If it becomes possible to develop Hyde's rocket, that fact probably will not be revealed publicly. It will be quite sufficient if some high-level people with clearances can come in and say, "Yes indeed, you've nailed it down, and it's too bad that you will never be able to publish exactly how you nailed it down. But we will agree and sign our names to a report to send in to the folks in Washington, saying that yes, you have done it, you have definitely established the feasibility of laser fusion for rocket propulsion." This may happen well before the end of the century. In the words of Lowell Wood, "We hope to do that in the very near term, in a few years rather than in a few half-decades."

And what of Nuckolls' and Wood's early and optimistic predictions, which held out the prospect of breakeven at one kilojoule? They appeared alongside other predictions, which stated that for laser fusion to serve such purposes as the practical generation of electricity, the laser would require an energy in the megajoule range. In 1970 this second prediction drew almost no attention. The reason was that in those days, no one could conceive of so powerful a laser; but a one-kilojoule laser was a device that might well be built with the technology then available. Thus, all attention was on the hope of one-kilojoule breakeven.

Today, by contrast, a laser in the megajoule range may well be more feasible than was a one-kilojoule laser in 1970. Even now, Emmett's designers are drawing up plans for a new laser, to be called Zeus. The name is appropriate, for it is to be the king of the lasers. Emmett hopes it will cost no more than Nova, but he expects it will reach the astonishingly high energy of five or even ten megajoules, making it brighter than a thousand Shivas. If Nova's pellets work, then Zeus will seek to ignite Nuckolls' high-performance pellets, demonstrating not merely a more intense hot spot but a propagating burn within the mass of the compressed fuel within the pellet. These pellets will be full-size, rather than being one-tenth scale models like those that will be tested with Nova. What is more, Nuckolls and his designers are inventing advanced pellets that may be suitable for a fusion power reactor while demanding lasers much smaller than Zeus. If Zeus fulfills its promise, it may stand as the most powerful laser that will ever be built. Future efforts in laser fusion then might treat Zeus as a point of departure and work downward, seeking to reduce the needed power.

This situation then represents a distinct turnaround from that of 1970. Even if the earliest lasers had worked as designed and had reached breakeven, that still would have been no more than a first step toward the vastly more powerful lasers that would have been needed for useful applications. It is easy to contrast the cheery optimism of 1970 or 1972 with the cautious, measured statements of today's laser-fusion leaders. But it is worthwhile to contrast the one-kilojoule lasers that couldn't be built in 1970 with Zeus, which could be built today, and with Nova, which is currently under construction. These contrasts are measures of the changes that have come to laser fusion as the laser-building art has advanced.

Nevertheless, that early optimism had its uses. In 1970, laser fusion was already more than a decade old, yet was plodding along amid low priority and lower budgets. Nuckolls and Wood were far from being the only optimists; similar hopes were rampant at KMS and at several other labs. But if the scientists had then appreciated how severe were the difficulties of laser fusion, that program might today be at nearly the same level as in 1970. After all, in 1962 Ray Kidder had predicted that breakeven would require a hundred kilojoules; eight years later, that prediction was lighting no fires. If today we are close to completion of Nova and can look ahead to Zeus, that is only because Emmett was able to build such precursors as Argus and Shiva. Those, in turn, stemmed directly from the surge of enthusiasm that came forth in response to the optimism of Nuckolls and Wood, of Brueckner, and of a number of others.

On its face, that optimism was ill founded. But without the spur of hope in achieving one-kilojoule breakeven, no one would have carried forward with the hard, necessary work needed for an era of megajoule lasers. After all, in 1970 the tokamak had already shown its high promise. The nation hardly needed a second and independent fusion program, founded on entirely different principles of physics. Yet from a different perspective, that early optimism was well founded indeed. The hope at that time, at both KMS and Livermore, was that laser fusion could be important to the nation, and that it might be achieved using lasers and pellets that even then could be built. Today we can still hope, and with much better reason than in 1970, that laser fusion will be significant and will be achievable with equipment that can be designed and built today. The hope has not changed, only the energy level and the date anticipated for its achievement. From this perspective, the goals of the 1972 *Nature* paper, the goals for which Kip Siegel worked and died, still are there, still are being sought.

7

The Entrepreneurs

W ITH the cautionary experiences of KMS Fusion before them, it would be entirely understandable if the nation's entrepreneurs and inventors were to leave fusion strictly alone. After all, there are plenty of opportunities in electronics and computers. Indeed, in California's Silicon Valley, four out of five new electronics firms succeed. There are opportunities aplenty to invest in condominiums, in shopping malls, or in oil and gas drilling, all of which offer enticing tax writeoffs. By contrast, if there is one lesson that might easily be drawn from the KMS affair, it is that fusion is best left to the Department of Energy. They can absorb the risks, pay for the big and costly programs that run for ten years and longer, proceed at their own pace without fretting too much about producing a product and earning a profit. Privately funded entrepreneurs can't. No Kip Siegel will ever build a TFTR, an MFTF-B, a Nova laser. Such projects will always rely on your tax dollars at work.

Yet on Torrey Pines Road north of San Diego, close to the coast, there is a two-story office building owned by E. F. Hutton. Its second floor has been leased by Inesco, Inc., whose chairman is Robert W. Bussard. Bussard is thoroughly familiar with the KMS story, but nevertheless he intends to succeed where Siegel failed. He aims to produce practical, commercial fusion power by 1990—and to be earning profits, big profits, not long afterward. Indeed, he wants to save the world, and much of his project has a definite Southern California go-go aura about it. To anyone from east of the Coast, the office surroundings would appear almost too pleasant to be easily seen as a place for serious endeavor. Bussard's most notable work, before he entered the fusion world, was in the invention of nuclear rockets for a manned flight to Mars. His chief financial backer has been Bob Guccione, founder and publisher of the magazines *Omni* and *Penthouse*. Yet Bussard just might make it. There is enough technical strength behind his efforts that he deserves to be taken seriously. What is more, he may be able to attract enough financial support to achieve his goals. If he stands among the darkest of dark horses in the energy sweepstakes, he ranks high indeed among those inventors whose ideas may turn the world around, if only they can be realized.

As with virtually everything else in the modern world of fusion, his story starts with Robert Hirsch. Bussard was working at Los Alamos, helping to build up their laser fusion program. Then, early in 1973, he recalls, "Bobby Hirsch called me one day and said, 'Why don't you come join me here and we'll do something with magnetic fusion?' " Soon he was installed as one of Hirsch's three assistant directors. During the next year and a half he helped push fusion budgets toward the tenfold increases that they experienced during the 1970s.

Another of Hirsch's assistant directors was Stephen Dean, who recalls vividly what Bussard did: "He came up with the grand ideas of why we needed more money." He could invent a great many reasons to justify more money, and he helped the others develop reasons why they needed a lot more. He also helped in the strategy. He knew that fusion didn't have a big constituency, but he had ideas about how to get around that. He said that the Indians had had a practice with the settlers: if there were only a few Indians, but they wanted the settlers to think there were a lot of them, they'd circle the wagon trains faster and shoot off more guns, to make the settlers think there were millions of them. He said they had to do the same kind of thing for fusion: raise such a ruckus, such a storm of pressure, of ideas and demands, that the people from whom they were trying to get support would think there were thousands of people out there demanding fusion, ready to get on with the job. Actually, at that time only Hirsch's core group really wanted to expand the program. Hirsch and his staff weren't getting this pressure from the laboratories. They created this pressure themselves. Bussard thus was a major influence in having Hirsch and his people think more expansively than they might have otherwise.

In mid-1974 Bussard left Hirsch's staff—"I figured I had sort of done everything I could to help make the program grow"—and set up his own firm to do energy consulting. One of his clients was Hirsch. By early 1976 he was working on a question Hirsch had put to him: What would it take for fusion to succeed economically in the real world, what would make it attractive to bankers and financiers? All apart from the tokamaks and mirrors then planned, what direction would fusion follow if it could be guided by economics and not physics? Bussard wrote a paper arguing that tokamaks would always be too expensive, unless they could be made small and compact. Then he set out to show what a small tokamak would require. In doing this, he was able to take advantage of an important advance achieved only a few months earlier, under the leadership of his good friend Bruno Coppi.

He had first met Coppi in Varenna, Italy, on Lake Como, in 1961, where he was lecturing at a seminar on nuclear propulsion. Coppi was finishing his Ph.D. studies at the Polytechnic of Milan. From the start, he was fascinated with Bussard's vivid imagination. Bussard helped him come to the United States, and during the mid-1960s he worked at Princeton. Then he went to MIT and was there when Lev Artsimovich gave his lectures on the tokamak, early in 1969. Soon MIT was on board the tokamak bandwagon, preparing to build one based on Coppi's ideas. Coppi wanted to build it small, which required that he use the highest magnetic fields attainable. He went on to work with Bruce Montgomery, the leading magnet designer at the Francis Bitter National Magnet Laboratory at MIT. Montgomery liked to say that when inventing anything, "Build it simple, build it stout, out of things we know about." One thing Montgomery knew about was the Bitter magnet, a high-field design invented by Francis Bitter himself. Reaching back to his Roman heritage, Coppi christened their tokamak Alcator, an acronym for "high-field torus" rendered into Latin. They built it out of Bitter magnets, guided it through early troubles, and then in 1975 achieved success.

What they found that year was that Alcator could have its cake and eat it, too. As their experiments progressed, not only were they getting unusually good plasma confinement, but they were getting high values of density as well. What was best of all, as the density went up, so did the confinement. This was quite unexpected. Hirsch was delighted, and that summer he asked the Alcator leaders to push the confinement parameter $n\tau$ as high as possible by Thanksgiving. Late in October the project leader, Ronald Parker, phoned Hirsch at home with the good news: Alcator had reached $n\tau = 10^{13}$. This was five times higher than any tokamak had previously reached, and was well within the range of what would eventually be expected from TFTR. But Alcator was no TFTR, the size of a mini–space station. It was less than four feet across, and barely the size of a truck tire. In the tokamak world, this achievement was the most significant advance since Artsimovich's original results with his tokamaks in Moscow in 1968.

This heartening relation, that confinement improved with increasing density, would soon be called "Alcator scaling." It was only natural that Bussard would soon be thinking about building small tokamaks along the lines of Alcator. Such a small machine would need the highest possible magnetic fields, but these fields could be attained with water-cooled Bitter magnets, like those that had been used on Alcator. It might then produce, in a very small space, a great deal of power, most of which would come off as neutrons. These neutrons would make the machine radioactive in a short time and would weaken its metal structures to the point where they would lose their strength within a month or so. But with its small size, the machine might be cheap enough to be used for power and thrown away. There would be no need for expensive and complex robots or other equipment for remote maintenance. The key to it all, however, would be how powerful the magnets could be. That, in turn, would depend on how well they could be cooled, for electric currents in the magnet coils would produce fierce heating. There would also be a serious heating problem in the walls of the tokamak itself, which would have to confine an intensely reacting and highly energetic plasma within its small space. But if there was one thing Bussard knew well, it was designing cooling systems for highly compact, very powerful nuclear systems.

Prior to entering the world of fusion, he had spent quite a few years working on advanced rocket engines. As early as 1952, he had proposed that it would be possible to build nuclear-powered rockets. His idea was to build a nuclear reactor out of uranium in graphite, run it at extremely high temperatures and power levels, and force hydrogen through it at high pressure. The hydrogen would carry away the reactor heat, and in the process would blast to the rear, producing a rocket exhaust with better performance than could be had with any chemical rocket fuels. His ideas led directly to the nation's program in nuclear rockets, which went on for seventeen years and spent $1.7 billion. This program produced rocket engines that were well tested and reliable, and it reached the point where Lockheed was to build a flight-test model to fly to orbit. But in 1973 the program was canceled. It was promising engines of far more capability than NASA could use on any reasonably foreseeable mission: indeed, engines the most appropriate use of which would be to take astronauts to Mars. But a manned Mars program would cost $100 billion, and simply was not in the budget. Despite being a technical tour de force, there was no way these rockets would be built. Still, this project made Bussard one of the world's leading experts on heat-transfer problems his compact fusion reactor would face.

The real question was whether his plasmas would ignite. A small tokamak would have no room for big, bulky neutral beams. Instead it would have to rely mainly on having the plasma heated by means of the powerful electric current flowing through its center, which would heat the plasma the way a current in a coil of wire heats the burners on an electric stove. This was the method used on Alcator, which still was very far from ignition. The problem was that he couldn't simply plan on using whatever current it would take to do the job. Too much current would make the plasma disrupt, dashing itself against its chamber.

Then in June 1976, Bussard went to Princeton for a meeting of Hirsch's lab directors and senior consultants, which was being held in a small auditorium in Lyman Spitzer's astrophysics building. There he ran into his old friend Bruno Coppi, and they went off to lunch. Coppi was a man who could easily bubble over with ideas, and right then he was particularly ebullient. He wanted to follow up his Alcator success by building a new tokamak, which he called the Ignitor. He had come up with an idea whereby, with a high enough magnetic field, he hoped to heat the plasma by means of its central current and reach ignition. He wanted to build Ignitor as a physics experiment, whose plasmas would ignite only momentarily and then go out. But Bussard had other ideas: "It just struck me, as we talked, that if we applied aerospace technology to the physics that Bruno was alleging could get you to ignition, we would actually have a machine that could not only ignite, but survive. It could make net power, if the numbers came out right. If you could get it to ignite. If, if, if."

They decided to work together, and Bussard soon was proposing that to get real power, the tokamak should run at two or three hundred times the power at ignition. Coppi said, "That won't work, Bob." Bussard replied, "Why not?" "Because the machine will melt." "No it won't, Bruno, not if we design it and cool it right." The question was whether all of it would work. To find out, a lot of detailed calculations would be needed, and these were beyond the range of what Bussard and Coppi could do on their own. Still, over the next few months they did what analyses they could, meanwhile taking time away from profitable consulting opportunities. By fall their ideas were still looking good, but they had no more than their own work, plus a handful of computer calculations they had been able to cadge from friends of theirs. Then Bussard went off to get other opinions, most notably those of Lawrence Lidsky, one of MIT's leading plasma physicists, and Alvin Trivelpiece, who had been the third of Hirsch's associate directors and who had taught plasma physics at the University of Maryland for twenty years. Bussard took the idea to Lidsky just to get his opinion. Had they missed something; was their plan crazy? Then he went and talked to Trivelpiece, whom he knew very well as a friend, to see if he'd look at it and say, "This is really nuts, you forgot some important thing." Nobody Bussard talked with could find anything crazy or missing. In November, he and Coppi went off to see Hirsch.

Hirsch intended to give them only fifteen minutes, but he got interested and had them stay for forty-five. His secretary kept interrupting him, but he said, "Go away, go away." Then he went to his boss, Robert Seamans, who was President Ford's energy czar. Hirsch told Seamans that here was an idea that had come in from the outside; it didn't come from the labs; it looked promising. Seamans said that by all means Hirsch should look for a way to support it. A few days later Bussard and Coppi were back in Washington, this time to meet with Hirsch's deputy, Ed Kintner. Kintner came in skeptical, but soon was thinking their ideas made sense, and hinted that they should submit an unsolicited proposal. He couldn't come right out with that—he couldn't solicit an unsolicited proposal—but Bussard knew that preparing such a proposal was the thing to do. That would give Kintner's staff something to review, and then they could see if there was a way to support it.

That December, Bussard and Coppi filed for a patent on their mini-tokamak, and then proceeded to incorporate their firm in Maryland. Bussard wanted to call it Enesco, for Energy Systems Company, but when he went to the office of the commissioner of corporations in Baltimore, he found the name was already taken. Within fifteen minutes he picked Inesco, for International Nuclear Energy Systems Company, and he then finished the legal formalities. When he told Coppi what had happened in Baltimore, Coppi clapped his hand to his forehead: "My God, Bob, do you realize what you've done!" "No, Bruno, what have I done?" "There is an Italian verb *innescare,* spelled with two *n*'s, not one, and the first person singular is *Io innesco,* which means 'I ignite.' "

They needed a name for their invention, also. Coppi had been talking about an Ignitor; he had worked on the Alcator; Lidsky had invented the Torsatron—they had to call it something. They had a *reactor-ignition* machine, and they called it Riggatron.* But that wasn't the real origin of the name. To qualify for a government contract, they needed to demonstrate financial stability, and the way to do that was to get a line of credit from a bank. They got their first credit line, for $50,000, from the Riggs National Bank in Washington, D.C.—and they named their device after, not before, they got the credit line. "I've never been sure whether Riggs liked it or not," said Bussard. "They're aware of this, by the way."

Their eventual contract was for $637,000, to cover nine months' work beginning in July 1977. This money gave them the chance to do some proper studies, including computer simulations to show how Riggatron plasmas might be expected to behave. Right at the start, Bussard had his eye on the man he wanted to lead those studies. His name was Ramy Shanny, and he had had quite a life. Shanny had been born in Palestine in 1935, before the State of Israel was founded, and had spent his early teenage years there on a kibbutz. He had gone to high school in England, then returned to Israel and had fought in the 1956 war with Egypt. Then he came to the United States and had done his college work at Tri-State College, an unaccredited school in the Midwest. He did quite well there, but that lack of accreditation meant that his choices of graduate schools were quite limited. In fact, only two universities would accept him: Princeton and Caltech.

He picked Princeton, and proceeded to study plasma physics. Then he went to the Naval Research Laboratory near Washington, and helped build up a major plasma group there. That was what attracted Bussard to him. In 1973, while with Hirsch, Bussard had found that Shanny headed the best plasma-physics group he could find, better even than any of the plasma groups Hirsch was funding. Although Hirsch was in the Atomic Energy Commission, he wound up getting Shanny's people in the navy to do some of the needed plasma studies; his own AEC physicists weren't nearly as good. To Bussard, that was impressive.

By 1977 Shanny had left the navy and was a free agent. Two weeks after Bussard's contract went into force, on a Saturday night in mid-July, Bussard picked up the phone and called him: "Ramy, I need you. The fusion machine is now funded; we're going to do a study. I need you to help me run it." That sounded interesting to Shanny, so Bussard went on, "Can you be here on the first plane tomorrow morning?" Bussard was near San Francisco; Shanny was in San Diego. He caught a 7:00 A.M.

*The term "Riggatron," used here and elsewhere, is a copyrighted trademark or trade name describing a compact fusion reactor concept invented by Dr. Robert W. Bussard of Inesco, Inc. All rights to this trade name are held by Inesco.

flight, and they spent the day talking about Bussard's work. By evening
Shanny was convinced. Bussard already had Coppi carrying out plasma
studies at MIT; now he would have Shanny running independent computations in San Diego. Thus, he would be able to check one set of
computations against the other.

Soon, however, there were problems on the way. By that time, Kintner
had moved up to take over Hirsch's position. A month later, in August,
Kintner called a meeting in which people from all the fusion labs were
asked to come up with ideas that would help push the program more
aggressively, achieve fusion more quickly. Bussard and Coppi came—
and soon found themselves being raked over the coals. Bussard had
expected merely to present a summary of what they were hoping to do.
Instead, Harold Furth got into an argument with Coppi about the
Riggatron. Then Furth started arguing with Bussard. Ron Parker, the
head of the Alcator project, complained that the ideas of Bussard and
Coppi had originated at MIT and they should give proper credit. Eventually Kintner himself had to stand up and tell everyone to calm down.
Afterward Bussard caught up with Parker, whom he had never met
before, and said, "It's hard to see how I could have stolen anything from
you when I've never met you. It's hard to see how Bruno could have
stolen anything, since he invented the whole Alcator line." But this
meeting was Bussard's first inkling that the fusion community might not
think kindly of him and Coppi.

Next, Bussard found himself caught in a crossfire between Kintner and
the Office of Management and Budget, which had control over all federal
spending. Within the OMB there was an official named N. Douglas
Pewitt, who was in charge of the fusion budget. Bussard had all along
made a point of touching all bases in Washington, so he knew Pewitt and
his associates. By the fall of 1977, Pewitt had decided that he wanted
Kintner to spend some money to look at a broader range of fusion
options. Pewitt phoned Bussard: "Bob, I'm going to run you up the
flagpole with Ed Kintner. Do you have any objections?" "Well, Doug,
what does that mean?" "I'm going to use you as an example to Ed, try to
force him to diversify the program, by guaranteeing to put money behind
your project and any other alternate concepts he can find." "Well, Doug,
what's this going to do to me; what are you going to do to me?" "This is
probably going to make Kintner mad at you." "Is there any way I can stop
you, Doug?" "No, I just wanted to let you know."

A week or two later, Bussard got another phone call. This was from
Don Repici, a young budget examiner who worked with Pewitt. Bussard

was just then in Arlington, Virginia, and Repici wanted to talk with him about his contract. Bussard came out through a blinding rain to a nearby Holiday Inn in Tyson's Corner, where they proceeded to sit down over coffee. Although Repici had a Ph.D. in physics, he did not feel he could assess Bussard's ideas. The most he could think was "My heavens, if ten percent of this is true, this is a spectacular program." Repici phoned Kintner and said, "This sounds like a great idea; what do you think of it?" Kintner replied, "Oh, it's an interesting idea; there's a certain amount of risk to it. I wish we could fund it. You know, we are funding it, but we just don't have enough money to put it all together, because it would cost five million dollars."

Repici, however, had enought clout to solve that problem. He arranged for $5 million to be added to the fusion budget, and a week later phoned Kintner again: "Say, Ed, I've got the problem solved. We'll give you the five million." Repici then went on to argue that Kintner needed to have a certain amount of risk money in his program, a few millions to spend on ideas like Bussard's, even if such risky projects turned out to be a crapshoot, as he put it. But Kintner was upset at being bypassed: "You can't do that, you're circumventing the decision process. It would show that if you go outside of the system, you'll be successful, and I'll lose control of the program." Repici replied that he appreciated that, but "What's more important, you losing a certain amount of control or a key idea getting funded?" At that, Kintner hit the ceiling, and it was obvious that he intended to have things done his way or not at all. Bussard would get extra money only by his personal say-so, and not by way of an end run around him and through the OMB.

Much the same happened in 1978, when Bussard tried to get $7 million in additional funding by way of a congressional amendment to the appropriations bill of the Department of Energy. Congressman Manual Lujan of New Mexico, a Bussard supporter and a member of one of the main energy subcommittees, described the lobbying against his amendment as "hysterical" and said that it "was very well orchestrated and most probably had an ulterior motive." In fact, by the time the contract was completed, Coppi, disenchanted by the politics involved, had returned to full-time research and teaching at MIT, leaving the project solely in the hands of Bussard and Shanny.

By then, Bussard's main concern was not so much whether he could get more money out of Kintner's office, but whether his work would merely be treated fairly and reviewed without bias. When Bussard and Shanny finished their studies, they prepared a three-inch-thick report, which was reviewed by outside experts at the direction of Kintner's office. This was standard procedure; these experts would assess the quality and significance of the work. Their opinions, in turn, would strongly influence Bussard's chances of being able to pursue it further.

These expert reviewers were not told, "Please critique this report." Instead they were told, "Go through it and note everything that's wrong with it." One of MIT's leading magnet designers, who was told precisely that, nevertheless prepared what he felt was a fair review. He noted shortcomings, but noted also that they were not fatal or insolvable, and added a number of favorable comments. When he read the official Department of Energy review document, which was based on his report and those of the other reviewers, he saw that all his favorable comments had been edited out.

Bussard was not without influence in Washington, and with the help of supporters within both Congress and the OMB, he prevailed upon Kintner to arrange for a second review. This one was more to Bussard's liking. The new panel arranged for some outside consultants to spend about two months running computer simulations of Riggatron plasmas, using their own plasma-physics programs. Shanny and Bussard didn't run the simulations; they merely stated what parameters should be used to represent conditions at which Riggatrons would run. The reviewers concluded from their calculations that, barring some undiscovered disaster in plasma behavior, Bussard's machines were almost certain to ignite. As for the heat transfer, the first review had said that Riggatrons would melt. For the second, in Bussard's words, "We had an honest, unbiased, non–bought-and-paid-for reviewer in the form of a gentleman from NASA's Ames Research Center. This man had built a high-temperature water-cooled heating facility, the Ames Jupiter arc, for re-entry testing of the space shuttle. It's a tremendous machine. It has run three years without a failure. The heat loads that he can take on that arc are three times the highest heat loads that we have ever seen required for any version of Riggatron reactors. And he sat on that panel and said that heat transfer is no problem."

Why should Bussard have had such trouble? What they were doing—trying to invent a quick road to low-cost fusion power—simply went against the grain of a federal fusion program committed to big machines, long schedules, and a very strong emphasis on physics research. To the people in the DOE, Bussard and Coppi were reenacting the roles of Kip Siegel and Keith Brueckner. Once again here were entrepreneurs who had spent years working within the fusion program, and who were now presenting ideas that they regarded as original but that were closely related to work the government had been funding all along. Moreover, they were taking these ideas to the Patent Office and setting up a private firm that was proposing to achieve commercial fusion in only a few years. This goal was to be achieved long before the government's fusion program would be ready with such a project, and then, to add insult to injury, they were asking the DOE to back them.

What was more, their Riggatron was completely at odds with the conventional wisdom about how to go after fusion. All the leading fusion physicists were thinking in terms of neutral beams and superconducting magnets, things they knew about. Bussard was thinking in terms of heat transfer. All his experience in aerospace engineering, his background in nuclear rockets and the cooling of rocket motors, applied to the Riggatron. And these were things the fusion physicists didn't know about. To be sure, plasma experts could no more be expected to understand the engineering problems of high-performance rockets than could rocket experts be expected to understand plasma physics. But Bussard was working in areas that he understood and his critics didn't.

In addition, there was just no easy way to bring Bussard's activities into the overall fusion program. Inesco was not like KMS Fusion, which could carve out a niche as a pellet research lab and attach itself as a tail to Livermore's dog. The Riggatron could be thought of as an advanced Alcator, but the Plasma Fusion Center at MIT was a rather small operation to begin with, and inevitably the MIT scientists would wind up dogfighting with Inesco over a limited pool of funds. Thus, the best Bussard could hope for was to obtain what he called "the government's USDA prime beef stamp," an official recognition that his ideas had merited funding at least for a while and that nothing in them was obviously wrong. Bussard would not find his future through government support; he would have to try his luck with private industry. Still, with industrial support he might make far larger profits.

Even before the second review, Bussard had been talking to officials at Litton Industries, and in the fall of 1978 they gave Inesco its next contract. This was not to do further work on the Riggatrons themselves. Instead, what these Litton executives wanted to know was, if Riggatrons would work as well as Bussard was saying they would, what could they do in the real world of energy applications? They asked Bussard to study the use of Riggatrons in a wide variety of situations that could make use of the cheap steam or cheap neutrons they would produce. For a year and a half he and Shanny immersed themselves in Litton's world, the world of power and industry. They concluded in the end that inventing Riggatrons amounted to inventing a new fuel that was the equivalent of burning oil at one dollar a barrel.

"The reason it's so cheap," Bussard said, "is that you're burning copper instead of oil." When mass-produced, the Riggatrons would be built of copper alloys and would cost about $1 million each. Each one would have a service life of about a thousand hours; then it would be used up and have to be disposed of. Because of this short life, it would actually be cost-accounted as if it were fuel, being continually purchased and used. But during that life, it would put out so much energy, considering its cost, that it would offer the eye-popping economics of one-dollar oil. And even if it were to fall short of Bussard's hopes, the difference between its cost and that of oil at $30 a barrel would offer plenty of opportunity for Riggatrons to turn the energy picture around by offering energy at the equivalent cost of oil at only $3, $4, or $5 a barrel.

An early application could be in replacing oil-burners in existing industrial plants. A Riggatron's small size would allow it to fit in easily. In Bussard's words, "You don't build the whole plant, just put Riggatron units next to the fuel tanks. Don't cut off the fuel, just tap into the steam lines, so in case something goes wrong you can still run the plant on the oil. You just don't burn any oil. It's about a ten percent add-on cost. Quickest, cheapest way to get off burning oil in power plants." Then, in building new plants powered by Riggatrons, only about 10 percent of the plant cost would be for items associated with these Riggatrons. The remaining 90 percent would be for heat exchangers, turbines, construction, and other things that plant builders and their bankers already know about and feel comfortable with. This would be very different from nuclear plants, where the nuclear aspect dominates everything. "You would have plants whose cost can be estimated fairly well, running on cheap fuel—copper—and that restores energy prices like forty years ago. But the manufacturer of these things isn't going to sell them cheaper. He's going to sell them slightly below the market and make more profit."

Bussard went down to Caracas and talked with people from Petróleos de Venezuela about their plans to develop oilfields in the Orinoco region. That area has three times the proven petroleum reserves of the entire Middle East, but most of it exists as heavy tars that are much too thick to flow into wells. Current plans call for part of the potential yield to be burned to produce steam at six hundred degrees. The steam would be injected into the oil-bearing formations to heat and thin the tars so as to make forms of oil that can flow. Steam made with Riggatrons would be much less costly than that made by burning oil, and in addition would save oil that would otherwise be burned for steam. The result would be a cut of $6 per barrel in the cost of producing Orinoco oil. Said Bussard, "They're not gonna wait for us; they're gonna build steam plants; they're looking at various kinds of nuclear reactors. But we can beat the cost of any of those. They're very interested in what we'll have—but we don't have it. When we have it, they'll be more interested and we'll put some plants down there."

He also went to Litton's shipbuilding yards at Pascagoula, Mississippi, to develop ideas about mounting Riggatrons on barges. He found that such portable sources of cheap steam could be very useful in improving the prospects for fuel alcohol made from sugar cane. Currently, it is made by squeezing the cane for its fermentable juice, then burning the squeezed pith and husks as a fuel to provide heat for the distillation. Bussard proposed to use a Canadian process that would turn the husks into plywood suitable for the building industry, and to use Riggatrons to produce the needed heat. The pith then could also be fermented, the plywood sold, and the result would be fuel alcohol at 35 cents a gallon. "You could take over the automobile market, make the cars run on alcohol. There are already a million cars running on alcohol in Brazil. Alcohol burns cleaner, solves the emissions problem—wonderful stuff. One Riggatron-driven plant could produce five thousand tons a day of fuel alcohol, from a canefield eighteen miles on a side in the tropics. A hundred and ninety such plants would run all the autos in the U.S. Those plants would cost four hundred and fifty million dollars each, on a barge mounting. Build 'em in Pascagoula; barge them anywhere in the world."

Bussard even looked at the prospects for desalination of seawater. "We found that because the steam we make is so cheap, if it all works, we can beat almost anyone else's method of desalination. We could produce fresh water from seawater at a dollar per thousand gallons, cheap enough that anyone can have it. As the price of oil declines, because it's driven under by things of this kind, the Arab states will be in terrible shape economically. But with this cheap water, and cheap electricity for fertilizer production, they can convert to agriculture. We ought to be able to stabilize the Middle East this way, if we do it right. I really believe it."

Not all potential applications turned out to be winners. District heating, producing steam to heat apartment and office buildings, turned out mostly to involve insulated piping; Riggatrons wouldn't make much difference there. But Bussard's study showed case after case where Riggatrons, if they worked as he hoped, would yield close to a 100 percent annual return on investment. After a year and a half of such studies, he said, "Look, now you know what it'll do; let's go develop this as a program." The men with whom he was working at Litton—Charles Bridge, the senior vice-president for corporate development, and Fred O'Green, the president—felt that developing the Riggatron would cost their company $100 million and would be a risky venture, difficult to justify to the stockholders. But they knew that fusion was going forward as a big government program, and they had the idea that Bussard could help them get in on that, help them win contracts to build things for the government. Bussard, however, wanted to put together a team and start developing the Riggatron. He and Litton then parted company, early in 1980. But by that time he already had other prospects in view.

In February of the previous year, he had been in Washington for the annual meeting of the American Institute of Aeronautics and Astronautics, a professional society in which he had been active for many years. These annual meetings were always lavish affairs, held in the ballrooms of the Sheraton-Park Hotel and featuring close to an acre of floor area filled with colorful displays of the latest in aviation and space. Bussard was in charge of putting together a symposium on the future of flight, and was looking for ways to stir up some publicity. A friend introduced him to Kathy Keeton, president of *Omni* magazine. Bussard didn't know much about *Omni*, other than that it was pro-space, but he thought she would be interested in his future planning. After a few minutes she said, "Why don't you come to New York and tell us about it?" He went up to their editorial offices on Third Avenue and talked with members of her editorial staff. While there, he thought, "Maybe Kathy would be interested in the fusion program I'm looking at." After the meeting, in which he had talked about his symposium, he said, "Kathy, I know this isn't why I came here, but you don't know what I really do. What I really do is, I've got an idea for developing a fusion device that might work, it might work quickly and cheaply, and if it does it might solve most of the world's energy problems." She said, "That sounds very interesting. Why don't you come over to the house and meet Bob, tell us about it?"

He hadn't planned on such a follow-up meeting, but fortunately he was free that evening. He didn't know that "Bob" was Robert Guccione, the founder and publisher of *Penthouse* as well as of *Omni*, whose net worth was close to $200 million. He went there along with the editor of *Omni*, Frank Kendig, thinking that perhaps he could interest *Omni* in an article on his fusion ideas. That address proved to be a luxurious Manhattan townhouse, where Guccione lived and where he conducted much of his business. He was upstairs in the middle of a meeting when they arrived, so Bussard talked with Keeton and Kendig for about an hour and a half. Then Guccione came down and joined them. They talked for about fifteen minutes. Bussard had brought with him some technical material, describing the Riggatron and its prospects, which Guccione proceeded to speed-read. He couldn't follow all of the physics, but he could understand what Bussard was trying to do and why it was important. Then he started asking questions. Whey weren't Bussard and his people doing this right now? Why wasn't American industry behind it? Why wasn't the government funding it? Why wasn't everyone unanimous in their technical opinion of it? He was straight, blunt, and to the point, and Bussard answered him directly.

Then Guccione said, "Look, this thing is so important, and I'm so committed to wanting the United States and the Western world to survive as entrepreneurial and free-enterprise nations. I'm very interested in supporting things like fusion, which the U.S. will be needing." Bussard replied that his firm was working for Litton and he was happy with Litton, going forward with them. Guccione went on, "Look, if for some reason things don't work out—I'll do it." All within fifteen minutes he had reached that decision.

When their meeting ended, it was nine in the evening, and Bussard still hadn't had dinner. Kendig called his wife, and the three of them made their way to the Princeton Club in midtown Manhattan. Bussard could hardly believe what had happened with Guccione. He asked Kendig, "Was that all real? Is that man real?" Kendig replied, "He's the most real man you ever met. He doesn't say things unless he means them."

Bussard and Guccione kept in touch, and early in 1980, Inesco and Litton parted company. Bussard went back to Guccione and said, "It looks as though we and Litton are not going to make the program go as we'd like it to. Is what you said real; do you want to pick up on it?" Guccione said, "Sure, let's go." In March 1980, they formed a partnership. Bussard's program called for spending $65 million to develop five prototype machines over five years. Guccione started by giving $400,000 to underwrite nine months of planning, throughout the rest of 1980. Then, beginning in January 1981, he started putting in more, a total of $4 million that year. In July 1981, he was quoted as saying, "I am convinced that the project will hit paydirt within five years for the modest price of $100 million. The real payoff for me would be to own the world's largest corporation. ITT and Exxon together would not make up the economic potential of the company that owned the world's first fusion prototype." During 1982, Guccione pumped in another $6.5 million.

With this, Bussard was able to make Shanny his executive vice president, and tell him to go ahead: "I want you to run the technical work, like you did at Naval Research Lab. I don't want to run it; I've got about ten thousand other things to do. My contribution will be to come around and kick the tires now and then." Shanny proceeded to pick their technical staff. Inesco started in 1980 with five people on the payroll; a year and a half later it had 88. Shanny lured several scientists away from General Atomic in San Diego, which was running an extensive fusion program under contract to the Department of Energy. Then others were hired from the plasma-physics group at Science Applications, Inc., a consulting firm whose managers specialized in bringing in a few good individuals and giving them lots of responsibility. Since SAI was well known for the excellence of its small groups, Shanny, in being able to recruit some of their scientists, was showing the quality of his organization. One of these was the plasma physicist Carl Wagner. When he was fifteen, back in New York, Wagner had been a chess prodigy who had played against twelve-year-old Bobby Fischer, another chess prodigy, who eventually became world champion. At Inesco, Wagner would be in charge of studying all the plasma issues.

Wagner had gotten his Ph.D. from MIT, and it was also at MIT that Bussard landed a prize catch. Seeking someone to head up his magnet-design group, he asked Bruce Montgomery if he'd be interested. Montgomery had his hands full right there in Cambridge, but he told Bussard, "I have a design engineer who's going to waste." This was Carl Weggel. His brother Robert, who was also at the National Magnet Lab, had built the world's most powerful magnet, a feat that won him mention in the *Guinness Book of World Records*. Carl himself had worked on the magnet designs for Coppi's original Alcator and had gone on to hold the main responsibility for the magnet design on MIT's next tokamak, Alcator C. Then he had designed magnets for the larger Alcator D, but the Department of Energy decided not to provide funds to build it. When Bussard went into Weggel's office, he needed to use only a little persuasion to convince Weggel that working on five different Riggatron designs, which he expected would be built, would be more interesting than working on an Alcator D that wouldn't. When Bussard got back from Cambridge, the other people at Inesco could see that he was walking on air. As he said to one of them, "You can't guess who I just talked to—and he's interested in joining us!"

Guccione's money also allowed Bussard to build offices and laboratories. He rented a large two-story brick house with a spacious lawn, in the Washington suburb of McLean, Virginia. This would be Bussard's home base and office on the East Coast, where he could meet with financiers and attorneys, while staying in touch with people in Congress and elsewhere in Washington. He leased the second floor of E. F. Hutton's office building in La Jolla, California, close by the Torrey Pines seacliffs where hang-glider pilots swoop and soar with their brightly-colored wings. His own office faced in the other direction, toward low hills; but to help him enjoy the view, directly outside his office was a red-tiled sundeck with lounging chairs. That sundeck was a good place for private conferences with Shanny, whose own office was on its other side. They made an unlikely pair for these meetings, Bussard soft-spoken and always in a business suit, Shanny gravel-voiced and dressed in denims, as though he were a telephone installer. Still, they had such meetings frequently. As Bussard put it, "We're like folksingers. I sing, and he plays the guitar."

Shanny had picked out that building himself, and for a reason: "We had to hire people, and they kept asking how solid is the backing, are you sure this is for real. Those questions evaporated and were gone, the minute we moved here. Also, if you provide people with a nice environment in which to work, you can probably get away with paying them less, and you have an easier way of attracting them." Bussard festooned his own office walls with his framed diplomas and certificates, and then complained that with its large glass patio doors and windows, there was no room on the wall for a blackboard or marking board. Shanny's taste in decorations ran mostly to Art Deco nudes and to pinup girls modeling swimsuits. ("They're too young for me," he said.) He liked to keep a radio tuned to light classical music. Most of their employees worked in individual offices, with carpeting on the floors and computer terminals on the desks. The terminals were linked to a VAX 11/780 computer, one of the most powerful of its kind, worth about half a million dollars. Soon Wagner and Weggel would be making heavy use of it, simulating the behavior of plasmas, calculating away on magnet designs.

Visitors were welcome, and Bussard and his associates liked to be hospitable. It was easy to get invited to lunch at a seaside restaurant like the Poseidon in nearby Del Mar. The Poseidon featured a terrace right on the beach and only a few yards from the blue surf. What was better still, it was full of leggy blond California mermaids. Del Mar itself was one of those picture-postcard seacoast towns that might have been taken directly from the Isle of Capri. Its main street, Camino del Mar, was lined with palm trees and pastel-colored houses trimmed with stucco. Along that same street, downstairs from an investment firm, was still another outpost of Bussard's empire. This was a suite of offices and labs given over to metallurgical research, studies of the copper alloys from which Riggatrons are to be built. The lab was quite well equipped, and even featured a small machine shop, to fabricate parts and make sure there would be no unusual problems for the machinist.

Bussard patched up his relations with MIT and with the rest of the fusion community, and began bringing in leaders in the field as consultants. One of them was Marshall Rosenbluth. Often called "the pope of plasma theory," Rosenbluth had spent a number of years at Princeton's Institute for Advanced Study, the research center Albert Einstein made famous. Then in 1980 Rosenbluth was lured to the University of Texas with a salary of $85,000 and the promise of his own physics center. It was all part of an effort by Texas to recruit enough scientific superstars to build a faculty its football team could be proud of. If you were meeting Rosenbluth for the first time, you could easily take him for a plumber, or perhaps an old-line union boss. He has a thick neck, a mass of barely combed hair, and a big pipe. What's more, it wouldn't hurt him at all to lose thirty pounds, or even fifty. At a conference of fusion leaders in 1982, where most of the audience was wearing business suits, he showed up wearing baggy blue pants and a faded old shirt. Still, there was no need for him not to be comfortable. By consensus he was, and is, the world's leading expert on plasma physics.

He came out to Inesco to look over Bussard's work. In his oracular way he gave his opinion. "I think that's a quite promising line of approach, which should be getting pursued probably more than it is. I think as best we understand the physics, that one could be pretty confident—maybe fifty percent, one always has surprises in this game—that one could make such a device ignite." Another consultant was Ron Parker of MIT. He said that if Bussard could keep his plasmas from leaking away too quickly, he would "be first in line to buy Bob's stock, if and when he [went] public." Parker was also quite willing to cooperate with Inesco, rather than regarding them as somehow taking unfair advantage: "We at MIT have very good relations with the people at Inesco. In fact, one could even think of us as a research and development program which, while it's not directed in any way by Inesco, certainly exists for their benefit. We're supported by government money, and Bussard has as much right to our research results as anybody else." Carl Weggel was quite prepared to return the compliment: "We are unabashedly patterning our approach on the MIT approach, using Bitter magnets, using extremely high magnetic fields."

Still, none of this was building Riggatrons. Their employees might be calculating and designing for all they were worth, but before Bussard and Shanny could build and test anything, they first would need to spend $50 million to buy a motor-generator set. They wanted a thousand megawatts of power, more than in the TFTR motor-generators: "We'll have to buy more than we need, so we don't run ten percent short. And we won't know what we need till we actually run the machines. Once we run a Riggatron machine, we'll know what it really takes." This motor-generator set would serve for the power supply at a test site Bussard was negotiating to build in Arizona. Such a site would include machine shops, computers, a control room with diagnostic instruments, even a cafeteria.

However, by mid-1981 Guccione was having second thoughts as to whether he wanted to carry all the risks of trying to underwrite the next ITT or Exxon. He told Bussard to look for additional partners, who would put up more of the money. This set Bussard flying off on a two-year quest for funds. In 1981 and 1982 he logged over a thousand hours traveling back and forth in commercial jets, totaling some half a million miles in the air. By early 1983 he was willing to say that he was engaged in serious negotiations with financiers, both in this country and overseas. Indeed, for a while things looked so promising that Bussard and Shanny believed that by the spring of 1983 they would have all the funding they would need.

However, the negotiations bogged down. This made Guccione unhappy, for Inesco's activities were costing him some half a million dollars a month. In May, he cut their funding, and Bussard was forced to lay off thirty employees. The layoffs included one of his senior aides. Moreover, it was clear that Guccione might very soon drop the other shoe and cut their funding even further, perhaps to zero. But Bussard and Shanny were prepared to face this. In late spring, they decided they would take their company public.

For nearly two years Shanny had been talking with Asher Schapiro, a Wall Street financier. Now Schapiro joined Inesco as one of its senior officers, and proceeded to put together the public offering. This would be a sale of shares on the over-the-counter market. In preparing the prospectus for this offering, Bussard could point to opportunities far closer at hand than the Riggatron. Inesco's metallurgists had invented a new copper alloy, of potentially wide application, to which Inesco held all rights. Also, Bussard was in a position to seek contracts from the Department of Energy, to carry out studies in support of the DOE's own programs.

A successful public sale of stock then would establish Inesco's independence, as a small research-and-development firm having its own block of capital. With this funding, Inesco would at last be able to avoid relying on Guccione as its sole source of money. Moreover, this independence would enhance Inesco's attractiveness within the rest of the financial community. Rather than being an adjunct to Guccione's publishing empire, Inesco would be able to negotiate on an equal footing with financiers and industrial firms, and to draw them in as partners. Bussard then would seek to have these partners provide the additional funds needed for a full-fledged program of Riggatron development. Guccione might well continue to be involved, but he would be only one of these partners.

With Inesco's financing in place, Bussard and Shanny will be able to go ahead and order their motor-generators, build their test site. They have their eyes on a spot of land near Red Rock, Arizona, thirty miles northwest of Tucson and just off Interstate 10. A little to the northwest is the jagged gauntness of Picacho Peak, the eroded basaltic core of an ancient and vanished volcano. Not far away is an old and nearly obsolete electric power plant, and Bussard has looked wistfully at it, remarking to friends that perhaps it will be the place where he will install his first commercial Riggatron power units and generate electricity from fusion for the first time.

Even when he orders those motor-generators, however, the fusion community will have to take Bussard very seriously indeed. He then will no longer be dealing in plans and designs only, or even in metallurgical research. He will be running a tokamak program larger and more advanced than the one at MIT, and his operation will be fully on a par with the work of Princeton or Livermore. He will be going ahead with the design and construction of five different versions of the Riggatron, differing among themselves in magnet shapes or choice of materials, to permit a broader range of comparisons. Then by mid-1988, while Dale Meade's TFTR has merely achieved $Q = 1$, while Ken Fowler's MFTF-B is doing no more than show good plasma parameters, Bussard may be breaking out the champagne on some warm Arizona evening to celebrate the attainment of ignition. "We will inch our way up the temperature curve," he says. "You'll never do that by theory. You'll do that by real-life experiments going another million degrees at a time, shot after shot after shot, to see what nature gives you."

Still, the real question is what nature indeed will give them. Bussard may be seeking to found the go-go growth corporation of the 1990s, but he has not spent his life coping with plasmas, and his plasmas may have other ideas. In contrast to his high hopes is the cautiousness of a dour man from Holland, Kees Bol of Princeton, who has spent many years seeing how plasmas actually behave. Bol does not have a large, sunlit office in Southern California with a sundeck outside. He has a small, cramped one, piled high with papers and reports and with a blueprint tacked to the door, just off a second-floor corridor. He worked with Dale Meade to build the PDX tokamak, and then took over as head of the PDX experimental group. Because he is the director of one of the world's principal tokamak experiments, his views naturally have been tempered by harsh experience, but there is more to it than that. Bussard may hold to the cheery optimism of a man who has spent his career leaping lightly from success to success. Bol, by contrast, has spent much of his life wrestling with adversity.

An accident in infancy cost him his left forearm. Then in 1954, when Senator Joe McCarthy was terrorizing the nation's scientists, he lost his security clearance and his job at Sperry Gyroscope, all within a half-hour. The FBI had charged him with being a dangerous radical. During his college days, he had been president of Students for Wallace. (Henry Wallace had been Franklin D. Roosevelt's vice president, and ran in 1948 with a left-wing splinter party.) Worse than that, however, was that he had spent some summers working as a counselor at an interracial camp run by a Quaker woman. Such an interracial camp was certainly subversive; the FBI was continually investigating that woman. And what sealed his fate was that his father (no, not he himself) had been a member of the American-Russian Cultural Institute, during the war. At that time, Russia and America were fighting as allies, but somehow that didn't matter. When Bol's supervisor fired him, Bol wasn't even given time to clear out his desk. He couldn't qualify for unemployment assistance. It was winter, and his wife had just had their third baby.

He survived by getting a teaching job at Adelphi University; it paid $3,500 a year, but his wife knew how to make do. His lack of a security clearance didn't matter there. Eventually he won a fellowship to study plasma physics at Harvard, and he was holding that fellowship when Lyman Spitzer recruited him to join his group at Princeton, in 1959. He has been there ever since.

"People are scared of studying confinement," he said. "We had the Model C Stellarator, which had confinement that was very poor. It was the so-called Bohm diffusion. We spent four or five years, really trying to find out what in the devil Bohm diffusion was. And we still don't know. All we know is that our machines don't show it any longer. This just gave studies of confinement a bad name. I mean, here we spent all this time, all this effort, and what did we get for it? Well, nothing. What did we do? We built a bigger machine and it all went away. We wanted to understand what was driving Bohm diffusion. The favored mechanism was diffusion caused by drift waves, but that was never really established. In fact, I don't think it's true. I just got back from Washington, and their attitude down there still is that confinement is not a real problem. I think they're scared of devoting time to confinement; they hope it'll go away the way Bohm diffusion did. The trouble is, there is reason to think that confinement on TFTR will not be very good. And if that's so, the Q equals one is likely to be compromised. I mean, if you haven't got the confinement, you won't get the temperature and you won't get Q equals one, either. And if TFTR fails to meet its goals, it will be because of the confinement problem.

"We don't understand what goes on in the plasma. It may be just too damned complicated. At one level of description we understand plasmas very well, and that is when you treat it as a fluid. At that level, plasmas do exactly what the theory says they ought to do; it really works. When you get one step beyond that, they constantly surprise us. There is an enormous amount of work that's been done, of course, on all kinds of plasma instabilities. They're always lacking something. I think that studying plasmas may be like predicting the weather. One of the problems with predicting weather is that small perturbations can amplify. And anything with that kind of a situation, you're probably just out of luck. You may find a statistical description, but you will probably never be able to predict in detail. A plasma may be just the same way. For instance, a little bit of impurity comes in at the edge of the plasma. It changes the local density profile, which influences the way the impurity comes in, and this may be an exponentiating effect which will end up with the plasma in one particular mode. It'll affect the current profile, which will affect the instability profile, which in turn will feed back on the way the plasma interacts with the wall, back on the impurities. So there are all kinds of these very complicated feedback loops that may be going on, involving not just the plasma itself, but the plasma in its whole environment, out to the wall. And I can well imagine that those interactions may be too much like the weather for us to ever really get full control."

Even if Bussard attains ignition, that will not be the end of the story, for he then will be up against the problem of plasma beta. Beta, again, measures how efficiently a tokamak uses its magnetic fields. Bussard has put great emphasis on the results of experiments made with a small tokamak at Columbia University called Torus II. It has attained a beta as high as 12 percent. However, the plasmas in Torus II have been much less hot or well confined than in the major tokamaks, and outside of Columbia, no one has gotten values as high as 5 percent. To achieve ignition, Bussard needs less than 2 percent, which should be no problem. But to achieve the Riggatron's full economic potential, he has been very forthright in stating that he needs 20 percent or more, "which you have to get." Can it be achieved? "Who knows? Nobody knows, no one will know till you try. The theory says you can get over thirty-five percent, some theories say fifty. Second regions of stability, the study at Oak Ridge that said thirty-three percent in 1979. But who knows? We'll never know till we try. If you can't get twenty or twenty-five percent, pure fusion by this route is not going to work economically."

How good are these theories? Again, Marshall Rosenbluth knows: "Second stability regions represent the great white hope. There's a problem in going through this unstable intermediate region, to get from the ordinary low-beta stability region to the high-beta stability region, with beta of forty or fifty percent, almost arbitrarily high. If you can reach this high-beta stability region, then of course all sorts of good things happen. But I think one has to be rather skeptical about whether this second stability region is real or not. There's no experimental evidence at all."

If Bussard can't get high beta, still not all may be lost. He claims that in that case, he could still build a successful power-producer by wrapping a low-beta Riggatron with a thick blanket or tank filled with natural uranium. Neutrons from the fusion reactions then would amplify their energy by splitting these uranium atoms. Of course, this amounts to reinventing the nuclear reactor, and making a success of that would call for much more than just getting good results with his plasmas. Still, it would be a reactor built according to new principles, and Bussard has long experience in inventing new kinds of nuclear reactors. Just possibly, he might get it right.

Bussard likes to compare his Riggatrons to a light bulb. An electric light has a filament inside a glass bulb, or envelope, and then on the outside there may be a lampshade. If the bulb burns out, you don't take apart the bulb to fix the filament; you throw it out and screw in a new bulb. Riggatrons will have a very similar arrangement. The compact tokamak will play the role of the filament, giving off not light but neutrons, and will be housed in a bell jar, a bell-shaped metal container somewhat resembling a light bulb's envelope. This bell jar will be hooked up to vacuum pumps. When a Riggatron burns out, tokamak and bell jar will be removed as a unit, thrown away, and replaced with a fresh unit; a quick-disconnect mounting will make this easy. Then outside the bell jar is the "lampshade," a large cylinder filled with lithium. The lithium will absorb the neutrons and convert their energy into heat. Also the neutrons will transmute some of the lithium to tritium, to keep the fusion reactions burning.

If it all works as planned, Bussard will certainly stand high among the Tom Edisons of the late twentieth century. The Riggatron then will rank as an invention on a par with the computer, the airplane, the automobile. Even if Bussard reaches only as far as ignition, he still will have gone farther than anyone else in the fusion world. As one of his Washington supporters has stated, if he gets that far, "he'll have to fight off the Brink's trucks loaded with gold." Financing his work will be no problem, and he will have plenty of money with which to try to go beyond ignition, into the realm of real economic promise.

Forty years ago the New Mexico desert was the place where the first atomic bomb was detonated. Now we may hope to see the Arizona desert as the place where another and vastly more hopeful form of nuclear energy will first be released. Certainly the Inesco story will stand as one of the fascinating tales of risk and entrepreneurship in the 1980s. But for Bussard there is very little middle ground between failing ignominiously and succeeding magnificently. He need not gain full success on his first try, or with his first Riggatron. He will be building several of them in a series, and will make improvements as he goes along. However, it will not be enough for him merely to get good plasma parameters. Too many labs will be doing that, and more, by 1988.

If he gets only that far, to the point of good parameters, he still will be running a major fusion-research center, with powerful motor-generators and a world-class group of scientists. That might make them look interesting to the federal fusion program, and put Inesco in line for a switch to government sponsorship. But if that happens, it will be on the

Department of Energy's terms, which will not necessarily mean that they will want Bussard to keep working on trying to build commercial Riggatrons. To keep his independence and freedom of action, Bussard must achieve at least ignition, and thereby seize the position of world leader in fusion development. If he can do that, the world will be his to win. And if he can't? Well, probably he will turn Inesco over to Ramy Shanny, and go back to energy consulting. Inesco itself may then settle into some comfortable niche within the DOE program, perhaps working closely with General Atomic, whose main lab is just down the canyon from Torrey Pines. And as for Bussard's financial backers—well, Guccione has all along regarded Inesco as a fine tax shelter, and an excellent source of tax writeoffs.

Note added: Early in November 1983, the American Physical Society held a meeting in Los Angeles. With hundreds of physicists assembled, MIT's Ron Parker electrified the gathering with news of a spectacular achievement. On November 3, his experimenters had gained a new record with Alcator C, a confinement parameter $n\tau$ of 8×10^{13}! This was twice the previous record, and for the first time represented an $n\tau$ good enough to yield net fusion power, if the plasma temperature was right. Parker thus had come close to fusion's Holy Grail, an $n\tau$ of 10^{14}. The betting was he'd have it by Christmas.

He gained this by using a new method of injecting fusion fuel into the plasma. Rather than puff it in as gas, the usual procedure, he shot the fuel in as tiny pellets of frozen deuterium. Alcator C used no tritium, mounted no neutral beams. Its plasma temperature was a modest 18 million Celsius degrees. Still it produced so much fusion energy that some people got antsy and ducked out of the control room. They were afraid that they'd receive more than their allowed dose of neutrons, from the fusion reaction.

Ramy Shanny was at that meeting, in a light-beige suit. He was ecstatic and was boasting of the new results to all who would listen. To him this $n\tau$ was a vindication: "It means we have a much stronger chance now of getting ignition!" His Riggatrons would be so similar to Alcator C that now they had appeared much more credible. Moreover, just within the previous month Bussard and Shanny had completed arrangements for their first offering of stock. This meant that during a few short weeks, they gained both financial stability and a dramatic new promise of technical success. Thus the people at Inesco were in high spirits indeed.

8

The Breakthrough

ONE of the spectacular photos from the early days of World War II showed the destruction of the destroyer USS *Shaw* at Pearl Harbor. A Japanese bomb reached the magazine, and the *Shaw* blew up like a fireworks display. Long thick streamers of smoke traced their arcs high in the air, driven by the explosion. Pieces of metal and machinery were hurled to great distances by the force of the white-hot fireball that tore the ship apart. Fortunately, there were few casualties. The ship was in drydock, it was Sunday morning, and most of the officers and crew were ashore.

On that particular day, a young midshipman in the Naval Academy at Annapolis was studying for his final exams. He was Edwin E. Kintner, and he was about to graduate and receive his commission as a naval ensign. His orders called for him to report for duty aboard the *Shaw*. Since Admiral Yamamoto had made that an impossibility, Kintner was reassigned to a different ship, the light cruiser *Trenton*. The *Trenton* was an old four-stacker built just after World War I, and had about as antiquated a power plant as was still in commission. It spent its time doing miscellaneous odd jobs, convoying troops to Australia by way of Guadalcanal, then cruising off Cape Horn to look for commerce raiders. Home base was in the Panama Canal Zone, and the *Trenton* never went back to the States. All in all, this was not Kintner's idea of how to fight a war, so he began to look for a way to get off that snafu bucket and into a real fighting ship.

An AlNav came through, a general order to all the ships in the fleet. It announced that the navy was looking for officers to enter a marine-engineering program at MIT, and anyone selected would be transferred for a year's engineering duty at sea, to gain experience suitable for the program. Kintner's spirits perked up when he saw that word *transferred.* The *Trenton* was so old that its machinery was completely obsolete, and Kintner was sure he would be sent to get his experience aboard a modern ship, so he applied. Sure enough, the navy selected him to go to MIT. However, they had a little surprise for him. They sent orders to his captain, and he was to be transferred, all right—transferred into the engine room of the *Trenton.* From duty on deck, Kintner went into the hot, noisy, greasy engine room, to serve there for an additional twelve months. All in all, it was mid-1944 before he emerged, and in his thirty-one months aboard the *Trenton* he had never once got back to the States. That was what he did to help win the war.

After nearly three years of sea duty, with never even a single day of leave, he finally got to MIT. He had not been there long when the bomb fell on Hiroshima and the war ended. Atomic energy was a deep dark mystery to Kintner, but he decided this was something he should learn about. He was allowed to take elective courses, so he went for MIT's two-semester course in nuclear physics. His Naval Academy background hadn't taught him much that would bring him to the level of the physics students who were his classmates, but he managed to pass, just barely. The course got into his academic record. Also while at MIT, he took up skiing, which he dearly loved. Upon graduation, then, he wanted to work at a naval base close to some good skiing, so he asked for assignment to Portsmouth, New Hampshire. He went there primarily for the skiing, but it happened that Portsmouth was a submarine yard. Soon he was involved in submarine engineering.

It was the late 1940s, and in the meantime, Captain Hyman Rickover was setting up a nuclear-engineering program at MIT. He soon began looking for young naval officers who might be good candidates. Among those recommended to him, Kintner led the list. Then Rickover found that Kintner had voluntarily taken those nuclear courses and, what was more, was already working on submarines. Kintner was skiing at Stowe, Vermont, when he received orders to report to Washington for an interview with Rickover. Rickover began by asking why Kintner hadn't stood higher at the Naval Academy. Kintner replied, "Because I wasn't smart enough." That was a good enough answer, and Rickover then said that he wanted to send Kintner back to MIT to study nuclear physics.

Kintner replied that he didn't want to do that; he wanted to be a submarine engineering designer. Then Rickover said, "We're going to build a submarine."

No one had ever built any sort of nuclear power reactor, no one had ever used the atom to produce power for any purpose—let alone for the highly specialized purpose of propelling a submarine. Kintner went back to MIT, got a master's degree in nuclear physics, then went to Washington to serve as one of Rickover's top aides. He designed the engine room and reactor compartment for the nation's first two nuclear subs, and was with Rickover out in Idaho when the first test reactor started up. On the way back to Washington, flying in one of those propeller-driven airliners they had then, they were having breakfast over Pittsburgh when Rickover said, "If the *Nautilus* makes two knots on nuclear power, it'll be a success. Nobody's ever done anything with fission energy, and if we can just get a vessel moving, it'll be a tremendous success." The *Nautilus* and the rest of Rickover's nuclear navy did a great deal better than two knots, and Kintner went on to head Rickover's Advanced Design Group. That group developed the early ideas for much of the rest of the nuclear navy, including the Polaris submarines as well as the aircraft carrier *Enterprise*.

That was how it was for Kintner, in the 1950s. Eventually, though, he decided to leave the navy and return to civilian life. He took over as president of a firm in Maine that did heavy engineering construction, and had been there a year and a half when the telephone rang. It was Rickover, and he needed help. Rickover had never asked any of his former associates to come back after they had left, but he wanted Kintner again. Kintner came on down to Washington to talk about it, intending to say no, but Rickover knew how to be persuasive. He convinced Kintner that both for Rickover's sake and for the good of the country, he should come back. They went back to working closely together, which in particular meant taking part in Rickover's bull sessions.

Rickover liked to sit down with his senior aides and just talk about things—navy regulations, politics, religion, education: whatever interested him. One day he had Kintner in his office, and was talking about how senior navy brass were trying to ease him out in favor of a replacement. This was a continuing problem for Rickover. He frequently had had to rely on his supporters in Congress to keep himself in charge of his nuclear navy, in the face of moves within the Pentagon aimed at ousting him. Kintner could certainly understand that, for as he proceeded to tell Rickover, he had himself been touched by one such move.

On his last day in the navy, just before leaving for Maine, he had come back from lunch to see that his secretary was white-faced: "The Secretary of the Navy is on the line; he wants to talk to you!" Kintner picked up the phone, and the Secretary was there: "I just learned that you're retiring. We need you. You ought to stay around." Then he went on to offer Kintner the opportunity to take over Rickover's job, once Rickover was eased out, and to be promoted to admiral in the bargain. Kintner didn't want to do it. He knew nobody would replace Rickover, and besides that, he had made his commitments in Maine and had even bought a house there. He made his point to the Secretary, and went on to Maine. In his mind, at least, that was that. He had never told Rickover of this, however.

Rickover immediately flew into a blind rage: "Why didn't you tell me!" "Because I didn't see that it made any difference." But it did. Had he known of this episode, Rickover never would have invited Kintner back. As it was, Kintner was right there working in Rickover's own organization, ready to step in as a potential replacement. If anyone wanted to follow up on the idea of replacing Rickover, then Rickover's replacement wouldn't be up in Maine; he'd be right there in the office down the hall from Rickover. Kintner had worked for Rickover for fourteen years, but he now was a threat and had to be got rid of.

That was in 1963. Fortunately, Kintner had a friend, Milton Shaw, who headed the program in breeder reactors within the Atomic Energy Commission. He invited Kintner to come work for him at AEC headquarters, twenty miles outside Washington. Kintner worked for Shaw for over ten years, and was heavily involved with the largest and most advanced breeder the United States has yet built, the Fast Flux Test Facility in Hanford, Washington. While working for the AEC, Kintner met Robert Hirsch. They were in a group of experts who were sent on a seventeen-day visit to the Soviet Union. The most inexpensive way for them to fly was on a twenty-one-day excursion to Europe, and they were all told to spend an extra four days in Europe, rather than come home early. Kintner had never been to Copenhagen, so he picked that city. By chance, Hirsch did the same. They spent an extended weekend there, had a fine time in Denmark together, and became good friends.

Then, early in 1975, there was a move under way to slow down the development of breeder reactors. Kintner wrote a lengthy memo to Robert Seamans, the head of the Energy Research and Development Administration, outlining his views about how the program should be managed. Seamans was interested in his recommendations, and let Kintner know he was glad to have the memo. The trouble was, Kintner had sent it to Seamans directly, bypassing two levels of his superiors. That was rank insubordination, and when Kintner came back from lunch the next day, he was no longer deputy director of reactor development. He had been fired.

Hirsch soon heard about this, and invited Kintner to talk about coming in as his deputy. Kintner demurred: "I don't know a damn thing about fusion." Hirsch said he didn't need to, that he was interested in Kintner's background as a manager of large nuclear projects. That was the kind of man Hirsch wanted as his deputy, rather than a lab director or plasma physicist. Kintner took the job in March 1975. By December Hirsch had turned over to him the day-to-day management of the fusion program. The following March, with Hirsch's promotion to Assistant Administrator within ERDA, Kintner formally took over Hirsch's job.

Once in office, he was faced with a worrisome prediction from theory. Two Soviet plasma physicists, Boris Kadomtsev and Oleg Pogutse, had predicted that as temperatures rose within a tokamak, increasing into the fusion range, the plasma would become more unstable and leak more rapidly. At fusion temperatures, their theory predicted, the leakage could be severe enough to bring back the bad old days of Bohm diffusion, setting all the good work of the last few years at nought. This prediction had not been made using the most solidly established and reliable theories, the ones that treat the plasma as a fluid. Rather, these Soviet physicists had gone to a higher level of detail, using a theory that considered the orbits traced by particles within the plasma. Such a "kinetic theory" was not so well tested or reliable, but the arguments behind their prediction were nevertheless quite persuasive. Kadomtsev and Pogutse called their predicted rapid leakage a trapped-particle instability, and among plasma physicists in 1976, it stood as potentially the most significant barrier standing in the way of success with tokamaks. Kintner and his directors decided to check that theory. The way to do this would be to install powerful neutral beams on the Princeton Large Torus, shoot for high temperatures, and see if they could be achieved.

"We sat around one day in my office," said Kintner, "and we asked ourselves what was the biggest, most significant thing that could be done to speed up the program. We concluded that the most important thing was to test this theory, and the way to do that was to put a lot of neutral beams on PLT, in a hurry. So we pressed—there are letters in the file that I personally wrote, to Oak Ridge to get on with the neutral beams, to Princeton to put them on and find out. The real force and motivation came from my office, to make a specific objective—get neutral beam power on PLT and try to breach that barrier. We put additional funds in. It was the kind of thing you would try to do in program management: pick an objective and go for it."

Then came the presidential election; Gerald Ford was out and Jimmy Carter was in. And all of a sudden, Kintner could see that the fusion program was heading into some turbulent weather. Since the 1968 tokamak breakthrough this program had been sailing forward rapidly, amid budgetary sunshine and blue skies. Now, however, there were storm signals. To this former navy captain, the budget was his weatherglass; a rising budget was as heartening as a rising barometer. But only a few weeks into the new administration, Kintner was presented with the first bad sign. The Office of Management and Budget cut his funding.

One of the budget examiners at OMB, who had responsibility for ERDA's energy programs, was Doug Pewitt. Kintner did not know Pewitt well, at least not then, though those two would come to know each other all too well a few years later when they would again cross paths, and swords. Early in 1977, Pewitt reviewed the TFTR program and decided it was going ahead too rapidly. He had reason to think this way, for the problem of trapped-particle instability made him think that TFTR should be slowed down until this theory had been tested on the PLT. He convinced his boss that here was a budget worth cutting, and these people then let ERDA's directors know that they intended to cut that budget request. Kintner did not know of these things, but one day in February he went up to Princeton to give a pep talk to the several hundred people who were working on TFTR. When he had finished, Mel Gottlieb, director of the Princeton plasma lab, came up to him and told him he had a phone call from the ERDA controller, Mel Greer. Greer got to the point immediately: "How would you like to take a $60 million reduction in your program for fiscal 1978?"

The magnetic-fusion budget had zoomed from $118 million in fiscal 1975 to $316 million in 1977; as recently as 1972 it had been $36 million. The outgoing Ford Administration had left behind a request for $371 million for 1978. If Greer could be convinced that this cut would do real damage to the program, then there might be a budget appeal, in which James Schlesinger, Carter's head of ERDA, would appeal directly to Bert Lance, the new director of OMB. But such appeals were not to be undertaken lightly, and there was no chance of one here. There was nothing for Kintner to do but take his cut. In the end, Congress restored some funds, and the 1978 budget came out at $332 million. But considering inflation, in real dollars this meant a budget cut. The cut came out of TFTR, and that program was delayed an extra year as a result. The next bad sign was that Hirsch resigned from ERDA.

Hirsch still had much of the idealism that he had originally picked up in the days when his mother was heading the local League of Women Voters. He still had a strong sense of wanting to serve the country. And he was shocked at how the new Carter aides had treated Robert Seamans, the head of ERDA, whom Hirsch greatly admired. Seamans had been deputy administrator of NASA during the Apollo program, he had been Secretary of the Air Force, had headed the National Academy of Engineering, and Hirsch regarded him as one of the really fine officials who had served in government. After Carter's election, Seamans tried to work with the new administration, to help make the transition go smoothly while Carter prepared to put his own appointees in charge. He wrote letters, made phone calls—and no one would respond to him. Carter's aides simply ignored him. Moreover, few of the Carterites had any extensive experience in government service. On the contrary, they took pride in being untainted by contact with Washington.

Hirsch found this treatment of Seamans deeply disturbing: "To me that bordered on being sinful. These people were so myopic and small that they didn't have the decency to respond to him. I began to have a bad feeling about these new people relatively early; my gut told me something was really wrong. I didn't know what they were doing, but this kind of behavior was inexcusable. It was narrowness, and a lack of class. They were very strongly motivated by ideology, they had very strong feelings about the rightness of their cause."

Following his inauguration, Carter chose as his energy czar James Schlesinger, the same man who in 1972 had named Hirsch to head the fusion program. Hirsch had been appointed to his present job by President Ford, and thus he had to meet with Schlesinger if he was to land a job in the new administration. Hirsch drove to the White House and went up to Schlesinger's office in the West Wing. The two had not met in several years, but they were ready to pick up where they had left off. Schlesinger started by saying, "Seems like the fusion program has gotten out of hand." Indeed, since Schlesinger had named him to take over that program back in 1972, its budget had increased tenfold. Hirsch replied, "That's your fault." Then they grinned at each other, having taken each other's measure. Schlesinger went on, "You want to know if you have a job." "Yes, sir." "Yup. I can't tell you exactly what it is. We're not going to be keeping very many people from the last administration, but I would like you to stay on." However, Schlesinger did not offer Hirsch a specific job. What Schlesinger seemed to be saying was that if Hirsch decided to stay on, they'd find something for him to do. Meanwhile, Hirsch had been talking to executives at Exxon about a job; they had contacted him after President Ford's defeat. Soon they made him a very good offer. That, together with Hirsch's bad feelings about the Carter officials, led him to make up his mind. In March 1977 he left government service and joined Exxon.

Meanwhile, Kintner still had a fusion program to run, and once again the spotlight was on Princeton. The immediate task was to get those neutral beams mounted and operating, and the man in charge of this was Harold Eubank. In a different life, Eubank could easily have been a barefoot boy from Mark Twain's Mississippi. Actually, he had grown up in the tidewater country of Virginia. His hometown had a population of a thousand, fifty years ago; today it still has a thousand. It is a country of sandy back roads and weathered frame houses with porches, of land cut by broad tidal rivers; fifty years ago, there was no railroad, and people pretty much got around by boat. Eubank finished high school there in 1942, without having any definite plans. Then later that summer, a girl he knew suggested that he come on down to William and Mary College. In those days people could decide in August to enter a class that was starting in September, which was what he did. He was drafted into the army during his freshman year, though, and spent three years in the war, but then came back to William and Mary. A physics teacher got him interested in the subject, and soon he was following the main line to a physics career, spending two years at Syracuse for a master's degree, then going on to get his Ph.D. at Brown. He was teaching at Brown in 1958 when he learned about the fusion program, which had been classified until then. That interested him, so he decided to try for a sabbatical year at Princeton. He wrote to Lyman Spitzer, got his invitation, and soon realized he had found a new home.

For Eubank's colleagues, in 1977, the work with neutral beams was the usual mix of problems, glitches, and impossible hours. One memorable glitch stemmed from a mistake made by Eubank himself. Parts of the beam systems had to be cooled with liquid nitrogen, at hundreds of degrees below zero. To make sure the nitrogen wouldn't run short, there was a 9,000-gallon container just outside, supplying it on demand. One Saturday morning, Eubank turned a valve to fill a chamber in one of the beams, then went off to have lunch and quite forgot about it. About half

196 THE MAN-MADE SUN

an hour later one of his colleagues stuck his head into Eubank's office: "Why is liquid nitrogen bouncing all over the platform out there?" Consternation! Eubank roared in and found it had been overflowing for twenty minutes. An entire corner of the PLT looked as if it had been taken out of Antarctica; the floor was completely covered with frost. A rubber seal had frozen hard as a brick; air was leaking into the beam-box vacuum, which was making matters worse. It took everybody about two hours to get everything thawed out, but they were fortunate. The seal might have shattered in the cold, requiring a lengthy replacement job, but that seal held.

As usual, they were running double shifts, which meant there was plenty of opportunity for people to work sixteen hours a day. It was rather like following one of the training rules of George Allen, head coach of the Washington Redskins: "Nobody should work all the time. Everybody should have some leisure. Leisure time is those five or six hours when you sleep at night. You can combine two good things at once, sleep and leisure." One physicist, who actually could work double shifts for four days a week, week after week, came to be called Iron Man. He didn't get paid extra, though, for as Eubank said, "We're professionals." Still, the usual work week ran about seventy hours, Monday to Saturday.

Monday was always maintenance day, a day for making good, solid repairs to things that had gone on the blink during the previous week. A key data channel might go out at nine in the evening, requiring that you just limp along till midnight, the end of the experimenters' work day. The next morning you would have time for a temporary fix, wiring things together to get through the week, but the following Monday would be when you would make the real repair. So Monday was always a full day, ten hours or so. The next four days would be double shifts, and then Saturday would be another eight hours. During the week, you might fudge a bit by coming in late in the morning rather than promptly at eight, but you really would stay till midnight, week after week, particularly during the experimental runs. Midnight, after all, was traditionally the time when the best data would be coming in, the time when the Responsible Person would finally put the beams up to full power, if he ever did so at all. But Sunday at least was free. As Eubank said, "I guess people might occasionally have looked at Sundays, but nobody ever had the courage to say, 'You really ought to run it Sunday, too.' I think there would have been a revolt."

By the fall of 1977, Eubank had two of his beams fully installed and working. Serious experiments in plasma physics could now begin. It soon became obvious that the plasma temperatures were being held back; the beams were not heating the plasma as they should. The problem, though, was not in the trapped-particle instabilities, the severe plasma leakage predicted by those Soviet scientists, which Eubank's experiments were seeking to study. Rather, the problem was in a flood of impurities which entered the plasma and which radiated away so much energy that the plasma was rapidly quenched.

It was not hard to tell where the impurities were coming from. As in all tokamaks, the plasma did not simply sit within its doughnut-shaped chamber, but rotated quite rapidly as if running around a racetrack. This allowed an easy way to keep the plasma from expanding all the way to the chamber walls. They built the walls with limiters, heavy bars of a heat-resistant metal like tungsten or molybdenum. As the plasma flowed past these bars while whirling rapidly, the bars would scrape off or absorb whatever plasma particles hit them, and these particles then would not reach the walls. However, when energetic plasma particles hit these limiter bars, they knocked atoms from their metal into the plasma. These were the impurities that were causing the trouble.

By December, with the two beams pouring in 1.1 megawatts of power, the plasma struggled up to a new temperature record, 25 million Celsius degrees. The previous record had been 23 million, but had been set with only half a megawatt of beam power. Clearly, all those impurities were making the plasma very reluctant to heat up. To make progress, Eubank would have to bring on his next two neutral beams. He also would have to replace the limiter bars with new ones made of graphite.

Why did he pick graphite, or carbon? Atoms of carbon, knocked into the plasma, would quench the plasma far less severely than would the metal-atom impurities. The quenching resulted from hot atoms radiating energy away from the plasma, cooling it. The radiation, in turn, came from the fact that the metal-impurity atoms had held on to some of their electrons; these electrons were what made the atoms radiate. The way to get rid of the radiation, then, was to get rid of these attached electrons in the atoms; and the best way to do this was to switch to carbon for the limiters. Carbon atoms, being light, would be completely stripped of their electrons in the hot plasma, and thus would not radiate at all. Light atoms like carbon would lose their electrons much more readily than would heavy atoms like tungsten.

Putting on the new graphite limiters and installing the other two beams would be work for 1978. As the new year began, however, the Carter administration was preparing to drop the other shoe. Throughout 1977,

Carter and Schlesinger had been working to elevate ERDA to the cabinet-level Department of Energy, which went into operation that October. Carter, a former nuclear engineer, had come into office with very definite ideas about energy. Even now it is hard to forget his nationwide speech of April 20, 1977, in which he proclaimed that dealing with our energy problems would be the "moral equivalent of war." During 1977 Carter had had to operate under a federal budget prepared by President Ford; he had carried forward the energy programs he had inherited, and had taken few new initiatives. But this was about to change.

Carter and Schlesinger wanted to shift funds away from nuclear technologies, most notably the breeder reactor, and instead to emphasize solar energy, conservation, and coal. They also wanted to put the main emphasis on near-term energy needs, rather than to give priority to long-range approaches such as fusion, which would not be ready for decades to come. In addition, Schlesinger wanted less emphasis on technologies for generating electricity in large power plants, which again included fusion. Finally, Schlesinger was not pleased with the way the fusion budget had mushroomed since 1973, when he had left the Atomic Energy Commission. He felt it would be possible to cut it by at least $100 million, possibly as much as $200 million.

Within the $10 billion Department of Energy, fusion stood as quite a significant program. Schlesinger started by naming as one of his top assistants a man he knew he could count on. This was John Deutch, whom he picked to fill the DOE's most powerful scientific post, that of director of his newly created Office of Energy Research. Deutch was a high-level Washington insider who had spent years moving easily amid the worlds of government, universities, and the military. In the early 1960s, aged twenty-three, he had been one of the whiz kids working for Secretary of Defense Robert McNamara. While in graduate school at MIT, he had spent his summers and vacations working on strategic-arms problems at the Pentagon. His close friends included Harold Brown, Carter's secretary of defense; Frank Press, Carter's science adviser; and Schlesinger. Indeed, during the Ford administration, when Schlesinger had been defense secretary, he had appointed Deutch to the prestigious Defense Science Board. Prior to being named to this energy post, Deutch had been chairman of the MIT chemistry department, and he was well known in the scientific community. Within that community, his appointment to the energy post drew widespread praise, the more so since he was known as a strong and forceful leader who was likely to get things done. As

Deutch said to William D. Metz of *Science,* soon after being appointed: "Whatever you write, don't say that nothing will change."

Schlesinger was certainly no fusion enthusiast. He had called it a "scientific sandbox," into which hundreds of millions of dollars had been poured so that scientists could amuse themselves. Deutch was not well acquainted with the fusion program and saw no reason to challenge Schlesinger's views. Only a few weeks after Deutch took office, early in 1978, Schlesinger told him to prepare a rationale for cutting back the fusion program. Deutch went to talk it over with Kintner's boss, Robert D. Thorne. They decided that the way to proceed was to set up a panel to review the program. Both Deutch and Thorne agreed that this review should find that the prospects for fusion were quite uncertain, so that it would be appropriate to cut it back to a less expensive level. Deutch then proceeded to set up his review committee, taking care to keep the choice of its members in his own hands.

Kintner was sitting in his office soon afterward, when his secretary brought him a letter signed by Deutch and Thorne. The letter was addressed to members of this committee, and stated what would be their job. In effect, they were to provide an intellectual basis for removing $100 million or more from the magnetic-fusion and laser-fusion budgets, the cut to be divided between these two programs. Naturally, Kintner got upset. As he later said, "This was just an unfair way to get at the problem, to get external people to come look at you and tell you what you're doing wrong." However, one thing was working in his favor. He knew the committee chairman, John Foster, and felt he could work with him.

Many people called him Johnny Foster, and he had a lot in common with Deutch, having also spent his career in high positions in industry and government. A physicist, he had started as a manager at Lawrence Livermore lab. One of his fellow managers, Harold Brown—"*the* Harold Brown" to those who had known him when—had been John Nuckolls' boss when Nuckolls had been inventing laser fusion. Foster went on to become director of Livermore from 1961 to 1965, then went to Washington to take on probably the most demanding position in American research: director of research and engineering at the Pentagon. Just then, in 1978, he was vice president for energy research at TRW, Inc. He was known there as a master of his profession, who could sort out winners from losers early in the game.

Kintner had met him during Foster's Pentagon days. They had lived about a mile apart in Annandale, Virginia, though Foster had much the bigger house, with more luxurious surroundings. Foster's son was in the

same junior-high class as Kintner's son Peter, and they got to be good
buddies. Kintner would sometimes ferry Peter back and forth, so Foster
knew who Kintner was. When Kintner put in a call to him at TRW near
Los Angeles, Foster was open to meeting with him. What Kintner wanted
was to fly out and discuss the fusion program with him, presenting his
position in private discussions between the two of them, off the record.

TRW lists its address as One Space Park, Redondo Beach, California.
It is a cluster of low white labs and office buildings, many of which show
an almost futuristic architecture, set amid sweeping lawns. Kintner flew
out and spent most of a day closeted with Foster. There he laid out his
whole program, from his viewpoint as a program manager: what had been
accomplished, what they were trying to achieve, what the effects would
be if the budget cuts went through, how the planning went. Foster didn't
agree with Kintner on all his points, but he understood Kintner's point of
view. Kintner came away feeling that he had given Foster a good deal of
insight into the real workings of his program.

The formal briefings to the committee took place in a lavish Department of Energy conference room within their main buildings on Massachusetts Avenue. The room featured individual microphones and
speakers for each participant, as well as a backlighted screen for project-
ing slides and viewgraphs from behind. Kintner led off the presentations
with a lengthy statement, and was followed by Harold Furth of Princeton
and Ken Fowler from Livermore. Kintner had put quite an effort into
preparing his statement, he really had his heart in it, and he came away
feeling he had been persuasive. The committee met for only three days of
briefings in mid-March, followed by two days of discussions among
themselves. To Kintner this was far from adequate to guarantee that they
would go deeply enough into his program or understand its true merit. He
asked Foster for the chance to meet with him again, before Foster would
write his committee's final report.

They agreed to meet back in Washington, in a suite owned by TRW,
within a building across the street from the Statler Hilton. Kintner and his
wife, Alice, drove in to meet Foster on a Sunday night. Foster was
traveling along with two other committee members, John Dawson, a
fusion expert from UCLA, and Mike May, former director of Livermore.
However, Foster's plane was delayed, and the building where the meeting was to have taken place was locked. The Kintners waited outside for
about two hours, then decided that since it was nearly 11:00 P.M., they
might as well go home. Ed said to Alice, "I'm gonna go over to the Hilton
and call the suite, see if there's anybody there. You stay and if anyone

comes to go in the building, you accost them." He went over, tried to call, and got no answer. Meanwhile, two people came by where Alice was sitting in her car. They opened the door to that building with a key, and went right in, but Alice didn't accost them. As she later told Kintner, "They just looked like two ordinary people; they just opened the door and went in." They happened to be Foster and Dawson.

Fortunately, Mike May was staying at the Hilton; as Kintner walked out from trying his phone call, he recognized May coming in. "Where's Johnny Foster?" "He just went in over there." With that, Kintner got hold of Foster and spent the next two hours with him. Kintner talked with him till one in the morning, discussing his reactions to the briefings, trying to sway him. As Foster later told Kintner, a major reason why the committee came to the conclusions it did was Kintner's personal intervention on those two occasions.

The resulting "Final Report of the Ad Hoc Experts Group on Fusion" was not even printed. A secretary typed it on a Selectric, with math symbols being drawn in by hand—at least one such symbol got left out in the proofreading—and it was reproduced by offset machine, like a community newsletter or announcement of a supermarket sale. That fact did not detract from its significance, for in Washington, such blue-ribbon committee reports are a specialized and highly influential form of politics. This one, however, was not what Deutch had expected. The report came out early in June of 1978, and as it stated on its first page,

> The first objective of the program must be to determine the highest potential of fusion as a practical source of energy. . . . The best partners in a marriage of scientific and engineering promise must be selected. . . . The Group believes that there is an urgent need to answer the questions concerning feasibility and that the momentum of the program developed over the past few years is a major asset and should be maintained. Obviously, the time required to determine feasibility is influenced by the level of funding made available to the program.

To say that "the momentum should be maintained" was tantamount to saying that the budget that supported that momentum should be maintained, the more so since the efforts then under way had been described as "urgent." The report pointed with pride to achievements of both tokamaks and mirrors. There was irony in the three tokamak achievements cited: two of them had been obtained at the Plasma Fusion Center at MIT, which was virtually an orphan by comparison with the far larger and more lavishly funded Princeton lab. These achievements were the discovery of Alcator scaling, and the achievement in Alcator of a confinement parameter, $n\tau$, of 3×10^{13}. Alcator scaling was the heartening law by which plasma confinement time was found to increase with increasing plasma density. That $n\tau$ was close to the value to be achieved eventually in the much larger TFTR. The third major tokamak achievement was that of PLT, which by then had reached plasma temperatures of 2.3 kilovolts, nearly 27 million Celsius degrees. These values of temperature and $n\tau$ were "within modest factors of those required for reactors." Therefore, the TFTR work "should be pushed aggressively."

The report also praised the high temperatures and values of beta in the mirror program's 2XIIB, described the tandem mirror as an "ingenious idea," and stated that "the TMX and MFTF deserve an aggressive effort." On the program management side, Foster's committee gave kudos to what Hirsch and Kintner had accomplished: "Clearly headquarters should be responsible for development and approval of program objectives, strategy and plans. . . . It was clear that such a headquarters group is established and functioning well."

All this was a victory for Kintner. It meant his budget would not be cut after all. In the face of Carter's energy priorities and of Schlesinger's stated preferences, the fusion program had vindicated itself by appeal to a commodity infrequently invoked in Washington: technical merit. Kintner's arguments to Foster may have been persuasive, but the program's actual achievements were more persuasive yet, and gave the foundation on which Kintner had been able to make his case. Nevertheless, Foster and the committee members had been talking to Deutch as well as to Kintner, and Kintner did not get all he wanted. His long-term strategy called for pushing ahead toward a new and very large tokamak, to follow TFTR as TFTR was to follow PLT. This had been Hirsch's idea all along: to move forward with a succession of increasingly powerful tokamaks, at last reaching the stage of a prototype commercial fusion plant. But following Deutch's lead, Foster's committee rejected this strategy. In the words of the report, "This strategy . . . places undue emphasis on a single approach, which constitutes an unnecessarily high risk. The commitment at this time to the construction of a next generation Tokamak project would further increase the risk. . . . Commitment to construction of a next generation Tokamak beyond TFTR, should not be made until results from TFTR and other related experiments justify it. . . . Before commitment . . . a convincing case should be made that Tokamaks can be engineered into attractive energy producers."

The report's phrase, "highest potential of fusion," actually was a code. It meant holding tokamaks to the level of TFTR, while giving mirrors the chance to catch up. Foster, after all, was a former Livermore man, a fact that was not lost on the tokamak people. His sympathy for mirrors tied in neatly with Deutch's determination that any new tokamak beyond TFTR be put off till well after President Carter would have served out his years in office. In the end, this was just what would happen. The next big project indeed would turn out to be MFTF–B, a mirror, while a decision on starting the next large tokamak would be put off to some future administration.

At Princeton in this summer of 1978, Harold Furth could see which way the wind was blowing, and he was not pleased: "We perceived that what the Foster Committee did was to blunt the forward march of the to-kamak, by a rather clever ruse. Instead of saying 'This is all very good but we don't have the money for you, in fact we want to cut you,' they said, 'This tokamak stuff is wonderful but perhaps we should raise fusion to its highest potential rather than just proceeding with success, and we really should be bringing in these alternate approaches.' Previously the argument was simply, does the world want fusion power or don't they. Now it's always, well, wait a moment, don't be in such a rush, wait until these other things have come along. So as soon as you introduce the idea that you have to wait to make sure this is the best of all possible things, it's a wonderful mechanism for waiting forever."

But Furth still had a trump card he could play, for Harold Eubank finally had all four beams on the PLT. The new limiters were installed too, water-cooled graphite limiters to replace the earlier ones of tung-sten. Everything finally was in place to make the critical tests of the theory of trapped-particle instabilities. As with other such major experiments, this one had been years in preparation, from the earliest negotiations with the Washington office to the final installation and debugging of equip-ment. But with everything ready at last, the actual tests might take no more than a few days. That was how it was in the Princeton control room, that July.

The control room in those days looked just as described in chapter 1, and the people did their experiments the same way. It was not that some Physicist in Charge ordered up a set of dial settings and presto, in half a dozen shots the problem of trapped-particle instabilities was gone for good. As always, the Responsible Persons took their sweet time in increasing the beam power each night, the more so since the arrangement of four beams on the PLT was quite new and had to be handled with particular care. But during a week or so in July, toward the end of each night, they were shooting close to two megawatts of beam power into their plasmas, and were pushing into regimes where the trapped-particle instabilities would be expected to make their presence known. Yet these instabilities simply were not there.

As the scientists kept pushing to higher beam power, the plasma was not deteriorating in its stability or leaking more rapidly; instead, its temperature was going up. First they got 4 kilovolts, 45 million Celsius degrees. This in itself was a milestone, nearly twice the previous record, and sufficient to produce actual fusion energy if the $n\tau$ was right. Then, pushing up the beam power, there were hints of 5 kilovolts on one diagnostic, followed by confirmation with others. Finally, on the night of July 24, they were shooting a full 2 megawatts into the heart of the plasma. The plasma responded beautifully, gave up all the reluctance it had formerly displayed, and soared to a peak of 5.5 kilovolts, over 60 million degrees.

This was it; this was the success everyone had been seeking, the reason why PLT had been built in the first place. It might well have been an occasion for celebrating, for passing around the champagne. But if you had been in that darkened control room with its racks of equipment looking like so many stereo back-panels, if you had been there among those racks with their blue cables and green TV screens, you would not have been celebrating. Instead, like everyone else, you would have been busy checking your instruments, trying to think of ways the data might be wrong, studying the data intently as the Responsible Person ran the shots over and over again. You would have wondered if it was for real, wondered if somehow you were being fooled. Only later, after the PLT had gone even farther with 2.5 megawatts, would you get your yellow T-shirt silk-screened with the achievement of which you had been a part: PLT NEUTRAL BEAMS—75 MILLION DEGREES.

Several months before this, Kintner had been in Mel Gottlieb's office, and had said in a half-facetious way, "Mel, what'll you do if you get five kilovolts?" Gottlieb demurred with a comment about trapped-particle instabilities: "No, we'll be lucky if we get three and a half." Kintner wasn't satisfied with this: "But, Mel, you ought to be thinking about this from a strategic view; what if you get five," and Gottlieb had demurred again.

That summer, Kintner and his family were vacationing in the Green Mountains of Vermont. His daughter Mary was in charge of maintaining some huts along the Long Trail, which ran down the spine of the mountains. He and his wife, Alice, took a week to hike along that trail, spending some nights with Mary in her huts. Then they came down out of the mountains to stay in Stowe. When Kintner checked in, the desk clerk told him that he had an urgent call from Gottlieb. Kintner didn't know what to expect. Had something blown up, had someone been electrocuted? He went up to his room and returned the call. Gottlieb could hardly contain himself: "We have six kilovolts! And we haven't finished yet; we've still got more power; we can get more out of it!"

Kintner understood full well the significance: "The theorists thought that they knew what they were talking about, trapped-particle instabilities. It was an absolute barrier, no one knew how we were going to get through it. Here was something that looked like the sound barrier in aviation—and suddenly it wasn't there. It just wasn't there! The PLT results were very fundamental because after that it was possible to say that there was no longer any question you could make thermonuclear energy. You might not be able to use it effectively, but you could sure as hell burn a plasma on the earth in reasonable-sized equipment. Here we had been working for all these years, spent all this money, and we had gotten, really, this fundamental, final step—the goddamn thing would fly! It might not fly more than a hundred feet, it might not carry human beings, but it would fly!"

The PLT results offered no guarantee that tokamaks could be made into attractive energy producers, or that they would be the form in which fusion would be offered to the world. But in sweeping away the trapped-particle barrier, these physicists removed the last serious doubt that TFTR would work as planned, and with it, that fusion energy would be produced in accordance with their programs and schedules. Any prediction that TFTR would fall short of its goals then might be based on Murphy's law, or on the general perversity of things. It could not be based on known results from experimental physics. From that July onward, in plasma physics, theory and experiment no longer would point to continued barriers or to severe and intractable problems. Instead, they would point to success.

These results had come just too late to influence the Foster Committee. They were very much in time to influence the policy statement on fusion that Deutch was preparing even then. But these political considerations were far from the minds of the Princeton physicists. Their eyes were on the next major fusion conference of the International Atomic Energy Agency, to be held in Innsbruck, Austria, during the last week of August. Artsimovich's tokamaks had been the featured attraction at the 1968

Novosibirsk conference, while tandem mirrors had stolen the show at Berchtesgaden in 1976. Now these PLT results stood to be the hit of the coming meeting at that other alpine city in its high valley, surrounded by mountains. There would be plenty of opportunity for other scientists to raise questions and criticisms. Then, amid such comment and debate, these results could be announced to the world in a proper scientific atmosphere.

Kintner agreed that the news should be held for Innsbruck and expected that Schlesinger would hold a press conference of his own at the same time. Kintner had Stephen Dean, one of his senior directors, consult with Gottlieb and Furth on a press release. But while they hoped to keep the story under wraps, the news was spreading rapidly through the fusion community, by way of the grapevine. On July 31, the Washington-based newsletter *Energy Daily* ran a short item on the front page, reporting "persistent reports of a major breakthrough in the U.S. program in magnetic fusion." On that same day, Kintner's boss, Robert Thorne, sent a memo to Schlesinger. He stated that the PLT results represented a most significant development, unique to the United States; that the success of TFTR was now assured, and that the scientific feasibility of fusion was virtually certain. He then pointed out that it was likely the national press would get hold of the story soon, and recommended that the Department of Energy hold its press conference on August 15, rather than wait for Innsbruck.

A few days later, the Princeton press release was ready, and was sent to the DOE press office to be approved. By mutual agreement, no one at Princeton was to release the news in advance; everyone was to keep mum till the press conference. One of the chief attachés within that press office was Gail Bradshaw. She was a tough and cynical old bird, wise in the ways of the bureaucracy, and she had the ear of Schlesinger. She saw the Princeton statement and then told him, "I always find it uncanny that we have these breakthroughs just before the appropriations cycle." Indeed, not only was Deutch preparing his policy statement just then, but early August was the time when the various program officials, in charge of solar, nuclear, fossil-fuel, and fusion research, were sending their fiscal 1980 budgets to Schlesinger's office for review.

Schlesinger needed little of Bradshaw's prompting to decide that she was right. He might be Carter's cabinet secretary in charge of energy breakthroughs, but fusion stood to be the wrong breakthrough. It had nothing to do with synthetic fuels from coal, energy conservation, or solar energy—any of which were far more likely to receive a presidential smile.

Worse, fusion was little understood in the energy world and was rather mysterious, but potentially quite glamorous. A report of a fusion breakthrough could easily be made to sound as though practical fusion power was just around the corner, instead of still being decades away, and that the energy crisis was over; we didn't need to worry about coal or conservation; all our problems would soon be solved. Schlesinger didn't need any of that, so he decided there would be no press conference.

At this point, another player entered the act. This was Morris Levitt, executive director of the Fusion Energy Foundation, a small but well-financed offshoot of the U.S. Labor Party. It operated out of offices on West Twenty-ninth Street in Manhattan. Its magazine, *Fusion,* was circulated mostly by being handed out in airport concourses. It featured articles by people like Lyndon LaRouche, sometime fringe candidate for the Democratic nomination for President, articles with titles like "Poetry Must Begin to Supersede Mathematics in Physics." Nevertheless, leaders of the fusion community were willing to talk to Levitt and his staff, and Levitt himself was no splinter-party malcontent. Eventually he would go on to become editor of a well-respected physics journal, *Laser Focus.*

On Thursday, August 10, Levitt put in a phone call to Gottlieb. No one at the DOE had passed the word to Gottlieb on Schlesinger's decision, so he told Levitt about the press conference set for the following Tuesday, although he had to keep the lid on the actual news. Levitt was quite interested, and even sent a note to the White House, urging that Carter himself take part in the press conference. But when he called the DOE press office the following day, he was told quite grumpily that there would be no press conference. This was the sort of thing that would set him off like a hound dog on a scent. Evidently there had been an important breakthrough; equally evidently, the DOE was trying to cover it up. There was little love lost between Levitt and Schlesinger, and Harold Furth knew the Fusion Energy Foundation well: "They were always ready to see a conspiracy to crush fusion, and they felt it was their duty to leak this out."

As the recipient of his leak, Levitt picked Dave Hess, a reporter for the Knight-Ridder wire service, which served such major newspapers as the Miami *Herald,* the Chicago *Tribune,* and the Baltimore *Evening Sun.* Hess began making phone calls to check out the story. When he called around to find out about it, he formed the correct impression that the scientists thought it was great stuff, and that the DOE didn't want it out. Finally, an unnamed official in Kintner's office "very reluctantly" substantiated the reports of a fusion breakthrough. This was just the combination that would make the press extremely interested, much more so than if the DOE had announced it as originally planned. The result was entirely the opposite of what Schlesinger wanted. The next two days happened to be slow ones for national news, and suddenly the unlikely and esoteric subject of fusion was the topic of a full-blown media flap. During August 12 and 13, days that those involved still call the "PLT weekend," fusion was probably the hottest news story in the country.

It started with Hess' story, carried in Saturday's morning Miami *Herald:* "Scientists at Princeton University have produced a controlled thermonuclear fusion reaction that experts are hailing as a major technical breakthrough." The news soon spread along the entire Knight-Ridder chain, and the Associated Press wire service put out a bulletin based on Hess' story. That started the phones ringing. Early in the afternoon at Princeton, Tony DeMeo, head of public relations for the plasma lab, was at home polishing his car on what looked to be a quiet day. Then he got a call from the Princeton University press officer: "I'm receiving lots of calls from the press regarding some breakthrough on PLT. They're asking if it's true that we've achieved control of fusion, that we're there, all our energy problems will be solved from here on out. I'm not up to speed on your program, so I want to refer all calls to you." From that point, DeMeo's phone didn't stop ringing till midnight.

During that same afternoon, Bill Peterson, a reporter for the Washington *Post,* got on the story. He called the DOE, and was told merely that industry sources were promoting the story. He called DeMeo, who was glad to tone down the wild rumors of an end to the energy crisis, but who still was under orders to keep the lid on the news till the DOE released it. Peterson thought he was stuck, but then he found some literature from the Fusion Energy Foundation. In it he found the name of Stephen Dean, the man in Kintner's office who had direct responsibility for the PLT work. Dean, meanwhile, had been sitting at home watching television, when his phone rang. It was the duty officer at the DOE press desk. She said her phone was ringing off the hook with reporters wanting to know about a breakthrough at Princeton; it was on the wires, too. She hadn't been able to reach Kintner, Thorne, or Schlesinger, so would he take the calls? Nobody else available at the DOE was technically qualified to answer the reporters' questions. Dean said he would, and soon was fielding his own share of calls.

Just after 6:00 in the evening, CBS News telephoned him: could he come into downtown Washington to be on the 6:30 news? Dean replied that it was already 6:05 and it would take him at least a half-hour to get there. The response from CBS: "Please try." He actually arrived about 6:40, while the news was already on. They ushered him into a separate room and did a three-minute taped interview. By 6:55 it was on the air, and Dean was able to see it go out over the CBS network. Then, later that evening, Peterson of the *Post* phoned with the draft of an article, which he wanted to check for accuracy. Dean was amazed at how well he had been able to piece the story together. He even had a quote from Furth saying that this was the most important development in fusion in ten years. Nevertheless, Peterson wasn't claiming that fusion was right around the corner; instead, his article was stating that there was still a lot to be done. By that point, Dean had concluded that the story was breaking wide open, it was completely out of his control, and the best thing he could do was to answer questions in a straightforward way. He gave Peterson all that he wanted, including the specific statement of the 60-million-degree temperatures. Peterson went on to quote Dean extensively, and was the first to report the achievement of these high temperatures.

Meanwhile, where was Kintner? It was all taking place entirely without his knowledge. Right along he had insisted that the news be held under wraps till it was formally released at a press conference. A few days earlier, before any of this had started happening, he had flown out on a visit to Livermore, and was thoroughly unaware that the news was available on any TV set. Saturday afternoon he got back to the East Coast on a flight to Dulles Airport. He got in his car and drove homeward, his radio tuned to WTOP, the Washington news station: "Here I heard this goddamn news release about a fusion breakthrough!" Once he was home, it was Kintner's turn to start fielding phone calls. One of them was from Dean, who had been trying for hours to reach him and finally succeeded toward ten in the evening: "Ed, I gotta tell you what's happening!" Kintner replied, "I know."

For Dean and Kintner, by then it was mass confusion. They were getting calls not only from the press, but from people at Princeton as well as from others in the fusion community, who wanted to know what was going on. Dean's phone kept ringing late into the night, and he finally decided not to talk to any more newsmen. By then it was past midnight, and he had even had calls from overseas. People from the Voice of America and the British Broadcasting Corporation called for telephone interviews, as late as three in the morning. Then after a few hours of sleep, Dean got up and brought in the Sunday morning Washington *Post* from his porch. He was astounded to see the banner headline "U.S. Makes Major Advance in Nuclear Fusion." Down low on the page was a much less prominent story, related to the death a few days earlier of Pope Paul VI. That headline read, in considerably smaller type, "Pope Is Buried."

By then, however, Dean and Kintner were in trouble. Another of the *Post*'s readers was Jim Bishop, the DOE's press spokesman. Like Gail Bradshaw, Bishop stood close to Schlesinger, and he proceeded to phone Kintner and blast him with both barrels. He accused Kintner of having been responsible for the press coverage, of having manipulated the media for the sake of putting Schlesinger in a box, and that all of Kintner's plans for a press conference had merely been a ploy to trap Schlesinger, stirring up publicity in ways that would keep Schlesinger from dealing with fusion on its merits. Then Bishop started to get personal. He insisted that the whole thing was just a budget ploy, to generate publicity at budget time, and that his people knew the fusion community was prone to do that sort of thing. Kintner tried to tell him how he had worked to handle it properly, insisting that everyone keep the lid on the news until it would be released at a DOE press conference. But Bishop simply wouldn't listen, and there was nothing Kintner could do but listen to Bishop's abuse and take it.

Next, Bishop phoned Gottlieb in Princeton and proceeded to dish out more of the same. Gottlieb told him he had been working with DOE people on preparations for a press conference. Bishop responded very angrily that there was to be no such press conference. From that point, matters rapidly got quite unpleasant as Bishop accused Gottlieb of having invented the PLT story and blown it up for his own purposes. Gottlieb by then was becoming very upset. Still, he held on to his temper as he told Bishop, "I'm going to get the hell out of it. If I get any more phone calls about this, I'll just refer them to you and refuse to make any statement." It was obvious to Gottlieb that Bishop simply didn't know what was going on and had leaped to conclusions without bothering to check. Nevertheless, Gottlieb was a proud man, and to be treated this way was deeply distressing.

He knew how hard everyone had been working all along on PLT, the sixteen-hour days, the months of struggle before achieving their 60 million degrees. Moreover, while working on the press release, he had consulted with his fellow lab directors to make sure he would not overstate anything. He had been particularly gratified when Ken Fowler of Livermore told him this was the most significant result yet in the U.S. program. On top of that, he was sure that Deutch didn't really understand the PLT project. Gottlieb had never been able to have Deutch willingly sit still for more than half an hour, to be briefed on it. The one time Deutch had sat for a longer briefing was when he had been stuck at Princeton during a snowstorm. Bishop's phone call thus was a direct challenge, not merely to Bishop's fellow DOE directors like Kintner and Dean, but to the professional integrity of Gottlieb and of his entire Princeton lab. Certainly Gottlieb's people hadn't been working their tails off merely so that some high-level Washington ignoramus could abuse him by telling him that his people had done nothing.

He just wasn't going to take it. He picked up the phone and called William Bowen, the president of Princeton University. He was feeling very angry, but Bowen had always been very supportive, and what was more, Bowen knew Schlesinger personally. Gottlieb then told Bowen how his lab had gained a legitimate achievement but that all he was getting was abuse, and if these were the rewards he was going to get, he would do two things. He would hold a press conference on his own, and then he would resign. Bowen said, "Look, Mel, just take it easy for a little while and I'll see what I can figure out about this." Then he called Schlesinger and talked with him about what was happening. Afterward he called Gottlieb back. He did not pass on the details of his talk with Schlesinger, but he told Gottlieb that things would settle down now.

Monday morning, Kintner and Dean arrived at work early, at 7:30. Kintner's boss, Robert Thorne, was still out of town, and filling in for him was Eric Willis, Thorne's assistant. Willis had phoned and told them to meet him in his office, early. When they arrived, he looked at them with a grim face and said he had been specifically ordered that they should be off the payroll and out of the department by the end of the day. Schlesinger was convinced that Kintner had leaked the news and that Dean had then blown the story wide open, and he wanted them fired.

Kintner had to try to help Dean: "I was between Steve and the fire and I was trying to protect him while meanwhile I was trying to save my own neck." He told Willis his story. At that same moment, early in the morning, Deutch and Schlesinger were conferring. Willis had been meeting with them, and he now went back into Schlesinger's office to rejoin them, leaving Kintner and Dean to wait outside. As befitted a man of cabinet rank, Schlesinger's office was as spacious as a large living room, beautifully draped, carpeted, and furnished with the best the General Services Administration could offer. It featured a huge desk flanked with the flags of the United States and the Department of Energy, the DOE flag being blue and white with the department's seal emblazoned on it. There were couches, a large mahogany conference table, walls lined with bookshelves, and several large framed certificates, ornately lettered, which were the documents whereby Presidents Nixon, Ford, and Carter had appointed Schlesinger to high office.

Willis started by arguing that Schlesinger and Deutch should hold up until they could get the whole story. Then they started debating whether Kintner and Dean actually had been responsible, and whether the Princeton results were significant after all. Eventually, Deutch convinced Schlesinger that the thing to do was not to do anything dramatic like firing people, but to hold a news conference after all and to downplay the results in a statement to the press. In the meantime, there was the matter of a letter from the Vice President, Walter Mondale. Mondale had also heard the news, and wanted a memo from Schlesinger on just what had happened. Kintner and Dean were still cooling their heels in the outer office when Schlesinger's meeting broke up. Then Deutch came out and told them to help in writing the letter to Mondale and in drafting the press release. They still weren't sure if their jobs had been saved, but at least they had a reprieve.

Meanwhile, in Princeton, Gottlieb had put on a pair of blue jeans for what he expected would be an ordinary working day at the lab. Shortly after noon he got another phone call. It was from Jim Bishop, but this time Bishop had a completely different attitude and was speaking in a much more respectful tone. He said there would be a press conference at 3:30 P.M., and could Gottlieb come to Washington immediately to take part? Gottlieb wasn't dressed for a press conference, and he could have driven to Washington in four hours, but not in three. Also, there was no scheduled airline service that would get him there in time. Fortunately, there was a small airport nearby with an air-taxi service, featuring two-seater planes. He rushed home, changed into a suit, and zipped on down to the airport. There he chartered a flight, and had the pilot fly him to National Airport, just across the river from downtown. A quick taxi ride brought him to the Department of Energy on Independence Avenue, and he arrived in time.

The fact that the DOE was to have a press conference meant that Tony DeMeo could also schedule one at the same time at Princeton. DeMeo had had phone calls from ABC–TV in New York, as well as from two Philadelphia stations, so now he could invite their camera crews and reporters to come out. By then the only important part of the story that had not been officially released was the specific number, 60 million degrees. A reporter from WCAU–TV in Philadelphia struck up a conversation with DeMeo, then pulled out a microphone and said,

"What was that temperature again, Tony?" He then added that his story wouldn't go on the air till after the press conference anyway, but DeMeo stuck to his guns. He wouldn't confirm the temperature record till he was told to, even if everyone knew it anyway.

At the Washington press conference, Deutch told Kintner, "Go sit in the back of the room and shut up." By all rights, Kintner should have been up at the front making the announcement, since he was in charge of the fusion program. But Deutch was taking this prerogative for himself. Some 75 reporters were there as Deutch delivered the DOE's party line: "This was something we had planned for a long time, it was in our program plans, we had expected it all along, that's why we had built PLT in the first place. Yes, we're doing a lot of good things in fusion, we're doing a lot of good things in our other energy programs too. It was basically a routine result. It came sooner and in stronger form than we anticipated, but it was no more than an item that bears on the first step of a very lengthy process, in looking toward the practical use of controlled fusion."

Then it was Gottlieb's turn. He said that the PLT results didn't mean that fusion would be providing energy to people's homes this year, but nevertheless they had indeed done something that was significant and far from routine. He explained how they had believed there was a fundamental barrier in the trapped-particle instabilities, which would block the temperatures from increasing. But when they ran the experiment, lo and behold, the plasma heated right up and the barrier wasn't there. Therefore, although they had not achieved breakeven, they were on schedule and he was confident they would achieve breakeven with the forthcoming TFTR.

By the end of the press conference, both Gottlieb and Schlesinger were satisfied. Gottlieb had not been held to a party line, forced to read from a DOE-approved statement, but had had the chance to describe the PLT results as he saw fit. Gottlieb as well as Deutch, however, had emphasized that in no way should these results detract from the need to push ahead with other energy programs, including conservation and synthetic fuels. The reporters wrote one more round of stories about fusion, then went back to their more usual stories of the day, including the merits of supposedly noncarcinogenic cigarettes and the appearance on Capitol Hill of James Earl Ray, convicted assassin of Martin Luther King.

Following the press conference, Kintner still wanted to meet with Schlesinger, and he went back to camping in his outer office. Schlesinger was back inside—meeting with Gottlieb. Gottlieb told him that Kintner had all along been acting properly and did not deserve to be punished. Schlesinger assured him that there would be no problem. Meanwhile, just outside, Kintner was waiting and asking the secretary if he could see Schlesinger, only to be put off: "No, he's busy." Finally he gave up and walked on down a long corridor to the elevator. As it arrived, Schlesinger came out of a private corridor from his office and got in the same elevator. Once the doors had closed, Kintner introduced himself: "I've been trying to see you all day. I think you've got some bad information and I want to correct it." Schlesinger replied, "Oh, *you* did it!" Down on the ground floor, then, they walked off for a short talk. Kintner then said that this was not a budget ploy but an honest-to-God scientific achievement of a high order, that he had handled the news in ways consistent with DOE policy, that it had gotten out through channels he had tried to block, and that by no means had he or anyone else been trying to trap him. Schlesinger listened, then said, "Well, don't do it again." Kintner replied, "If we ignite the oceans of the world, I promise you we will not hold a press conference." Schlesinger said, "Okay, don't do it again."

And that was the end of the PLT weekend, except that Eric Willis caught up with Dean and gave him one final tongue-lashing: "Now look, I hope that you've learned a lesson out of this. Don't talk to press people. You may be a hero for a moment, but they will do you in later. I've saved your ass on this one but I hope it never happens again."

Meanwhile, the PLT story was continuing to spread overseas. During that week, *Le Monde* and *Le Figaro* in Paris ran articles, as did *La Repubblica* in Rome and *O Globo* in Brazil. In Moscow, *Pravda* reported the news in characteristic fashion:

> It would be incorrect to think that the advocates of "cold war" are taking the upper hand everywhere. News of an entirely different type is also being reported these days. . . . Scientists at Princeton University have achieved a major success in the area of thermonuclear fusion. They succeeded in obtaining a temperature of 60 million degrees C in an experimental tokamak reactor. This was accomplished thanks to cooperation with Soviet scientists. . . . Comrade Leonid Brezhnev has said, "The vital interests of workers of all countries require that everything good accomplished internationally not be permitted to be erased."

By then it was the last week in August, time for everyone to get together for the conference in Innsbruck. It was natural that the Princeton results would be questioned very closely. The man who gave the paper was Rob Goldston, who a year earlier had gone into Harold Eubank's office to warn him when the liquid nitrogen had been flowing all over the PLT. He knew he would be sitting in the hot seat, so he took good care to bone up. When he got there, so many European scientists wanted to question him that the conference organizers set up a special evening session. Goldston later said that nothing like that had happened to him since taking his Ph.D. at Princeton in plasma physics, and defending his thesis in his oral exam. For one thing, they wanted to know how real were the temperature measurements and why they should be believed. Goldston showed that four separate and independent diagnostics had all given the same answers. They also wanted to know if this was a true plasma temperature, even if correctly measured. The plasma density had deliberately been kept quite low, while the beams had come in with an energy of 40 kilovolts. Were these energetic beams really heating plasma, or were their atoms merely bouncing around like so many high-energy pingpong balls? Again, though, Goldston had data from his diagnostics to show that the plasma energy had shown a "thermal distribution," and that in no way had the beams been merely mimicking the appearance of hot plasma. By the end of the evening, Goldston had made his points.

Afterward, Madame Rasumova, one of the attending Soviet scientists, went up to Gottlieb and gave him his reward. The DOE might have been reluctant to acknowledge what his lab had achieved, but she knew, and she presented him with a handmade Firebird. In the lore of her country, the mythical Firebird had brought fire down from the sun in the manner of Prometheus, and this myth had inspired Stravinsky's *Firebird*. It was a very fine trophy, eight inches tall, green and orange-red with gold spangles. She was presenting it to him for having overturned a theoretical prediction by her own countrymen, Kadomtsev and Pogutse. Gottlieb treasured it, the more so because she had made it herself, and because he knew her as a very warm person whom everybody liked. For years afterward he kept it on display in the PLT control room.

Then in September, Deutch presented the DOE and Congress with a formal policy statement on fusion. As expected, he largely followed the recommendations of the Foster Committee. He rejected any notion of an early start for a new tokamak project to follow TFTR, and called for further emphasis on mirrors as well as on alternate fusion approaches. But he recommended that the fusion budget not be cut. Instead, it was to be maintained at its current level, with cost-of-living adjustments for inflation. Also he emphasized the importance of demonstrating breakeven with TFTR, and gave that project high priority. All this was a vindication of Robert Hirsch's bold leadership, in pushing for TFTR long before there was a proper scientific basis for predicting its success. Also, Hirsch's success in winning tenfold increases in the fusion budget now would be consolidated into policy. It would not be reversed with huge budget cuts, only a year and half after Hirsch had left Washington.

These two events, the Foster Committee's report and the PLT results, occurring as they did within a few weeks of each other in 1978, together constituted the breakthrough. After that summer, it was clear that Schlesinger had been defeated in his attempt to cut the fusion budget. Also, it was no longer possible to argue convincingly that trapped-particle instabilities would prevent the TFTR from succeeding. On the contrary, all indications now pointed to success. Moreover, this was the last time during the Carter administration that Kintner would have to face down the budget-cutters. Instead, the groundwork was now in place for a serious effort to launch a new set of fusion initiatives, to give Kintner a new round of budget increases—and to push for the demonstration of practical fusion power no later than the year 2000, a full fifteen years before Deutch's policy statement said it would be achieved.

9

Plasmas on the Potomac

I N the state of Washington, the Fourth Congressional District sprawls across the eastern part of the state, extending from the Cascades across some of the nation's finest farmlands. Its farmers have always boasted that their crops are grown in rich volcanic soil, and rightly so, for when Mount St. Helens blew up in 1980, its ash cloud drifted over this region and brought a fresh supply. During the 1970s, this district was represented by Congressman Mike McCormack, and when he speaks of it there is a certain huskiness in his voice: "Eastern Washington is lovely country, deserts and mountainous country and lakes. It's so wonderful because the sky is so big and blue, the water is so pure, and the people are so happy and relaxed. It's a lovely place to be in, a lovely place to represent. My district was primarily agricultural. The delightful part of the district, the part that was so much fun especially during campaign time, was the apple country. You could fly over almost endless miles of apple orchards, and come down and drive and walk in them, where they are so clean you could almost eat off the grass. The whole community would get involved with the apple harvest, picking or moving or packing the apples. There's one building in Selah that does nothing but take reject apples and squeeze them into apple juice. Hundreds of trucks a day bring in apples and empty glass jars, then haul away the pulp after the juice has been totally squoze out. They take that pulp to feed to the cows, by the way, and the cows are very, very contented. You see, the pulp is stored and it ferments, and I think the cows are half-swacked all year. That's where the expression 'contented cows' comes from, I think."

There is more to the Fourth District than apple blossoms, however.

There is also the sprawling Hanford Works of the Department of Energy, which date back to World War II as a production center for weapons-grade plutonium. Some 15 percent of the people in this district live in the nearby communities and depend on Hanford for their livelihoods. Thus, it was natural that when McCormack was elected to Congress in 1970, he would be closely concerned with issues of energy and nuclear power. McCormack himself had bachelor's and master's degrees in chemistry and had worked at Hanford for a number of years before entering politics. He thus was one of fewer than half a dozen congressmen with technical backgrounds. His first committee assignment was the House Committee on Science and Technology. After being reelected in 1972, he was named chairman of that committee's subcommittee on energy research. He also obtained a seat on the influential Joint Committee on Atomic Energy.

He quickly set out to make a name for himself in the field of energy legislation. During the next few years he wrote a number of bills that were passed as acts of Congress, dealing with solar energy, geothermal energy, nuclear safety, and electric automobiles. This work built for him a solid reputation in Congress as a leader in dealing with energy issues. Moreover, during the Carter administration, he played a key role in the years-long contention between Congress and the White House over the issue of the experimental Clinch River breeder reactor. Carter wanted to shut that project down, but Congress kept voting funds for its continuation. As McCormack later said of his relations with Carter, "We were eyeball-to-eyeball the whole four years over this issue."

McCormack was also very interested in fusion. He gained his energy-subcommittee chairmanship and his Joint Committee seat at nearly the same time that Robert Hirsch took over the fusion program. As he later put it, "Hirsch and I immediately found each other and started to work together." He soon became virtually a partner of Hirsch and worked effectively both within Congress and with officials of the Office of Management and Budget to help increase fusion funding. By 1979, at the start of McCormack's fifth term, he was one of the most senior members of the Science and Technology Committee. By that time, fusion had become the apple of his eye.

The 1978 breakthroughs had recently taken place, and McCormack was saying to his colleagues, "The controlled use of fusion, when we accomplish that goal, will be the most important energy event since the controlled use of fire." He made speeches on the House floor. He emphasized over and over that fusion was the fundamental, ultimate source of energy in the universe, that the sun and stars were fusion reactors, and that with an intelligent federal program, we could achieve the control of fusion. As he liked to say, "It would not only be one of the most important events in modern history, but it would provide an absolutely unlimited supply of energy, clean, cheap, practical, flexible energy, for all mankind, for all time." He did more than just talk. Hardly had the Ninety-sixth Congress begun its work than he began to lay the groundwork for action to boost the fusion budget.

Hirsch was out of government and was working for Exxon, but the two men had become close friends while Hirsch had headed the program, and they had stayed in touch. One day Hirsch stopped in to visit McCormack in his office in the Rayburn Building. McCormack raised the idea of setting up a fusion advisory panel for his subcommittee: "I'd like to do something like this; would you go think about it?" Exxon was not involved in fusion work. When Hirsch checked, he found there would be no conflict of interest and that Exxon's management would let him do it, as a public service. By early March of 1979, McCormack was ready to go ahead. As he stated in a letter to Hirsch, "The purpose of this panel shall be to advise the Subcommittee on the pace of development of . . . fusion, to evaluate various fusion concepts, to suggest budgetary approaches, and to generally provide expertise to assist the Subcommittee." Hirsch was given virtually a free hand to pick the panel members.

"They were people who had been involved in fusion and had what I considered to be a more aggressive approach," Hirsch said of his panel. There were thirteen members in all. More than half were people who either had been strong fusion advocates like Hirsch, or else were the directors of organizations that stood to benefit directly from any stepped-up fusion effort. These included the heads of the fusion programs at Princeton, Livermore, MIT, and General Atomic. Also on the panel were Alvin Trivelpiece, who had been Hirsch's assistant director for fusion research; Joseph Gavin, president of Grumman, which was one of the main contractors on TFTR; and Robert Conn, an engineering professor at UCLA, who had made his reputation with pioneering studies of designs for large fusion reactors.

In short, Hirsch's panel was a special-interest advocacy group, pure and simple. They all were very distinguished and of high reputation, but that qualification could not obviate a simple fact: that asking Hirsch to give advice on the proper pace for the fusion program was precisely the equivalent of asking Senator Barry Goldwater whether Arizona water projects should be increased or cut. As a leading molecular biologist had

written on a somewhat similar occasion, this was one of the few instances
in which the incendiaries were asked to form their own fire brigade. Thus,
the panel's conclusions could have been predicted in detail from the
moment Hirsch first set foot in McCormack's office: "It is none too soon
for the Department of Energy to plan a comprehensive strategy for
bringing a ... fusion electric demonstration plant on line by 1995."

McCormack next arranged for his subcommittee to hold hearings on
the subject of fusion energy. He set up the hearings so that Hirsch and
others could sing their hosannas and hymns of praise in a way that was
likely to lead to legislation. Also, Ed Kintner and Steve Dean could now
be invited to join the hallelujah chorus. When Kintner appeared to
testify, McCormack lost little time in getting to the point: "Do you see
any reason why we should not compress the schedule? Do you see any
reason why we must take our steps sequentially? Is it not realistic to take
them somewhat in parallel, and therefore not wait until we have every
detail from one project before we start the next?" Kintner picked up the
refrain: "You know my position. We want very much to proceed,
absolutely. We believe we are dealing with the future of the human race."

By early 1980, McCormack was ready to introduce his bill. He had
brought in a fusion expert for his committee staff, Allan Mense of Oak
Ridge, who helped write the bill and then lobbied for its support. On
January 28, he introduced what would eventually be called the Magnetic
Fusion Energy Engineering Act of 1980. It was sweeping in its audacity.
Thus, under "Findings and Policy," it declared:

> The Congress hereby finds that—
> the United States of America continues to be dependent on imported oil,
> and is faced with a finite and diminishing resource base of native fossil
> fuels ...
> the energy crisis can only be solved by firm and decisive action by the
> Federal Government ...
> every nation of our world possesses in the oceans and waters of our
> planet an easily accessible and inexhaustible supply of fuel for fusion energy
> which cannot be embargoed, is inexpensively recoverable, and is usable
> with minimal environmental impact;
> the early demonstration of ... magnetic fusion energy ... will initiate a
> new era of energy abundance for all mankind forever;
> the widespread use of fusion energy ... will help provide energy inde-
> pendence for all nations of the world ...
> the early development and export of fusion energy systems ... will
> improve the economic posture of the United States, and ultimately reduce

the pressures for international strife by providing access to energy abundance for all nations . . .
    it is contemplated that the programs established by this Act will require
the expenditure of approximately $20,000,000,000 during the next twenty
years.

The specific policy to be followed, as set forth in the bill, was vintage
Hirsch. The country was "to proceed immediately with all work necessary to construct and operate a fusion engineering test facility by 1986."
This meant that in six short years, the United States was to have the next
big machine beyond TFTR, which would produce actual fusion energy,
going well beyond breakeven. Then, the next goal would be to proceed
"with all steps necessary to construct and successfully operate a magnetic
fusion demonstration facility before the end of the century." This would
be the full-scale prototype, which would actually generate electricity. All
this was basically what Hirsch had been urging since 1971, and now his
friend McCormack had introduced the bill, H.R. 6308, to mandate that it
be done.
    Having introduced the bill, McCormack then proceeded to get his
fellow Congressmen to sign up as co-sponsors. He started by sending out
"Dear Colleague" letters to each of his fellow House members, featuring
statements in his florid style:

FUSION: UNLIMITED ENERGY FOR
ALL MANKIND. WE MUST START NOW!

Dear Colleague:
NOW—FOR THE *FIRST* TIME—WE CAN UNDERTAKE A FU-
SION PROGRAM WITH *CONFIDENCE* THAT WE *CAN* SUCCEED
WITHIN A SPECIFIC TIME.

Allan Mense then followed up with technical information for the congressional staffs. Eventually McCormack would sign up some 160 of his
colleagues as co-sponsors.
    All this was good clean fun, but did it mean anything? The answer was
that it could potentially mean a great deal. McCormack was no simpleminded fusion enthusiast. He was an experienced and sophisticated
Washington insider, who had learned his trade in the course of shepherd-
ing his earlier energy acts to passage. Now he was following a coordinated

strategy, aimed at launching a new set of fusion initiatives. This strategy had two major elements, the first rather obvious, the second considerably less so.

The obvious part of the strategy was the bill itself. McCormack was one of the ranking Democrats on his committee, as well as being chairman of his subcommittee. That meant he could count on getting his bill to the House floor for action. 1980 was an election year, and congressmen would gladly be helping to pass each other's pet bills, to help one another get reelected. Fusion, unlike solar or nuclear power, had no strong constituency either pro or con, but energy was a hot issue in 1980. As one lobbyist later put it, "Opposition was limited, because fusion is viewed as fairly benign; it's like motherhood and apple pie right now. Congress needed to vote for an energy *thing,* particularly one with the potential to save the world. Also, no extra money would be spent at first."

In drafting his bill, McCormack had deliberately gone for the most rapid pace and most advanced goals he could justify. At some later date, he might strike a compromise, backing off to less advanced goals—which nevertheless would amount to a fusion program that still would move forward faster and more boldly than previously planned. Moreover, his bill would provide an opening for action under the congressional budget process. The bill was an authorization measure; it did not provide that funds be voted or spent. But it could stand as precursor legislation, to be followed by congressional appropriations, the money that would actually make things go. Indeed, McCormack was already talking with Tom Bevill, chairman of the congressional appropriations subcommittee that would handle such funds.

That was the obvious part of his strategy. The less obvious part was that he wanted to move the administration's fusion policy off dead center. That policy had been set in 1978, in the wake of the Foster Committee's review. Its basic points were that the fusion program was to receive only cost-of-living increases rather than to be allowed to expand, and that no major new tokamak project should be started till the mirrors had a chance to catch up. With his House bill, with his own review panel headed by Hirsch, McCormack was deliberately challenging that policy. Rather than continue to stand pat with it, the administration would have to respond to McCormack's initiatives, by launching a formal review of that policy. That review, in turn, might provide an opening for a new and more advanced fusion policy, more in line with what McCormack wanted.

Still, there was danger. That review would be in the hands of the President's science advisor, Frank Press, and of the Secretary and Undersecretary of Energy. Press was quite satisfied with the existing fusion policy and saw no need to change it. The Secretary was Charles W. Duncan, who had replaced Schlesinger in 1979. He, too, was cool to the idea of a new set of fusion initiatives. His Undersecretary, in turn, was John Deutch. Deutch had not liked even the 1978 fusion policy; he had set up the Foster panel, which had led to that policy but had wanted actual cuts in the fusion budget. He would certainly prove no friend of a new policy. Thus, McCormack's problem was to push for this policy review without having it wind up in Deutch's pocket. Deutch might easily respond to McCormack by setting up a new high-level review panel and having them say, "We stand on and reendorse the 1978 policy." If that happened, McCormack would be left without a leg to stand on.

Still, Deutch's opposition was something that might be overcome. As part of the reshuffle in which Schlesinger went out and Duncan came in, Deutch had moved up to his post of Undersecretary, and Carter had picked a new man to replace Deutch as director of energy research. This was Edward A. Frieman, who, like Deutch, was an Ivy League science administrator, but who came from Princeton, not MIT. Unlike Deutch, however, Frieman was a fusion man. He had been a professor of plasma physics, and indeed had been Robert Bussard's thesis advisor when Bussard had worked for his Ph.D. twenty years earlier. Then he had spent a number of years as Mel Gottlieb's deputy, helping to run the Princeton lab.

For Frieman to replace Deutch could obviously be a stroke of luck, even though Carter had not picked him as a fusion partisan. But there was reason behind Carter's choice. Despite Carter's strong antinuclear stance, much of the DOE's energy research would continue to deal with various forms of nuclear energy. However, to pick a man with a strong nuclear background would raise controversy and would send the wrong signal to Carter's antinuclear constituency. Also, it might well be seen as a softening of his position on nuclear energy. Fusion, by contrast, offered people who had appropriate backgrounds and expertise, but who would not pose these political problems. Indeed, of the four people named to this energy-research post by Presidents Ford, Carter, and Reagan between 1976 and 1981, three had strong fusion backgrounds.

Frieman was the man McCormack wanted to see directing this review. Thus, McCormack's strategy was to strengthen Frieman's hand, by mobi-

lizing political support for a new fusion policy. Frieman worked for
Deutch; he could not oppose him, since Deutch was his boss. But with
enough support behind McCormack, Deutch would see which way the
wind was blowing, and would defer to Frieman rather than insist on being
strongly involved himself. Deutch might oppose McCormack, but he
would do so only to a point. McCormack was chairman of the congressional subcommittee that dealt with many of Deutch's programs, and
Deutch had to deal with him on a lot of issues. If McCormack's position
was sufficiently strong, Deutch would stand aside and give Frieman a free
hand.

Seeking such support, McCormack started at the top. In the last weeks
of 1979, he had been trying to arrange a meeting with Jimmy Carter
himself. They might be at loggerheads over the Clinch River breeder
reactor, but they were both Democrats and were both highly interested in
energy policy. McCormack exchanged correspondence with Stuart
Eizenstat, one of Carter's top aides, writing, "I hope you will help me
persuade the President now to an Apollo-like program for fusion."
Eizenstat, quite naturally, wanted to hear the views of the Department of
Energy. On December 31 John Deutch, acting for Secretary Duncan,
wrote a memo to Carter:

MEMORANDUM FOR: THE PRESIDENT
SUBJECT: Early Warning on a Proposed Meeting Between Congressman
        McCormack and Yourself
Congressman McCormack . . . is attempting to schedule a meeting with
you to discuss the Department's Magnetic Fusion Energy Program. His
position is that the program is technically ready to move into an "Apollo"
program crash mode. . . . This would entail, in his view, rapidly growing
budgets from the FY 1980 level of $353 million to $1 billion by 1983. The
Department's position is that the program is not prepared technically to
enter into an Apollo-like mode. . . . The current program level that you
have approved for 1981 for $403 million is adequate. . . . [We] fully support
the level you have approved for 1981.

Carter did not invite McCormack to meet with him. McCormack
obtained a copy of this memo, and asked Secretary Duncan to comment.
Duncan replied that he had not authorized Deutch to prepare that memo
in his name, and had not been aware that Deutch had done so. He
proceeded to raise hell with Deutch over this. Nevertheless, this took the

wind out of Deutch's sails for only a moment, and McCormack still needed support. On January 21 he wrote a letter to Carter:

Dear Mr. President:
   This letter is to formally request that you declare the development of magnetic fusion energy as a major national priority; and establish, as a national goal, the construction and successful operation of a . . . demonstration plant before the end of the century. . . .
   It is important to recognize that the decision to move forward now with an Apollo-like program . . . is a *political decision*. There is little doubt anywhere in the fusion community that an aggressive development program . . . will allow the country to reach this goal. There is no doubt that the concept of an Apollo-type program has already caught the fancy of Members of Congress, the press, and the public. . . .
   As you are aware, I have assembled a Fusion Advisory Panel of the finest scientists, engineers, and industrial managers in the world. We stand ready . . . to meet with you to discuss this matter. . . . May I hear from you soon?

His letter closed with two quotations he felt Carter would appreciate. From the Book of Proverbs, as he wrote, "Where there is no vision, the people perish." Also he gave a quotation attributed to a turn-of-the-century New York architect, Daniel Burnham: "Enact no little plans; they have no magic to stir men's blood."

Just then Carter had other things to stir his blood: the hostages in Iran, the Soviet invasion of Afghanistan, the inflation rate, which was approaching 20 percent a year, and the challenge of Edward Kennedy in the upcoming primaries. On its face, then, this letter looked like a quixotic effort. Actually, it represented a sophisticated means of eliciting a response. McCormack was sufficiently powerful within the Congress that this letter would receive more than a perfunctory answer, and whatever substance there was to the answer, McCormack would turn to his purposes. Moreover, McCormack was well aware of the procedure by which Carter would reply to this letter, and was quite prepared to use this procedure to his own advantage.

He knew that Carter would never see his letter, at least not for quite a while. Instead, it would be handled by some staff member who took care of the President's mail. That staffer would read the letter and see dollar signs in it: a congressman is asking for a new federal program. He therefore would send the letter over to the agency that deals with dollar signs: the Office of Management and Budget. At the OMB, someone would see that the letter dealt with an energy program, and thus would send it to the man in charge of energy budgets, Hugh Loweth. Loweth, in turn, would see that the topic was fusion, and would send the letter on down to the man who was responsible for the fusion budget, a budget examiner named Don Repici. Repici, in turn, would have the task of drafting Carter's response to McCormack's letter. And Repici was a strong fusion man. Back in 1977, he had intervened directly with Kintner himself, to try to win support for the fusion ideas of Robert Bussard.

By being clever, then, McCormack could negotiate with Repici, knowing that Repici's position might later be endorsed by the President himself. Repici's draft of a response would have to be approved by Loweth and by his superiors within the OMB, by Frank Press' office, as well as by Carter's own domestic-policy staff. But Repici was quite aware of what would and wouldn't be acceptable. Even before he sent the letter to Carter, McCormack had had his aide Allan Mense talk about it with Repici, discussing what McCormack would say in the letter as well as what Repici would recommend in response.

In addition, having written the letter and introduced his bill, McCormack proceeded to make the rounds within Washington, looking for support. He talked with Frank Press, the science advisor, and found he wasn't interested in a new fusion initiative. He met with Secretary Duncan, and got nowhere. He talked with Frieman, who was sympathetic but who had to follow the lead of Deutch, his boss. Then McCormack went to the OMB, where he received a pleasant surprise.

For McCormack, a trip to the OMB would be no casual matter but would involve a formal meeting with James McIntyre, the OMB director, arranged well in advance. McIntyre had moved up to take over the OMB in the wake of the Bert Lance affair, when Lance, Carter's hometown banker, had resigned as OMB director after being caught amid financial improprieties. McIntyre, in preparing to meet with McCormack, would rely on the number-two man at OMB, Bowman Cutter. Cutter was fascinated with technology. Not only had he been pushing for the OMB to support technical programs; he had also discovered, to his pleasure, that right inside his own OMB was a group of people with strong technical backgrounds. This group included Don Repici.

"There were seven people in the OMB group that had the Energy Department's programs," Repici recalls. "Five of them had Ph.D's. I had mine in physics; my boss had his in astrophysics. We were working in a budget shop; we were what they derisively call green-eyeshade S.O.B.'s.

But we really were a bunch of people each with a fair amount of technical clout in his own right. Cutter liked our whole group. He thought it was fascinating that among all his bean-counters was a bunch of people buried down there who had these technical Ph.D's. He even came to our Christmas parties." Thus, when McIntyre had to prepare for the meeting with McCormack, Cutter reached down within the OMB organization, past Hugh Loweth, all the way to Repici.

Budget examiners like Repici are the low men on OMB's totem pole. Ordinarily, when a subcommittee chairman from Congress meets with the OMB's director, no such low men would be present; but Repici was asked to come along. He, Cutter, and McIntyre were among the OMB people who met with McCormack. "We had a grand meeting," recalls McCormack. The discussion was about fusion, and the technical detail soon got too deep for McIntyre, at which point Bo Cutter took over the discussion. Then McCormack went into even deeper levels, and Cutter waved to Repici. "Don, please respond." After a little more of this, McCormack said, "Obviously you people are well informed on this, and there's no need to debate these issues."

"It was the funniest moment I'd seen," said Repici. "It was one-upmanship. It was delightful." McCormack had come expecting to spend most of his time explaining why fusion was important; but Repici and Cutter had been well prepared in advance. To McCormack, it was most unusual that people should have done their homework in this way. This meant he could quickly move on to the nitty-gritty: the question of future budget levels. At this, it was McIntyre's turn: "Yes, we'll entertain a new initiative. We want to be reasonable, we don't want to give the store away, but if the Department of Energy review indicates its merits, then, Mr. Chairman, we'll be willing to go along at a reasonable pace." That was a very positive response, especially from within the OMB.

The fact that McCormack had found support within the OMB, of all places, was quite unexpected. He might have found it in the upper reaches of the Department of Energy, among people who were eager to expand that department's programs. By contrast, the OMB would have been expected to hang back, since they were the guardians of the budget. In fact, nearly the reverse had happened. McCormack now had the support he had been seeking, and the consequences were as he had wished.

The news of McCormack's meeting at the OMB reached Deutch quickly. Deutch now could see that fusion was headed for a review, and that there was support for a new and more expansive policy in both

Congress and the OMB. Deutch still might have been able to rally support for his own position within the Office of Science and Technology Policy, the office of Carter's science advisor. But Frank Press had no strong personal views on fusion; he relied on the counsel of one of his staff members, Dick Meserv. Meserv was mildly skeptical about fusion but was a good friend of Repici and understood Repici's position. Moreover, Press rarely got involved in energy issues. In short, Deutch had been outflanked, and he would let Frieman, the fusion man from Princeton, carry the ball in the policy review.

An early question for Frieman was whether he could take advantage of the support he had within the OMB, by having them play a role in this policy assessment. Loweth at OMB could not simply offer his good services; that would have upset Deutch. The usual procedure in such policy reviews was that they be technical reviews, involving DOE and the science advisor only; the budgeteers would come in much later. Frieman went over to OMB and talked with Loweth and Repici. Then he went back to the DOE, and argued that Loweth had some smart people like Repici, who could make a contribution. An expanded fusion program would have budgetary impact; thus it was appropriate for OMB to take part. With this argument, Frieman won his point. What was more, Frank Press declined to take an active role, deciding to defer to Frieman in the review. The entire management of the policy review would now be in Frieman's hands.

Meanwhile, Repici's draft of a reply to McCormack's letter to Carter was making its way toward the President's desk. It finally got there in April.

THE WHITE HOUSE
WASHINGTON
April 22, 1980

Dear Mr. Chairman:

Thank you for your interesting and provocative letter urging the acceleration of magnetic fusion energy development.

I am aware of the promise that fusion energy holds for long range, relatively clean, and inexhaustible energy. I strongly support the development of a technology that offers such hope for meeting future energy needs.

The Department of Energy, the Office of Management and Budget, and the Office of Science and Technology Policy are assessing the recent scientific advances in the program to determine the best course for the future. Their effort should be completed by June of 1980. It will enable us to design an orderly and aggressive approach to the challenge.

I applaud your foresight and bipartisan leadership on this issue and welcome this opportunity for us to work together. The Administration is committed to the fusion option. I would urge that, upon completion of our examination, we strive jointly to make this option a reality.

Sincerely,

Jimmy Carter

It was just the sort of courteous, noncommittal letter any President might write to an influential subcommittee chairman of his own party. It was all very gratifying to have Carter write, "the Administration is committed to the fusion option," but that statement potentially meant very little other than that they were spending money on it. There was no mention of speeding up the program, let alone of an Apollo-style commitment. Moreover, there was no mention of McCormack's "finest scientists in the world."

Nevertheless, the letter contained something significant. It contained the President's written promise that the fusion policy review would be taking place in the way that McCormack had sought. Carter's list of participants in that assessment—the DOE, OMB, and OSTP—in practice amounted to Frieman, who was in charge of energy research at the DOE, along with Don Repici and his boss, Tom Palmieri, within the OMB. Repici and Palmieri were both quite supportive of fusion, and would carry forward with the position that McIntyre had put forth at his meeting with McCormack. This meant that McCormack now had the President's blessing for a policy review the results of which stood a good chance of being very favorable. Thus, McCormack could now send out another round of letters to his fellow Congressmen:

PRESIDENT CARTER ON FUSION ENERGY

Dear Colleague:
  You will be pleased to learn that the *President* has, in response to a letter from me, indicated his continuing support for the *development of magnetic fusion energy.* . . . In view of the great urgency of this matter . . . it gives us reason to believe that we may be able to join in a united commitment to an Apollo-like program for magnetic fusion energy.

Carter's letter wasn't the only thing McCormack now had going for
him. A few weeks earlier, Deutch had resigned. This act had had nothing
to do with the fusion question. Deutch, you may remember, had been
chairman of the chemistry department at MIT before being brought into
government, and had simply decided that he wanted to return to MIT.
His departure, however, came at an opportune time. Frieman now would
have no real opposition within the DOE.

As early as February, Frieman had put together his review panel. He
had started with one of the panelists from the 1978 Foster committee, Sol
Buchsbaum, a vice president at Bell Labs. This was the same Buchsbaum
who had been one of Hirsch's advisors, and who in 1973 had helped
persuade Gottlieb at Princeton to go along with Hirsch's plans for the
TFTR. Early in 1980, Frieman got in touch with Buchsbaum and asked
him to put together a new fusion review committee. His choice was a
natural result of the Princeton–Bell Labs connection, for Buchsbaum had
long been a leading science advisor to Princeton.

Buchsbaum's group would be no collection of fusion partisans whose
views could be predicted and hence discounted in advance. Indeed, its ten
members included only two fusion experts: Robert Conn, the reactor
specialist from UCLA, and the redoubtable plasma physicist Marshall
Rosenbluth. The other panelists included such luminaries as the president of Caltech, the chairman of Caltech's engineering college, the
director of the Stanford Linear Accelerator Center, and the recently
retired head of NASA. Also, since Buchsbaum had been on Foster's
panel, Buchsbaum now returned the favor by naming Foster to his own.
This was a shrewd move, since it would give Foster the opportunity to
duck away gracefully from his earlier recommendations, and would avoid
criticisms he might make if he were left out.

In sum, here was a group whose recommendations would carry weight.
The prestige of its members was one reason why Frank Press had elected
to defer to Frieman in the policy review; Press could hardly have put
together a better group. But Buchsbaum's group was independent and
would reach its own conclusions. If these experts came out against the
conclusions of McCormack's Fusion Advisory Panel, his own group
headed by Hirsch, then McCormack would be blown right out of the
water.

One person who was well aware of this possibility was Ed Kintner.
Frieman was his boss, but when Frieman set up this new panel, Kintner
was not pleased: "I didn't like that at all. Here were another bunch of
people, only two years after Foster, to look down our throats and tell us

we had tonsillitis. I was sure that when you had this number of people of such high caliber, you couldn't get them together without having them come out with some major criticism of what we were doing." This time there would be no hurried trips to TRW or midnight sessions in a VIP suite, but otherwise the process was thoroughly familiar to Kintner. Once again there was the matter of preparing the best possible briefings and presentations. Kintner went up to Princeton to make a presentation, after which he was questioned closely. Then Foster said to him, "Ed, I think I now understand that what you want to do is to make a machine which creates a burning plasma, and prove you can keep it running." As that panel proceeded in its review, Kintner occasionally asked Bob Conn, whom he knew, how things were proceeding, and Conn replied, "Oh, I think you're going to like what you'll see." Kintner kept on getting little driblets of information—"Yeah, they're really going to support what you want to do." Then, in June, Frieman called Kintner up to his office to look at the draft of their report. "How do you like this?" he asked. And Kintner was totally flabbergasted.

Ever since 1972, at the dawn of the modern fusion era, first Hirsch and then Kintner had been awaiting a certain day. That day would see fusion move out of being a program of science research, and turn into a real engineering program aimed at building genuine fusion reactors. Kintner's own background had been in managing just such programs, involving nuclear reactors. To Kintner and others, the distinction between science and engineering was critical. Science meant plasma physics, research that would lay the groundwork but would not yet build the large, complex facilities that would actually use fusion energy to light a light bulb. Engineering, by contrast, meant facing up to such facilities in all their complexity, just as Hirsch had looked forward to doing from the start. In particular, it meant starting a large new tokamak project, the goal of which, for the first time, would not be to study plasma physics. Instead it would serve as a testbed for developing the systems necessary to produce useful fusion power. Such a project would be a long step toward the day when the plasma physics would simply be there, understood well enough so that the needed knowledge would stand discovered and ready for use.

This was the day Kintner had been awaiting, and now this high-level panel was saying that this day was ready to arrive:

> The U.S. is now ready to embark on the next step toward the goal of achieving economic fusion power: exploration of the engineering feasibility

of fusion. . . . The engineering program that the Panel envisages is a long
and a difficult one. . . . A doubling in the size of the present program (in
constant dollars) in five to seven years must be expected.

A broad program of engineering experimentation and analysis should be
undertaken under the aegis of a Center for Fusion Engineering (CFE). . . .
The program we advocate should center around a . . . Tokamak-based
Fusion Engineering Device (FED) which should provide a burning, perhaps even an ignited plasma. . . . The device should be in operation within
ten years and cost no more than about one billion of 1980 dollars. . . . The
CFE should undertake the design and construction of . . . the Fusion
Engineering Device. . . . The U.S. Fusion Program is ready, in our view, to
embark on such an engineering program.

"The Buchsbaum committee came up with this idea of a Center for
Fusion Engineering, which went beyond anything we had talked about,"
Kintner recounted. "I was amazed, absolutely amazed, at how this
turned out. How did I feel? It was beautiful. It was the culmination of my
life. I felt that being head of the fusion program at that time—that all my
previous life, working for Rickover, working on breeders and so forth, all
my training at MIT, was a preparation for that. And I felt that no living
person in this country today could have done that job as well as I did it."

The Fusion Engineering Device, in turn, amounted to a wide loophole
in the Foster Committee's idea of "fusion's highest potential," in its
insistence that no major new tokamak be started until mirrors had a
chance to catch up. The loophole consisted of the fact that it might be
possible to build such a big new tokamak, but to argue that the fact of its
being a tokamak was really just happenstance. The FED would be a
testbed to develop technology for future reactors. Then, whatever form
they would take, whether tokamaks, mirrors, or something else, any such
reactors would need a great deal of complex apparatus: large superconducting magnets, lithium-containing blankets to absorb neutrons and
produce tritium, neutral beams, microwave plasma heaters, robots for
remote maintenance, fusion-fuel injectors, means to remove impurities,
tritium-handling systems, and much else. The hope was that these issues
would predominate in the FED, rendering its status as a tokamak little
more than a detail.

Harold Furth knew what was happening: "They hammered out an
ingenious strategy, which was to say that the tokamak should go ahead
and build the reactor—but not in the spirit that this was going to be the
sole reactor line. Just in the spirit that due to a fluke the tokamak

happened to be so successful, so why not go ahead and utilize it to build a reactor? But it's only to explore fusion technology. The word *generic* crept in there. The idea was, the FED was supposed to develop generic technology, not to prototype the tokamak reactor line, even though in fact it would very likely do just that. And the whole idea of generic technology was pretty stretched. As one could easily test by asking Ken Fowler, if you lay out your program to get to a mirror demo [demonstration plant actually producing electricity], do you think that a tokamak FED is a step that you would naturally include? And the answer would be 'No, of course not.' Nonetheless, for purposes of this exercise, everyone was persuaded to see the brightest side of it, and to agree that doing a tokamak FED would be of value no matter what reactor line you ultimately chose for the demo." But Fowler knew full well how the game was to be played. Any major boost for fusion would bring a rising budgetary tide that would lift all boats, including his own. Indeed, the Buchsbaum report carried a strong endorsement of his MFTF-B.

McCormack had been following these developments closely. The Buchsbaum report was released on June 23, and the next day McCormack predicted, in a speech on the House floor, that it "will turn out to be the Administration's formal declaration that it now agrees with [McCormack's] position." A week later, he started making arrangements to have his bill passed quickly, under a parliamentary procedure known as suspension of the rules. For this, he had to appear before the Democratic Policy Committee, headed by House Speaker Thomas (Tip) O'Neill. Suspension of the rules was a procedure to be used for routine bills spending no more than about $1 million. It certainly would not apply to any bill seriously proposing to bind the nation to a new fusion program that would cost $20 billion over the next twenty years. However, as McCormack stated in a letter to O'Neill, "The bill is noncontroversial and has wide bipartisan support."

McCormack's appearance before O'Neill was somewhat like an appearance in traffic court, with O'Neill sitting as a judge. The entire proceedings took only about five minutes. Previously, McCormack had sent information on the bill to O'Neill's staff, and when McCormack's turn came up, O'Neill could turn to the proper items in his folder and take it from there. There was no objection, either from O'Neill's staff or from other members of the Policy Committee, so O'Neill approved this expedited procedure. The key to it all was that the bill would appropriate no money whatsoever; that is, it would write no checks.

When Congress votes to spend money, it does so by a two-step process. First there is the authorization bill, which sets a spending level: "If we give you permission to spend money, it will be only up to this limit." Then, maybe, there is the appropriations bill: "Now you have our permission to write the checks." McCormack's bill authorized the same amount, no more and no less, as had already been approved in the Department of Energy authorization bill for the coming fiscal year. Thus, the reason the bill was noncontroversial was that in its practical effect, it amounted to a set of rhetorical ruffles and flourishes surrounding a reendorsement of a spending level that had already been approved.

Still, these rhetorical flourishes had their uses. To Frieman they meant the opportunity to have Congress go on record in favor of fusion aims that he himself wanted. A strongly pro-fusion congressional act, particularly if passed by large majorities, could be taken to mean that Congress would be willing to appropriate the funds that would bring the Buchsbaum recommendations to life. Frieman got his chance to become involved when McCormack started looking around for ways to get his bill passed in the Senate. McCormack's obvious choice for a Senate sponsor was Senator Henry Jackson, who like himself was a Democrat from Washington State, and who was chairman of the Senate Energy Committee. Getting Jackson on board in this way, having him introduce the bill in the Senate, would then leave little doubt where the Center for Fusion Engineering would be built: at Hanford, Washington. But Jackson declined to do it.

Why would Jackson decline to go along with one of his home-state Democrats? The reason was that he owed McCormack no favors. Jackson was close to being a political boss in Washington State. Over the years, he had handpicked political allies to run for Congress with his blessing, had provided funds and talent for their campaigns, and then had cashed in their political IOUs once they got to Washington. However, McCormack had gotten elected on his own; without Jackson's permission, as it were. Moreover, Jackson was jealous of McCormack. McCormack saw himself as the recognized expert on energy legislation in the House. Jackson claimed this role for himself in the Senate, and he would have liked to have extended this domain to the entire Congress. Thus, by supporting McCormack's fusion bill, Jackson would enhance McCormack's leadership in the energy field. Even so, Jackson might have helped McCormack if McCormack had been in his pocket. But McCormack was in no way beholden to Jackson, or vice versa.

Fortunately, there were other senators who could help, and the one who eventually did it was Paul Tsongas of Massachusetts. Tsongas had been interested in fusion for some time; a few years earlier, while still a congressman, he had phoned Kintner one day and arranged for them to have breakfast together. Tsongas counted MIT among his constituents, so after meeting with the voluble Bruno Coppi and two other leading MIT plasma physicists, he agreed to sponsor McCormack's bill.

Frieman got in touch with Will Smith, staff director for the Senate energy committee, and proceeded to rewrite much of McCormack's bill for what would be its Senate version. He dropped the wilder statements about "a new era of energy abundance," "energy independence for all nations," and "the energy crisis can only be solved by the Federal Government." He also dropped the reference to "$20,000,000,000 during the next twenty years," and put the completion date for the FED at 1990, not 1986. All this, however, was in line with McCormack's strategy.

Frieman and Buchsbaum had come up with the idea for the FED, as a machine less advanced than the one Hirsch wanted to build. The Senate bill, by accepting this, amounted to McCormack's backing off to a less ambitious program and compromising. Yet even so, this bill would provide for an enormous speedup of the pace of the nation's fusion program. This Senate bill thus amounted to a legislative endorsement, not of Hirsch's committee recommendations to McCormack, but of Frieman's own Buchsbaum Committee. In the end, this was the version that was enacted. It passed in both the House and the Senate by voice vote. On October 7, President Carter signed it into law.

Even before then, by the end of August, the groundwork was in place for Frieman to launch a new fusion policy. What happened was that the Buchsbaum report was formally accepted by Frieman's principal review group, the Energy Research Advisory Board. During the same week, the House passed McCormack's original bill, by a vote of 365 to 7. This meant that Frieman could recommend a new policy based on the Buchsbaum proposals, knowing that his congressional flanks would be protected. In addition, the President's April 22 letter to McCormack now could be taken as offering an opening to the White House. Frieman lost no time in moving ahead. He had Kintner prepare a revised budget proposal for fiscal 1982, boosting fusion from $479.6 million to $525.2 million. This was in line with the discussions Frieman had had with the people at OMB. He prepared a set of policy alternatives in the time-honored Washington style, grouping them by threes, with his preferred choices in the middle.

These preferred options he flanked by different-paced alternatives that
he could describe as "too fast" or "too slow." Then on September 9, he
sent a lengthy memo to the Secretary of Energy:

ACTION MEMORANDUM
SUBJECT: Prompt Action in Response to Recommendations to the Sec-
    retary in the Magnetic Fusion Energy Report of the Energy
    Research Advisory Board
Issue: Whether to accept the mission as recommended by the Energy
Research Advisory Board (ERAB) of the "exploration of the engineering
feasibility of fusion" by the early 1990's. . . .
Urgency: The ERAB recommendations . . . for significant and prompt
acceleration in the magnetic fusion program have received wide attention; I
accept those recommendations. The FY 1982 budget is about to be trans-
mitted to the Office of Management and Budget. This budget, which was
developed several months ago, does not recognize or support the strong
recommendations of the ERAB. Simultaneously, the House has passed a
bill and the Senate is likely to pass a companion bill soon calling for
acceleration of the magnetic fusion program. We expect that a bill will
arrive at the President's desk before the end of this Congressional session.
President Carter has sent a strongly supportive letter . . . to Representative
McCormack [this was his April 22 letter] . . . indicating Administration
action following the current review of the magnetic fusion program. . . .
Recommendation: Accept the mission of demonstrating the engineering
feasibility of magnetic fusion during the next decade by providing an
enhanced FY 1982 budget, the required staff positions, and the issuance of
a Department fusion policy restatement.

Secretary Duncan then arranged a meeting with people from the
OMB, the Domestic Policy Council, and the office of Frank Press. Once
he had their concurrence, he sent a letter to the White House. It was not
so different from the one McCormack had sent to Carter the previous
January, but this was the real thing.

MEMORANDUM FOR: The President
FROM: Charles W. Duncan, Jr.
SUBJECT: Establishment as a National Goal the Demonstration of the
    Engineering Feasibility of Magnetic Fusion Energy

    In April, you sent a letter (copy attached) to Representative McCor-
mack, strongly supportive of magnetic fusion energy development. In that
letter, you stated that further action would be taken based on the outcome

of an Administration review. . . . That review is now completed and the result is a recommendation to establish as a National Goal the demonstration of the engineering feasibility of magnetic fusion energy by the early 1990's. . . .

Achievement of the goal would provide the Nation, in about a decade, with the knowledge upon which to make an assessment of the ultimate practicality of fusion. . . . Having knowledge in the early 1990's of a new and vast energy supply to come will affect our energy management for the following two decades before fusion energy becomes widely available.

Striving for this worthy goal can be a great inspiration for the American people. Joseph G. Gavin, Jr., president of the Grumman Corporation, recently wrote to me . . . "I was involved in the Apollo Program and remember vividly the 'lift of a driving dream' that all Americans received from a great technical challenge undertaken as a national goal." That phrase can apply yet again—to fusion energy.

The timing of such an announcement is also appropriate from a political point of view. Broad bipartisan support is developing for the two fusion acceleration bills presently before Congress, the first of which, H.R. 6308, was passed with near unanimity. . . . In the coming weeks, establishment of a stirring National Goal could well have a beneficial supporting effect for your leadership.

My recommendation is to proceed with the establishment of this National Goal for magnetic fusion energy. We stand ready to help you in any way.

Carter soon gave his assent, and on October 15 Duncan signed his approval to Frieman's recommendations. The nation had a new policy for fusion energy, approved at the highest levels of government.

What did it all mean? In the years since 1980, McCormack has frequently been honored by leaders of the fusion community. Within that community the lore and legend have grown that he singlehandedly initiated his bill, led the fight for its passage, and ultimately succeeded in using it to change the nation's fusion policy. Yet as a substantive piece of legislation, McCormack's Magnetic Fusion Act was no more than a precursor to what might take place in 1981, after the presidential elections. Then, in the Ninety-seventh Congress, McCormack might be able to push for the appropriations that would begin to make the Act real. This, however, would be a matter for the vagaries of the regular congressional appropriations process, and in no way could McCormack be certain of success. The list of the 160 co-sponsors of his House bill told that story. They included the majority and minority leaders, Jim Wright and John Rhodes, which meant that McCormack would have few problems in dealing with the House leadership. Indeed, Jim Wright spoke for several minutes in support of the bill, during the floor debate prior to its passage in the House. But one name was conspicuous by its absence from McCormack's list of cosponsors. This was Tom Bevill, chairman of the appropriations subcommittee that would have to approve funds for the bill in 1981.

Major policy shifts do not flow from bills passed with virtual unanimity, under suspension of the rules. When a bill passes with no real dissent, it usually means that Congress is doing routine business, or is going on record in favor of apple pie and motherhood, or is caught up in a blaze of emotion over some sudden crisis. McCormack was fond of noting that the 1970 Clean Air Act had been passed under suspension of the rules. But that Act had passed amid just such a blaze of emotion, when the need to Do Something About the Environment was at its height. Fusion, by contrast, was not that sort of incandescent national issue or controversy.

McCormack never had to face any real opposition, either from the White House or from powerful interest groups such as the environmentalists. Yet such opposition surely stood to develop, once the fusion program began to grow bigger and more visible. Even so, McCormack never mustered more than a partial regiment. He never gained support from leaders of the House and Senate Appropriations Committee, which would have been essential if Congress were to mandate actual spending. His floor manager in the Senate, Paul Tsongas, was a first-termer who had not yet established himself as a power in the field of energy legislation.

McCormack thus would have needed far more than he had, to forge a solid phalanx that could establish Congress as a stronghold of serious support for fusion. Ironically, McCormack had been prominent in forging just such a congressional phalanx during those same years, in support of the Clinch River breeder reactor. Carter had opposed this project, but Congress had repeatedly voted funds for its continuation, over Carter's outraged objections and even in the face of his threats of a veto. The advocates of fusion never got close to having Congress develop this level of support and commitment.

This did not mean that the Fusion Act was on a par with having Congress declare July to be National Fusion Month. In addition to his efforts in Congress, McCormack knew how to move the administration. He was able to strengthen the hand of the Director of Energy Research, Ed Frieman, while undercutting his boss, John Deutch. He knew how to look for support within the offices serving the White House. He was shrewd enough to get out front on an issue, fusion power, that really was ripe for development, and he was able to use his position within Congress to push the administration in the direction of favoring its development.

McCormack's leadership thus went deeper than putting Congress on record with his Fusion Act. That was only the more obvious part of his work. More significant was his use of that bill, together with the support which he had discovered in the OMB, to create a pro-fusion climate. With this, he and Frieman could lead the administration into developing a new fusion policy through its own internal procedures of policy review and assessment. That was highly significant. With its factionalism and its fluctuating tides of opinion, Congress alone would be a very weak reed with which to sustain a shift in policy. An administration-backed policy, by contrast, might well stand for years and be followed by the Congress with little debate. By mid-October 1980, three weeks before the election, such a policy was in place, and its architect had been McCormack. The real significance of McCormack's activities lay in his work behind the scenes.

10

The Face-Off

THROUGHOUT the summer and fall of 1980, Ed Kintner was working as in a dream. He was like a farmer in May, planting a vineyard in black loam on a sunny hillside. He had the responsibility for planning the new fusion era, but his plans would be no mere wish-list, to be ripped to shreds by budgetary machinegunners the minute they got out of his office. Instead they would be his response to a mandate, one that went beyond what he could have reasonably sought only a few months earlier. He had the task of planning for the emergence of fusion as a new high-technology industry, which might well grow as the aerospace industry had grown in the 1960s, and might become comparably significant in the nation's economy. Few federal managers have ever had such an opportunity thrust upon them.

At the core of his thinking was the Center for Fusion Engineering. It would be a major research center, as large as Princeton or Livermore, and would build and operate the FED. Kintner knew that many people would want to design fusion reactors by applying what was well known in the field of nuclear reactors, but in his mind this would not be enough. "We had to have basic scientific efforts in terms of metallurgy, materials sciences, heat transfer, coolants—all the basic ingredients. Many of them had to go back to fundamentals." He wanted to set up the center in the fashion of California's Jet Propulsion Laboratory, which has developed and built the planetary spacecraft that have sent back such spectacular photos of Mars, Jupiter, and Saturn. Those spacecraft were built with electronics and computers, communications systems, astronomical instruments, and control systems; but the JPL people had not simply taken the needed equipment off the shelf. Instead, they too had gone back to fundamentals, and had raised these engineering sciences to new heights. Kintner had been present at the start of nuclear engineering. He now hoped to bring about the advent of the new and equally demanding discipline of fusion engineering.

The most difficult problems in this field would involve metallurgy and materials. Any fusion reaction would give off vast swarms of highly energetic neutrons, more powerful than in any nuclear reactor. In time these neutrons would shatter the crystalline structure of any metal, robbing it of toughness and resiliency, eventually making it no stronger than chalk. With its metals weakening and turning brittle, a fusion reactor would have to be shut down to have the weakened structure rebuilt. This would have to be done entirely by robots, since the irradiated metals would be intensely radioactive. Doing all this would cost a great deal of money. However, no one knew what the actual effects of the neutrons would be: how rapidly a metal would weaken, or whether existing materials like stainless steel could be modified to keep their strength longer. No nuclear reactor could produce neutrons sufficiently energetic to give appropriate tests. Thus, even before 1980, Kintner had been pushing ahead with a materials-testing facility, to be built at McCormack's old stamping ground of Hanford, Washington. This was FMIT, Fusion Materials Irradiation Test. It would feature a powerful accelerator somewhat resembling a neutral beam system, to produce an intense beam of energetic neutrons. Well before the FED would be ready, the FMIT would be answering basic questions about the metals that would go into it.

Also, Kintner had never stopped thinking about alternative approaches to fusion. It was all very fine to have tokamaks as the lead approach with mirrors as the backup, but he wanted a backup to the backup. In 1978 he had invited some 100 fusion experts to a meeting, which participants would remember as "the Gong Show." Members were asked to rate a number of fusion approaches, ranking them on the basis of their readiness and their attractiveness as reactors. The winner was a concept with the unlikely name of Elmo Bumpy Torus, or EBT. "Elmo" was an acronym for "electron magnetic orbits"; it also referred to the electrical discharge known to mariners as Saint Elmo's fire. The concept involved two or three dozen short magnetic-mirror cells strung into a circle like a chain of sausages. The plasma within this torus would have bulges or bumps; hence, "bumpy."

To Kintner, EBT was not only an important fusion approach but was also a project that would help draw industrial firms into fusion. They would help build a fusion industry, and they then would lend their political clout to help keep the federal money flowing. There was already a small EBT operating, at Oak Ridge. Kintner had wanted to proceed with a much larger EBT, the size of the Princeton Large Torus, to prove out the EBT principle, and he wanted its contract to go to industry. The Oak Ridge fusion director, Herman Postma, opposed this plan; he wanted his own lab to build it. He went over Kintner's head to appeal to John Deutch—and Deutch backed Kintner. "I could have kissed him when he did it," said Kintner. "It was the only time I ever felt like kissing the guy, and that's a hard thing to say if you know him, because Deutch sure isn't a kissable character. Maybe his wife loves to kiss him, but I sure didn't feel that way very often." The Buchsbaum report stated, "The present results and positive reactor prospects warrant strengthening of the EBT program." That September, Kintner's office gave a contract for the new EBT to McDonnell Douglas, the aerospace firm.

So, throughout much of the summer and fall of 1980, Kintner was in managerial heaven. Then came the election; Carter was out and Reagan was in. Also out was McCormack, swept aside in his reelection bid by the Republican tide.* Kintner had voted for the third-party candidate John Anderson, but he felt no forebodings when the election results were in. Indeed, he felt the fusion program might do even better under Reagan, particularly since, unlike Carter, Reagan was pronuclear. Kintner was surprised when a friend called and told him the news about McCormack, and was sorry for him personally. But he felt the program was in such solid shape that it would go forward. His view was widely shared. As Harold Furth put it, "It wasn't a matter that Carter was for fusion and Reagan was 'agin' ' it. It was more that Carter was for sunshine and against breeders, and marginally in favor of fusion. Reagan was for breeders and against sunshine, and marginally in favor of fusion. So this was very advantageous for fusion, in that it put us on a smooth road, rather than a rocky road of violent ups and downs. In fact it was in some sense the best possible situation, to be somewhat loved but not too much."

The first hint of a problem came when Reagan brought in his transition team to prepare to take over the Energy Department. That group was headed by Michael T. Halbouty, a self-made millionaire from Oklahoma,

---

*McCormack's district had always been strongly Republican, and McCormack knew that although he had kept his fences mended back home, he was likely to be vulnerable in a good Republican year.

a man of rather colorful views. He had criticized the tendency of oil drillers to seek new oil only near existing fields: "I call that drilling close to the belt, and I have been raising hell about it. We ought to drill more honest-to-God wildcats, and I mean in the boondocks." He had also grumbled at the standard seismological techniques used to find oil: "Trouble is, we rely too much on those damn black boxes to do our looking for us." Halbouty was very much in favor of Reagan's campaign stand on energy, which was that the best thing the government could do would be to get out of the way and let the producers produce. The windfall-profits tax had added personal venom to Halbouty's conviction that our energy problems had stemmed from government regulation and that the Department of Energy should be dismantled. These attitudes were entirely adverse to new initiatives in fusion, which would soon be needing more money.

Reagan liked Halbouty and wanted him to be Secretary of Energy. Halbouty was interested. But when he looked into the legal requirements, he found he would have to divest himself of his oil and gas properties, which he was not willing to do. As his Energy Secretary, Reagan picked James B. Edwards, a former dentist who had risen to be governor of South Carolina. Edwards, once he was installed, had to face the question of Ed Frieman. Frieman wanted to stay on as director of energy research, and there were a number of people who wanted him to. He was not politically partisan and was widely respected, both in Congress and in the Office of Management and Budget. In the end, Edwards went to the White House and said to Ed Meese, one of Reagan's closest advisors: "We want to keep this man." Reagan himself then made the decision: "We're not going to keep anyone in an appointive position who was a Carter appointee."

While in his DOE post, Frieman had had a deputy, Doug Pewitt. Pewitt's first name was Nelson, but everyone called him by his middle name. He would now move up to replace Frieman, in the post of Acting Director of Energy Research. This arrangement was to be only for the next few months, till Reagan could get around to naming his own man to fill that post. Pewitt was not keen about staying in government. He and Frieman had worked closely together, and Pewitt wanted to continue doing so. Frieman had gotten a vice presidency at the consulting firm of Science Applications, Inc., in La Jolla, California, and Pewitt had a standing offer to join that firm, at twice his government salary. But he didn't like the way things were beginning under James Edwards, and he decided he should stick around.

"The reason I stayed around," he said, "was to keep one of those South Carolina people from taking over as Director of Energy Research." These were the associates of Edwards, whom some called the "South Carolina thugs." They included people like Ben Rusche, who was known for grabbing the hands of Congressmen's wives at receptions and virtually crushing them in his grip. Rusche also happened to be the man with whom Pewitt had to deal on fusion issues.

There were other things about the new Edwards regime that bothered Pewitt. Early on, after Edwards had come in, he made up yellow-striped badges for his aides. If you didn't have a yellow stripe, you couldn't get into the inner sanctum, where Edwards had his offices. Pewitt did not appreciate this fortress mentality: "My grandfather was the only registered Republican in two counties; I have been a loyal Republican all my life. And I was excluded from that inner circle." Actually, Pewitt was an enthusiastic partisan of whatever party occupied the White House. During the recent campaign, he had pledged undying fealty to some of Jimmy Carter's top political aides. Now, however, it was time to turn Republican, and his exclusion rankled. He felt that Edwards himself was a fine fellow, but that Rusche and the others around Edwards had their own agenda. They wanted to position themselves so that if Edwards was able to be reelected as governor, then they could return with him and go back to running the State of South Carolina.

As acting director, Pewitt was now Kintner's boss. This was a situation that would quickly bring stormy weather. Pewitt was a budget cutter by inclination, and the leaders of the fusion community didn't like budget cutters, especially not right then, with the Fusion Act newly passed. Pewitt had no particular bias against fusion, but he had other energy programs under his purview as well. As he put it, "I do have a penchant for cutting budgets. I dissipated that penchant, in that year of 1981, by taking the fossil-fuels program from $1.6 billion to $400 million. I didn't need to take it out on fusion." Certainly, the fusion community could take little comfort from this.

Pewitt's relations with that community had not been good. He had all along been a critic, skeptical of their claims: "Fusioneers are their own worst enemies, frequently. When challenged, they do not hold up very well to legitimate inquiry by honest-minded individuals." He felt the program would need much more work in plasma physics before it could go ahead with the engineering work that would build a reactor: "Kintner's idea that you can just bully your way through the engineering, when you don't have the necessary understanding of plasma confinement, is without merit." He also was skeptical of the claims that the fusion community would gain that understanding with TFTR. TFTR was designed to approach the Lawson criterion, widely regarded as virtually the Holy Grail of fusiondom. This long-sought goal would achieve sufficiently good plasma confinement to give a confinement parameter, $n\tau$, of 10^{14}. But to Pewitt, "The Lawson criterion has less to do with building a successful fusion reactor than the boiling of water has to do with inventing the steam locomotive."

It thus was little wonder that many fusion people sniped at him when they could. In the words of one senior manager, "Many people referred to Pewitt as 'Deputy Dawg,' in reference to the TV cartoon character, who maintained his office because of his heart of gold, in spite of fouling up every job he attempted." Pewitt had been dealing with fusion for a number of years. He had been its budget examiner at the Office of Management and Budget, and had put through the $60 million cut in the fusion program in 1977, which had delayed TFTR by a year. "That was when Kintner learned to hate me," he said. "But the job of a budget examiner is to examine budgets, and the way you do that is to recommend a cut somewhere and see what happens."

He had some very definite views on Kintner: "Some people are good deputies [he was referring to Kintner's position in 1975 under Hirsch] but can't go farther. He did not have an army of admirers. He had tried to position himself to take over Naval Reactors, which was under Rickover. Nobody who ever had Kintner working for him, trusted Kintner; he had a track record of not being trusted. No loyalty up but he expects absolute loyalty from the people below. Ed just never understood that on occasion the real world requires you to take something other than what Ed Kintner says. The art of the possible never appealed to him."

Kintner was quite prepared to return the compliments: "He was the kind of fellow of whom Admiral Rickover would say, 'He lies.' Pewitt was the sort who would say one thing and do another, then turn around and deny he had said it. I didn't trust him, didn't respect him, and never had. I don't think he was ever committed to anything in terms of trying to help it, including fusion. He was always on the negative side, always pushing things back instead of furthering them, which is not my style. I don't cotton to people like that." Pewitt was a captain in the naval reserve, and on occasion had been seen wearing his dress uniform, with medals and gold braid. When he left the OMB to go to the DOE, the women in his office got together and presented him with an admiral's hat. But despite his first name, to Kintner he was no Admiral Nelson. He was more like Captain Queeg of *The Caine Mutiny*.

The trouble in all this was that as Acting Director of Energy Research, Pewitt now would be Kintner's boss.

The first issue they had to deal with, early in 1981, was the budget for fiscal year 1982, which was to begin in October 1981. The first move would come from the OMB, located across Pennsylvania Avenue from the White House, in the red-brick New Executive Office Building. The fusion budget was in the hands of Don Repici and his boss, Tom Palmieri. Repici was still the budget examiner, which was an ordinary-sounding title, but his domain included all the energy-research programs, making him the OMB's counterpart of Ed Frieman. He had flourished in the days of the Carter administration, and he expected to play the game by those rules: "In many ways, as a budget examiner, in dealing with your counterpart who was an assistant secretary or program director, you often knew just as much about the programs as they did. You cut through to the essence of things, in discussions. Plus, in the Carter administration there was always extra money. If you knew your way through the process, you could always find a few extra million dollars, if you needed it." Repici sent over to Pewitt a budget mark of $480 million for 1982, a hefty gain over the $350 million of 1980 and the $394 million of 1981.

Kintner and Pewitt worked out their own budget mark, to be sent back to OMB: $505 million. However, this would need Edwards' approval. Kintner was to go up for what Pewitt called the "Star Chamber proceeding," to make a formal presentation of this budget and its justification to Edwards and his aides. However, Kintner took the occasion to make a trip out of the country, leaving his deputy, John Clarke, to face the Star Chamber. This was the same John Clarke who in 1973 had directed Oak Ridge's effort to capture the big tokamak project that eventually went to Princeton with the name of TFTR.

In Pewitt's words, "Clarke's presentation was the sort of thing that if you were not an advocate for the fusion program, his arguments tended to put one's teeth on edge. They were the sort of thing that challenged the whole process of reviewing a budget. Rather than defending the basis for the program, his arguments were adversarial." Within the Star Chamber, Edwards' people decided that on the basis of what they had heard from Clarke, they would put the budget mark at $350 million, not $505 million.

This was well below the level sent over from OMB, which was entirely the reverse of what might have been expected. Ordinarily, the top people in DOE would be fighting for higher budgets, but these were not ordinary times. Edwards had been brought in to fulfill Reagan's campaign promise of dismantling the Department of Energy, and was not about to defer to David Stockman, Reagan's new director of the OMB. In Pewitt's words, "Jim Edwards was not about to allow Stockman to run the Department. His macho was up. Those were going to be his budget marks. Stockman wasn't going to put his money where Edwards didn't want it."

This was the sort of thing that Pewitt had stayed in Washington to fight against, and he decided he would personally go to bat for the fusion program: "John Clarke had made such a sterling presentation that I had to intercede. I told them that if the fusion office couldn't put together a program that made sense to Edwards, I could, and that I would shoot for around $450 million. Asking for a $100 million increase seemed to be, on the spur of the moment, the most I could do." On the basis of Pewitt's presentation, Edwards approved sending a recommendation of $440 million over to the OMB, which still was less than OMB's $480 million.

At OMB, Repici and Palmieri put the subtracted money back. Then a discussion developed as to how much money the DOE would let the OMB put back. This was a complete role reversal. Always it was the departments like DOE who were howling and screaming for money in the face of a recalcitrant OMB; now the OMB was virtually forcing extra money on them. The only reason the DOE position didn't prevail automatically was that Senator Baker, the new Senate majority leader, was telling his fellow Reaganauts that he wanted to protect the fusion budget. After all, Baker was from Tennessee, home of Oak Ridge and the proposed home of the new Elmo Bumpy Torus. Eventually they compromised at $460 million.

To Kintner, all this was thoroughly bizarre. Much of this budgetary back-and-forth had gone on over his head or in his absence, and the main thing he could see was that the budget mark had come in from OMB at $480 million and gone back to them from Pewitt at $440 million. Kintner called this the "Pewitt mark," and he didn't like it. Budget-cutting was what people at OMB did for a living. At DOE the standard procedure was never to look a budgetary gift horse in the mouth, but to take the money and go with it. To Kintner, fighting for higher budgets, it was thoroughly distressing to think of his boss as an OMB wolf in DOE clothing. In any case, however, Kintner never again had to deal with such a role reversal, for by now it was February 1981, and David Stockman was instituting his new regime within the OMB. For Repici, the old free-wheeling days would soon be gone for good. In his words, "George

Keyworth, the President's science advisor, took a very aggressive lead in deciding what was technically relevant and what was not. The OMB did not engage in technical discussions anymore. The group got de-fanged." This was the last time the OMB would ever try to add money to the fusion budget.

Then in late February of 1981, Pewitt was invited to appear at hearings before the House Science and Technology Committee. He brought Kintner along, to sit with him at the hearings table. Clarke came along, too, and sat among the spectators, as if he were there watching proceedings in a courtroom. Pewitt then gave his testimony: "Faced with the need for fiscal constraint, we are not proposing to take the steps that would imply large future-year expenditures. However, the program proposed will represent a significant expansion of the program effort." Afterward, a lobbyist who was also sitting among the spectators asked Clarke, "Just who is the fusion advocate in the DOE?" Clarke replied, audibly, "Fusion has no advocate in the DOE."

That was a bad move. Clarke might not have liked the Pewitt budget mark, and in private he might say what he wished. But in public—and a congressional hearing was certainly that—Clarke would be expected either to show loyalty to his superiors, or to resign from government. Clarke had spoken loudly enough for Pewitt to overhear. By himself, Pewitt could have ignored it. But shortly afterward, Joel Snow, one of Edwards' advisors, came up to Pewitt and told him what Clarke had said. This put a different face on the matter, for in Pewitt's words, "this particular individual was not above pedaling right down to Ben Rusche and saying, 'Look, Pewitt is countenancing insubordination, he's protecting Clarke and Kintner.' " Neither Clarke nor Kintner was from South Carolina, and Pewitt had to do something, or he'd be in trouble with Rusche and the rest of Edwards' people. Still, when Pewitt decided what action he would take, in his words, "it wasn't even a particularly difficult thing to do. John had been indiscreet in public, in a way no public official should be."

A few days later, Kintner had a meeting with Pewitt, and at the end of it Pewitt told him, "Oh, by the way, tomorrow morning at nine o'clock I want John Clarke down here. We're assigning him a special assignment, to review the personnel requirements for fission." This meant the DOE's programs in nuclear reactors, which had nothing at all to do with fusion. Kintner replied, "Well, what the hell is this all about? Why John Clarke?" "Well, it's a six-month assignment and he's available, he knows the right questions to ask, he's a bright guy." Kintner went on, "What do you mean he's available? Nobody told me about this. What am I supposed to do without a deputy for six months?" "That doesn't matter; we've already decided."

What really was happening was that Clarke was being sent to Siberia— but not to a conference in Novosibirsk. He was to be punished for talking out of turn, by being sent to manage a study that had nothing to do with his real job responsibilities and that Kintner regarded as entirely meaningless. It was clear to everyone that this was a punishment, and Clarke was so upset that he was on the verge of resigning: "There's no point in staying around if this is the way people are treated for stating their views."

Kintner was equally upset: "This was the worst piece of personnel management I'd seen in my entire career. Clarke was my deputy. If he wasn't doing right, I'm the one who was supposed to fix it, not those other guys. I'd never had anyone, not even Rickover, reach down below me and discipline one of my subordinates for not doing the right thing." He disagreed in writing with this action, and then in talking about it, Pewitt said, "Well, you know, we really had a question whether it should be you or Clarke." Clarke did not resign. He decided to go ahead and direct that manpower-requirements study, and to try to do so good a job that it would turn out to his credit. But meanwhile he would be off in Siberia, in exile from the fusion program, and everyone knew it.

Still, Pewitt was not the only senior administrator Kintner was dealing with. As he had predicted, the program structure he had put together really was turning out to be strong enough to stand. The Buchsbaum report and the Fusion Act had proposed budget increases of 25 percent in fiscal 1983 and in 1984, on top of an allowance for inflation. Kintner fell back to recommending increases of 10 percent, as the minimum with which a meaningful start could be made. In the spring and summer of 1981, his view was prevailing. Thus on June 26, Ray Romatowski, who was Pewitt's boss, sent a memo to Pewitt:

> The recommendation that DOE undertake to determine the engineering feasibility of fusion energy during the next decade is approved for FY 1983. This is with the understanding that the Department is committing to engineering work but is not prepared to commit to the construction of a Fusion Engineering Device at this time.

This memo set forth a rising budget profile for fusion, which in constant or uninflated dollars was to get $557 million in 1983, increasing to $773

million in 1987. With allowance for inflation, of course, the number of
dollars to be spent would be larger still. Then, within DOE's internal
reviews, this $557 million budget mark was worked over further and led
to a set of three budget cases for 1983. The "Request" of $532.3 million
was what they would seek from OMB. The "Target" of $498.3 million
was OMB's guideline as to what it would accept, while the "Decrement"
of $445.1 million was the largest budget cut they felt they could live with.
All this was a substantial cut below the numbers Kintner had been
working with at the end of the Carter administration, which were $525
million for 1982, $596 million for 1983. Even the "Target," however,
would provide $25 million for Kintner's materials-testing FMIT, $24
million for Elmo Bumpy Torus, and $5 million to make a start on the
Center for Fusion Engineering.

The summer of 1981 brought more good news as Reagan finally got
around to naming his Director of Energy Research. Pewitt would be
gone, but not very far. He took a position on the staff of George
Keyworth, Reagan's science advisor, who was a good friend of Pewitt's
and had asked him to stay in government; thus Pewitt would soon be
heard from again. Reagan's energy appointee was a good fusion man,
Alvin Trivelpiece. He had taught plasma physics at the University of
Maryland, coauthoring a leading text in the field, and had then gone on to
join Robert Hirsch as one of his research directors. And since in the
coming budget battles the hinge of decision would turn on Trivelpiece's
judgment and personality, it is worth digressing a bit to introduce him.

He could easily affect the slightly overbearing air of a senior professor,
but Hirsch well knew his other side. Hirsch had decided to try to recruit
him to be one of his directors, in 1972, but just then Trivelpiece was off
vacationing in Canada. A few days later Hirsch phoned again, just as
Trivelpiece was walking in the door of his house: "Hi Al, it's Bob. I want
you to come work for me in the Atomic Energy Commission." "What,
are you nuts? Are you crazy? I just walked in the house, what kind of a
thing are you talking to me about?" They decided to meet for lunch the
next day, at a Shakey's Pizza out on Rockville Pike in suburban Maryland. Trivelpiece walked in looking like a big shaggy bear; he had been up
in the north country for weeks, and hadn't shaved. Hirsch explained what
he had in mind, and finally got Trivelpiece to agree to come in for a visit to
the AEC's headquarters. Hirsch then proceeded to lay some groundwork
with a phone call to one of the AEC commissioners, William Doub: "Bill,
I have a man that I really want to hire, but he doesn't want to come. Can I
bring him up to talk with you when he comes in?"

When Trivelpiece arrived, Hirsch soon took him to Doub's office. After doing a very effective selling job himself, Doub said, "Tell you what, Trivelpiece, I think you ought to meet the chairman." He soon had Trivelpiece in the office of James Schlesinger. Trivelpiece's head was quite thoroughly turned. Here were these people, Doub and Schlesinger, whom he had only read about before, urging him to come join them. It all was an incredible vista that was opening before him, so he decided to take this new career direction. He didn't know what he was getting into, but Hirsch did: "It wasn't even a matter of hitting the beach running, it was like running while you're still in midair. It was really a head-twister for him, because we were off and running like crazy, and he had to sprint to catch up." After a few months, he remarked to Hirsch, "You know, when I interviewed here, Bill Doub and Jim Schlesinger gave me these great stories about how it was going to be—and, you know something? I haven't seen either one of those guys since then!"

That was in 1973. In 1981, on his second tour in Washington, he was now the very model of a modern presidential appointee. Tall, lean, elegantly tailored, with the silver-gray hair of distinction, he well fitted his office. And his office, in turn, was part of a suite with the floor space of a large ranch house. The office itself was as big as a large living room, but it was far better carpeted, draped, paneled, and furnished than most. Out of the picture windows that formed nearly an entire wall there was a dramatic seventh-story view of L'Enfant Plaza, with the Potomac and the Virginia shore in the background. Trivelpiece had learned his Washington lessons well. He knew exactly what to do when presented with a live mike, in our interview. This is not to say he clammed up; quite the contrary. His lips moved; he uttered strings of articulated sounds that could be identified as human speech. But he conveyed no information beyond what was available in standard DOE press handouts. He gave a virtuoso performance of granting a half-hour interview while keeping all his beans safely unspilled. Clearly, here was a man who would go far.

Kintner liked his new boss: "I thought that we had a friend. He might not be as forceful as some other people in carrying out what he believed for fusion. But after dealing with Pewitt, here was a guy I knew, I could talk to; he was an honest, human person." They proceeded to work together to develop 1983 budget plans that Kintner hoped would save the most important parts of his program: "These were the things I had worked for five years to put into place, and they were all there, and I didn't want to give them up." They included the TFTR, which now was nearly complete; the MFTF-B, which he wanted to push forward rapidly;

the materials-testing FMIT; the Elmo Bumpy Torus; and the Center for Fusion Engineering with its big new tokamak, the FED. He was willing to negotiate over their funding levels and pace, at least to a degree, but he was adamant that they all go forward. To get advice on these matters, and to establish a consensus among the leaders of the fusion community, Kintner convened a meeting of his lab directors and senior consultants, to be held at Los Alamos during the last week of September.

Then on Monday of that week, the day before he was to fly to Los Alamos, the roof fell in.

Over the weekend, Reagan had decided to cut the budget deficits. His decision was to impose a cutback of 12 percent in all federal departments, other than defense, whose spending was "discretionary": that is, not mandated by law. Kintner's program was definitely in that category. "I was in the office and my immediate response was—there wasn't any way to respond. It just said, the budgets are now this. You couldn't argue; it came down from the White House, bang, take twelve percent out."

He flew to Albuquerque and caught an air-commuter flight to Los Alamos. The laboratory there fills dozens of buildings sprawling on a high mesa, seven thousand feet up, and connects to the adjacent town by a bridge over a ravine. The country is ponderosa pine amid stark mountains, purple in the early morning. The road leading to Santa Fe winds down the mountain in a series of switchbacks, with sharp dropoffs and spectacular chasms on the other side of the guardrails. The weather in late September was clear and cool, even at midday. Here the Atomic Age had begun, during the war, when in great secrecy John Robert Oppenheimer had brought together the scientists who would build the first atomic bombs. Some traces of those days still could be seen: wooden barracks on Trinity Drive that had served as wartime apartments for scientists' families. These were just up the hill from the bridge. Los Alamos itself, however, had long since metamorphosed into just another fair-size New Mexico town.

But Kintner's mind was not on the scenery, which he had seen on many previous visits, nor on the history. The latest budget cut meant his 1983 budget would be the "Decrement" case, of $445 million. This meant that FMIT, Elmo, and FED would all go into a curious state of hibernation. Little driblets of money would keep their planning up-to-date, but in no way would there be enough to bring them to life. As he would say a year later, "FMIT isn't defunct; it's fast asleep."

He knew that this latest decision would cut the heart out of his program, and he was bitter: "A program needs bone structure, and that's the part that's been destroyed. A chicken without bones doesn't weigh much less, but it doesn't fly very well." As he met with his advisors, he felt he was fighting in the last ditch. This was his last chance to save the basic goals of his program. He had put together his program with care and love, over the past five years, and had seen it enacted into a national policy endorsed by Congress and by President Carter. He knew how hard it would be to win back what he was being forced to surrender: "When you gave up on that, you didn't reestablish that trench line easily." With his lab directors and other advisors, he decided that here he would make a stand. All his new initiatives might go into the budgetary deep-freeze, to be thawed out only in some more clement climate in the future. But he and his advisors all agreed that even if everything else was lost, there were two fundamental things that absolutely had to be held on to and kept on schedule: TFTR and MFTF-B. In particular, it was entirely essential that MFTF-B go ahead as planned, along with its supporting TMX-Upgrade, so that a valid comparison might be made between tokamaks and mirrors. In no way could MFTF-B be delayed. On its pace and schedule, Kintner would make his stand.

Kintner was not the only one concerned about the fusion budget. Don Repici had left the OMB, but his boss, Tom Palmieri, wanted to have some questions answered. Palmieri decided to call on Steve Bodner, who headed the laser-fusion group at Naval Research Laboratory. Bodner had got a Ph.D. in plasma physics at Princeton in 1965, but his work in laser fusion meant he had lost touch with magnetic fusion. Nevertheless, he knew he could get up to speed quickly: "I knew everybody in the program. I'd been a carpool-mate of Harold Furth; Ron Davidson, the head of the MIT program, had been at my wedding; I knew them all very well." Palmieri had him come in for an interview and told him what he wanted: "We're looking to cut the budget. Stockman wants all the budgets cut, and we need someone with technical expertise, now that Repici is gone." Bodner wouldn't agree to that: "All I'll agree to do is come in and review it. I may agree it can be cut, but I may argue for more money instead. I can talk up and down to people in the program, find out what's going on, but that's all I can do." That was exactly what Palmieri wanted to hear.

Bodner started quickly, and soon was spending a lot of money on phone calls from OMB. He was burning up several hours a day talking with people, and he filled up notebooks with comments from all of them as to what was going on. Right at the start, there was the matter of the 12 percent budget cut. Bodner felt that he should try to keep it from being applied to the fusion budget. The people at OMB had the leeway to do this because the cut was to be taken out of a large overall budget category, "Energy Supply Research and Development," within which fusion was only a small part. Hence fusion funding could be spared by making deeper cuts in programs such as solar energy, which were not in favor in the Reagan administration. Bodner convinced Palmieri that fusion should be spared from the cut, at least in the 1982 budget, and this view prevailed at OMB, early in October.

With that, he told Palmieri that he was going up to see Kintner and bring him the good news. Bodner put some budget recommendations on blank sheets of paper, no letterhead, nothing official. He didn't have that authority, anyway. He gave Kintner a document showing the revised OMB budget, by which he would be getting back the 12 percent. Kintner gave no reaction to it at all. He barely looked at the document, then he put it down. Bodner didn't know that Kintner had a low opinion of him: "He's a hatchetman; he's gonna do a hatchet job on the magnetic fusion program. He's in over his head. They brought him in so he could tell them how to cut the fusion budget." As their meeting progressed during the next hour, Bodner told Kintner of his high regard for the vigor and life in the program, the way people were dealing with the real issues and were doing good physics. But Kintner was not swayed.

Then Bodner made his first mistake. Having established what he thought was good rapport, he gave Kintner his sheets of paper with his suggestions as to where the money should go, the 12 percent he had saved. On the basis of little more than a week in his new job, he was recommending that Kintner should use the money to add support and flexibility to the ongoing plasma physics work, in what were called the operating programs: "I had found these complaints almost everywhere, in talking to people. There was a shortage of theorists, not enough money for diagnostics, the experiments were inadequate to treat some of the physics issues. They were talking about shutting down PLT and PDX at Princeton and that was wrong; we needed them both."

He didn't know that this was a sore spot with Kintner, a point on which the latter was quite defensive: "Bodner was talking to these people, underlings, two or three echelons down in the laboratories. He told me he couldn't talk with the lab directors because they were all giving him the party line. So he talked to the underlings, and they all said they didn't have enough money to do their jobs. My view is, I know they don't think they have enough money to do their jobs—but they *do* have enough money to do their jobs. If you're just going to give everybody as much money as he thinks he needs to do his job, he's not going to do his job very well. And in my judgment they have plenty of money to run those projects and run those experiments and get the results. That isn't the fundamental problem that's holding them back; it isn't money; they got it." But unlike these underlings, Kintner had not been working sixteen-hour days amid shortages of instruments and equipment. Bodner was certainly entitled to want to ease their situation a bit, in these operating programs. But he was talking to an old-line navy commander who ran a taut ship, and who insisted on full control of the magnetic-fusion program that he directed. In particular, while OMB might set his budget level, Kintner insisted on a free hand in allocating funds within what they gave him.

Then Bodner made his second mistake. He suggested that Kintner should set up a stellarator program and, what was more audacious yet, he suggested it be set up at a specific location, the University of Wisconsin. Stellarators were looking good. Lyman Spitzer of Princeton had invented the concept in 1951, and after two disappointing decades the United States had abandoned them for tokamaks, but Japanese and German scientists had been more patient. Near Munich, Günter Grieger and his colleagues had built a stellarator with all the modern improvements that Spitzer never had, including neutral beams for heating the plasma. They called their device by the good German name of Wendel-stein. In 1980 they had claimed plasma confinement five times better than had been attained by tokamaks of comparable size. At the Los Alamos meeting in September, Grieger had shown a cartoon with two people standing forlornly in the foreground, one holding an inner-tube sugges-tive of a tokamak, the other with a straight tube having pinched-off ends, representing a tandem mirror. Behind them, rising above the horizon like an enormous sun, were the spiral coils of a stellarator. Grieger then presented his results and said, "So you see, stellarators are not so bad after all."

Bodner was being wildly presumptuous to make such suggestions, the more so since he had been on the job only a few days. After all, these decisions were Kintner's to make, along with his advisors who had spent decades with the program. He could have easily dismissed Bodner's ideas, letting him know that his job as budget examiner did not include making such suggestions. Kintner could have said, "Look, you're new here; I know you're eager and you want to do a good job. I'm glad you like the program. But if a wrong decision is made about budgets within the fusion program, I am the one who will have to take the blame, not you or anyone at OMB. I therefore must have enough authority to fulfill my responsibility, which is why your suggestions just aren't appropriate. If they're wrong, you'll be safely back at your old job, while I might have my ass up before a congressional committee."

That wasn't what happened. Kintner accepted Bodner's papers without a word. Then, a day or two later, Palmieri started getting phone calls: "Bodner is using his position of power and influence at the OMB to take care of his old buddy at the University of Wisconsin, Leon Shohet."* In fact, Bodner had met Shohet for the first time only a few days earlier. Palmieri had to defend Bodner against such an underhanded and personal attack. He went to see his own boss, Hugh Loweth, who was only two levels down from David Stockman himself. Then Palmieri told Bodner, "You've been attacked and I'm not going to take this. We're going to join ranks on this thing, and we're not going to concede anything to them on this issue. We're not going to even talk to them any further. I won't stand for this sort of behavior, from anybody."

Over in the DOE, Kintner had been complaining to Trivelpiece that the OMB was trying to usurp his prerogatives by getting involved in the detailed management of his program. Naturally, Trivelpiece backed Kintner. The upshot was that good relations broke down between the DOE's energy-research programs and the OMB, and it was difficult for them to talk to each other. Bodner, for his part, had opened up channels by which he was talking to people in the labs and elsewhere in the fusion community, but he could not talk directly to Kintner. Kintner was still working to save his 1983 budget, and just at that moment, good relations might be worth literally tens of millions of dollars.

Meanwhile, Bodner had also been looking over the 1983 fusion budget. In it were two large items that stuck out and caught his eye: the construction programs for Kintner's holiest of holies, the TFTR and MFTF-B. There was very little he could do about the TFTR line item. That project was simply too close to completion, and any budget cut would damage it severely. But MFTF-B was scheduled to go from $47 million in 1982 to $71 million in 1983. Was this increase really justified? It was certainly a large item, and thus a potential candidate for the chopping block. After all, Palmieri had given Bodner his marching orders: "Do you believe that the current budget crunch is temporary, that it will last only one or two years, and then magnetic fusion will grow again? If you do, then you will conclude that the DOE program is well balanced and that it

*The plasma physicist who was in charge of stellarator studies there.

is appropriate to protect the big construction projects. But if you believe that this is not a temporary phenomenon, that it will last for many years, then it is not clear that the program is well balanced."

Bodner was bothered by this, and decided to find out what was the reasoning behind the rapid construction pace on MFTF-B. The original 1980 idea had been to push and have it ready by 1985, so its experiments could be run concurrently with those of TFTR. The FED, the next major tokamak, would be ready by 1990. By then there would be several years of experiments on both TFTR and MFTF-B, and it would be possible to choose between mirrors and tokamaks for the *next* big project: the electricity-producing demo plant, to follow FED. But if there would be no FED by 1990, then MFTF-B might readily be delayed and its budget cut. The cut, in turn, could go to shore up the operating programs, which Bodner was convinced were being held down for the sake of the big construction projects.

Then by means of one of his channels, from "some people from the labs who talked too much," he found out about a recent meeting of tokamak leaders. It had been held in Airlie House, an elegant lodge in Virginia. To Bodner, the people involved appeared very eager to keep that meeting in the family. There they had discussed the fact that tokamaks were in trouble, and might even have to be abandoned. The two issues that were worrisome were the beta of tokamaks, and their steady-state operation. In the early fall of 1981 there still were few good reasons to think that a tokamak could be run continually, rather than in a pulsed fashion, on-off. Stellarators, by contrast, were inherently steady-state, and, what was more, Wendelstein had operated without plasma disruptions. Also, mirrors were showing excellent beta, the measure of how high the magnetic fields would have to be in a practical reactor. The Airlie House participants concluded that if these matters could not be improved on, they might have to shift the tokamak program toward stellarators, or even do a complete switchover to mirrors.

"I found out about that meeting; I learned about it," said Bodner. "That was, of course, most unfortunate for the Department of Energy, for OMB to know that such a meeting took place. The managers of the program had met quietly to work out a backup strategy, what to do if. This meant to me that everybody was worried, that there was a real problem in this program, which was being shortchanged for the sake of construction." To Bodner, this meant all the more strongly that he was right in wanting to add money to the operating programs. Doing that would strengthen the physics groups who were studying things like beta and stellarators; it meant slowing down the big projects in order to do more research.

Palmieri was interested, but he wanted Bodner to find out more: "I know about the logic that Kintner gave for proceeding rapidly with MFTF-B, but there may be another logic. There may be a good reason to proceed with the mirror on the same pace, independent of whether you're building a Fusion Engineering Device. There may be a technical reason why you have to have it."

By now Bodner felt it was high time to meet with Ken Fowler and talk with him directly about his MFTF-B. Bodner had worked as a plasma theorist at Livermore, years earlier, so he knew Fowler well. He went up to Manhattan to a conference of the American Physical Society, in October. There he met with Fowler for lunch at an Italian restaurant, around the corner from the Sheraton Center. It was quite noisy there, which suited his purposes. He intended to lay down a smokescreen by asking a lot of technical questions, then ask about Fowler's construction schedule and its rationale. They were sitting at a small table, they had trouble hearing each other, and Bodner asked what was the logic of proceeding with MFTF-B if there wasn't going to be an FED. Fowler replied, "The basis for MFTF-B has nothing to do with FED." Bodner said that he didn't believe it. Fowler went on, "No, that's not the logic; it has its own internal logic." Bodner replied that he didn't understand it. Then Fowler made his mistake: "You don't understand these things, Bodner. You don't understand big programs, their momentum, the logic and how they work. You don't have any understanding of this at all."

Evidently this was a sensitive point for Fowler. In fact, Fowler's "logic" was not technical, but political. He felt the MFTF-B program had to keep pace with TFTR. If it fell behind, then Princeton might win success while he had no more than a big construction site, which could kill his program. But Bodner by now was leaning strongly to the view that the MFTF-B project was running way ahead of itself. That was a $226 million project, and it would depend on the complicated and unproven idea of thermal barriers. This idea would be tested out and proven on TMX-Upgrade, but not till 1982 and 1983. Slowing construction of MFTF-B would preserve needed flexibility and avoid prematurely locking in on a design that later experimental results might show to be premature.

Still, Bodner wasn't sure. He went to talk with "a senior person from one of the labs"—a person at the level of Harold Furth, though it wasn't Furth, someone whose identity Bodner prefers to protect. He explained what had happened with Fowler and asked, "Is it possible that Fowler is right, that I don't know what I'm talking about?" This person replied, "Yes, it's possible, you could be wrong. But then again, you could be right and Fowler could be wrong. You'll have to think it out for yourself." Bodner decided that since Fowler hadn't answered his question, he would go ahead with his own conclusions. These were that it would be correct to cut the MFTF-B budget item, and that there was no logic behind its going ahead so rapidly. Mirrors might look very promising, but basic research came first.

Bodner went back to Palmieri with these conclusions. Palmieri listened, made some phone calls, did some checking through sources of his own, and decided Bodner was right. He was also aware of the Airlie House meeting, and agreed that the money saved on MFTF-B should go to relieve the squeeze on the operating programs, by providing more experimentalists, theorists, and equipment. They decided that the thing to do was to recommend taking $42 million out of MFTF-B and distributing it among the operating programs. This could only be their recommendation; David Stockman would make the final decision. Still, their recommendation would carry weight. If it went through, it would mean that there would be no actual cut in the fusion budget.

The next thing for them to do was to plan their strategy, to sit down together and decide how to get Trivelpiece to go along. The standard, textbook procedure would be to invite him to submit a revised 1983 budget, switching the $42 million. This would carry a none-too-subtle hammer to hold over the heads of Trivelpiece and Kintner: the threat of a budget cut, of simply taking back the $42 million altogether, on top of their previous budget cuts. But this standard procedure carried risks. Trivelpiece would want to cooperate with Kintner on this, and Kintner's feet would be set in concrete so far as letting people from OMB tell him to reprogram that $42 million. In the likely event that Trivelpiece then would decline to revise his budget, Bodner and Palmieri could then go ahead and recommend the cut. And like chess players looking three or four moves ahead, Bodner and Palmieri could see that at that point, things might get sticky.

Faced with such a cut, Trivelpiece could choose to appeal, seeking restoration of the $42 million, either to go back to MFTF-B or to be put in the operating programs. He would appeal directly to Stockman and, if need be, to Reagan himself, on the ground that this budget cut would do major damage to the fusion program. In such an appeal, Trivelpiece would have an important advantage. He had much more technical information available to him about these programs than did Bodner at

OMB, and could put forth the stronger argument. Still, Trivelpiece would be taking a risk. He had several other large research programs under his direction, with budgets comparable in size to fusion. All these other programs were being well taken care of in the 1983 budget. He couldn't expect to have everything he wanted, so if he appealed, for that reason alone Stockman or Reagan might rule against him, and he would lose. Then the $42 million would be gone for good.

But Bodner and Palmieri were running a risk, too. They knew that such a cut really would damage the fusion program, and would not readily be made good in future years. They didn't want to do that. They were working to solve some problems within the fusion program, but they didn't want to set in motion a chain of events that would spiral out of control. Palmieri in particular was a seasoned professional, determined to have the best possible fusion program within the available budget limits. So, without anyone else knowing, they sat down together in a small office to discuss their options, just the two of them. Palmieri asked, "What if Trivelpiece calls us on it?" And they agreed that if Trivelpiece refused to submit a revised budget, then they would recommend that Stockman should approve the original budget, which included the $42 million. At the last minute, then, they would back off. But in Bodner's words, "No one knew what we were up to. It was all done very quietly."

In other words, Bodner and Palmieri were bluffing.

Having decided on their strategy, they had to line up support for their position. They went across the street to the Old Executive Office Building. This was a huge, rambling, rococo gray sandstone building with a mansard roof and forests of columns surrounding the entranceways, resembling an Austrian opera house. It had been built in 1870, just down Pennsylvania Avenue from the White House, and had once housed the entire staffs of the State, War, and Navy departments. Now, however, it was home mainly to some of Reagan's budget and policy-advice groups. Within this relic from the days of Ulysses S. Grant, they could meet with people on the staff of George Keyworth, Reagan's science advisor. One of these was Doug Pewitt. Soon they agreed all around that MFTF-B was going too fast and its budget should be cut. With this help from Keyworth's office, Bodner and Palmieri now had a solid phalanx of support.

On Monday, November 9, came the showdown. Bodner and Palmieri met with Kintner and Trivelpiece, in the latter's conference room. There they sat at his long conference table with its gorgeous seventh-story view. Palmieri started by laying it on the line: The mirror program could not be justified at its pace, and the operating budgets were too low. They had the choice of moving the $42 million into the operating budgets, or of facing the budget cut and appealing. But there was no way that MFTF-B would be kept at its original level of $71 million, because the people on George Keyworth's staff—and Kintner knew this particularly meant Pewitt—also felt that MFTF-B was to be slowed down and its budget cut. That would count for a lot at the forthcoming Director's Review, when David Stockman would personally approve each budget item. At that review, Kintner believed that Doug Pewitt would be sitting in the back row, and if OMB recommended a higher figure for MFTF-B than had been agreed to, Pewitt would object. In the presence of Stockman, that would be fatal. It was essential that OMB not recommend a higher figure than had been agreed to in Keyworth's office—again, because OMB was supposed to be cutting budgets, not raising them. All this was Palmieri's way of showing that Keyworth's people agreed with him. In Kintner's mind it raised the specter of Pewitt as an additional persuader.

Trivelpiece then had an idea: to call in a review panel of senior fusion people, let them look at the pace of MFTF-B and make a decision. Palmieri was all for it. There still were more than two months before Reagan would give his State of the Union speech and announce his 1983 budget, and Palmieri offered to hold back on setting a final budget number till just before then. He wasn't certain he had the authority to do that, but he could offer this as a negotiating ploy. Kintner liked the idea, too: "I've had a lot of reviews of my program, and I know they'll back MFTF-B as a good project." Bodner replied, "Wait a minute; that's not the issue. Everyone agrees MFTF-B is worthwhile. The question is the pace. The panel must be told the budget numbers, and then be asked if the money should go to MFTF-B or to the operating programs." Kintner said, "Hell, no. I make the decisions in this program. Those people cannot decide." He added that such a committee would probably agree with OMB and not with him, because "they'd see all this money sitting there and they'd all want it. Their personal desires would overcome their expert opinion and they'd choose to reduce MFTF-B." Trivelpiece's idea died right there.

At that point, out of desperation, animal instinct—or perhaps his past experiences in dealing with OMB—Kintner decided the OMB people were bluffing and that he could call their bluff: "Look, you make this decision; you don't need me any longer as director of the fusion program. But then you'll have to be prepared yourself to run it from now on, because if you tell me how to spend that forty-two million, that's my authority that you're taking away. Look, you're going to tell me you don't have forty-two million? Take the damn forty-two million! But don't tell me to take it from one place and put it in another." As he later told Trivelpiece in private, he didn't think they were brave enough to take it.

If Trivelpiece had felt as Kintner did, then right then and there Kintner would have won, although they wouldn't have known it immediately. But Trivelpiece instead went back to arguing that Kintner should have his $42 million for MFTF-B. Palmieri replied that within a few days they would have to decide what they wanted to do, and the meeting broke up. At that point, Bodner was worried: "I was afraid that Trivelpiece was going to reject any change and just accept the cut. Then we would lose, because we were going to back down. I was getting a little frightened at this point."

But Trivelpiece was concluding that the time was ripe to try for a compromise. Confrontation and heated arguments were not his style; he was a conciliator. He knew how emotional Kintner had been all along, and had often seen him that way himself. Once he had asked which project Kintner would prefer to save, the materials-testing FMIT or the Elmo Bumpy Torus. Kintner had replied, "That's like asking which of your children do you want to sacrifice!" Trivelpiece decided to use some of his own channels. He called the former budget examiner, Don Repici. Repici was out of the OMB, but he was working as a consultant to Hugh Loweth, Palmieri's boss, so he was in touch with the situation there. Moreover, as he said later, "Al and I were friends." Trivelpiece wanted to know what was really going on, and "Are they playing honestly with me at OMB?" Repici replied that Palmieri was playing square but that he didn't know about the role of Keyworth's people, though it was probably significant. But Trivelpiece was in a position to negotiate directly, not only with Palmieri, but with Keyworth. He had come to know Keyworth rather well, on a recent trip to Japan, somewhat as Hirsch and Kintner had long ago struck up a friendship in Copenhagen.

Kintner by now was prepared to budge, but only a little. He had asked Fowler if he could stand to take any sort of budget cut on MFTF-B, and Fowler had written back that he could keep that project on schedule even with a cut as large as $10 to $15 million. So Kintner was willing to go that far, though he would be adamant if it came to going farther. Trivelpiece accepted this, and backed Kintner up. He proceeded to argue for this as a compromise, in lengthy phone calls and meetings.

But Palmieri and Keyworth were firm: no, the cut in MFTF-B would have to be bigger. For one thing, there was the problem of a cost overrun on TFTR, amounting to $10 million. That would have to be covered in any case, and it would absorb most of what Trivelpiece was offering, leaving almost nothing to help the operating programs. But that wasn't the only problem. Palmieri and Keyworth really were determined to slow down the MFTF-B, and the way to do this would be to insist on a bigger cut than Fowler, and Kintner, could accept.

At that point, Trivelpiece had reached the crux of the matter. The choice was his to make: to continue to stand by Kintner, or to make a further move on his own initiative and over what would be Kintner's outraged objections. But Trivelpiece was in an excellent position to make such a move. He had been a leader in the world of plasma physics back when Kintner was still working on nuclear submarines. He had been working closely with Robert Hirsch when Kintner was still involved with breeder reactors. He knew personally all the key fusion people, and was quite prepared to make a judgment on his own. He made some phone calls, and then reached his decision: MFTF-B indeed should be slowed down. With this conclusion, he would step up his bid.

Palmieri, for his part, knew there was nothing sacred about $42 million as a budget number. He had all along expected to compromise, and he had come back with a counteroffer: $25 to $30 million. Trivelpiece decided he could go along with $25 million, and he then talked it over with Keyworth. Then Palmieri went across the street to the Old Executive Office Building and met with Keyworth, and they too settled on the $25 million figure. Keyworth, Palmieri, and Trivelpiece were now in agreement. Kintner was out in the cold, for these discussions had gone on at a higher level and had taken place over his head.

The elements were now in place for a deal, but Kintner still might try to sink it. Palmieri had set Tuesday, November 17, as the day when Trivelpiece would have to decide. He would have to choose whether to submit a revised budget, or to stand pat—and call their bluff, though he didn't know that would be the result. The amount in question was now $25 million, not $42 million, but still Kintner might persuade Trivelpiece to reject it. On Friday the thirteenth, this looked quite possible, for on that day Bodner found out that Kintner was about to take off for a meeting in Brussels, Belgium. On its face, this action looked entirely bizarre. If the one thing in the world that mattered most to Kintner was the MFTF-B, he should be staying in Washington to fight for it. Kintner had told Trivelpiece that the Brussels meeting was important, and that he had to go. But Bodner had heard of the times when Kintner had gone off at critical moments and left matters to John Clarke. Kintner's trip could actually be a ploy, to distance himself from a compromise he did not accept. Then he might come back afterward, denounce the compromise as one he was not a party to, and thereby try to overturn it. Palmieri told Bodner to try to call Kintner and resolve this, in a way that would keep OMB from losing. Bodner phoned Kintner the following morning and asked to meet with him. His reply: "You're killing the program, Bodner; you're destroying it." Bodner tried to be conciliatory: "Nobody's trying to kill the program; we all like the fusion program, but we have some differences. Why can't we sit down and talk about these things? If you and I talked and understood things, we could probably get a fair number of people to agree also. Why can't we get together? I know you're leaving this afternoon on a flight. Can I meet with you beforehand?" "Hell, no. I'm not going to meet with you, Bodner. It's not your fault, but you're being used by others, and you're damaging the fusion program."

On Monday morning, Trivelpiece phoned Kintner's office to talk about what to do. Kintner had left for Brussels, but John Clarke was there. By now Clarke had long since finished the assignment that Pewitt had given him early that year, and was back working with Kintner. With Kintner out of town, Clarke was to act in his place, so Trivelpiece drove out to meet with him and his staff. By now even Clarke had abandoned Kintner's damn-the-torpedoes approach to proceeding full speed with MFTF-B. He listened to Trivelpiece, then agreed that there were plenty of other worthwhile things that could be done with the money: "The operating budgets were being starved because of the needs of the projects. We were in the position of a business that lays down expansion plans and borrows a lot of money, with balloon payments. Then the economy goes bad, business goes sour, and you can't make the payments. So the operating programs had been taking the blow."

This was a dangerous view for Clarke to take, for Kintner had continued to argue that there was no problem with the operating budgets. As Kintner saw it, it was merely a matter of the labs being accustomed to flush times, in the wake of the tenfold increases in the fusion budget during the past few years: "Those fusion labs never came close to the tight budgetary constraints that the naval-reactor labs lived with year after year—and they produced results." Thus, Clarke had to tread carefully, because he might easily find himself in the position of a first mate trying to take over a ship from his captain. Kintner might easily fire him for what would be tantamount to a mutiny. However, Clarke could let Trivelpiece take the initiative and merely go along.

Kintner was to stay that night in the Hilton in Paris, on his way back from Brussels. When he arrived there was an urgent phone call from Clarke. Clarke said he had talked to Trivelpiece, and Trivelpiece was going to go ahead and make the deal with OMB: if we switch the $25 million from MFTF-B into operating programs, you don't cut our budget. He knew how Kintner felt, had done what he could, and, if Kintner wanted to, he might try phoning Trivelpiece on the overseas line. Kintner thought about it and decided not to: "How can I call from Paris and convince him on something he's already decided in his own mind to do?" He decided that when he came back, he would write a lengthy memo to Trivelpiece to set forth his objections. But Clarke came away from the phone believing that Kintner would not stand like a stone wall and block the deal.

The next day, Tuesday, Trivelpiece went over to meet with Palmieri and Bodner. He said he was ready to make the deal, but Bodner wasn't sure he could do it: "Kintner's not involved; is this going to mean anything?" Trivelpiece replied that he had asked Clarke to talk with Kintner in Europe, and Clarke had reported that Kintner would write a letter of protest, but would keep the matter within the family and out of the headlines. In particular, Kintner would not go to his friends in Congress, for what would amount to an end run around the regular budget procedures. So Trivelpiece was able to give his word that there'd be no double cross and that everybody would be locked in.

Kintner came back into town and wrote his four-page memo to Trivelpiece, urging that he change his decision. Nothing happened. Kintner then phoned and asked Trivelpiece what he would do about his memo. Trivelpiece replied no, he had already made his agreement; he didn't feel he could go back on it. And Kintner then said, "You'll have my resignation on your desk at nine o'clock tomorrow morning."

Trivelpiece then tried to talk Kintner out of leaving, asking him on what basis he would stay. The only basis he would consider was that the $25 million go back into MFTF-B. However, there was nothing Trivelpiece would do about that. Then Ken Davis, the Deputy Energy Secretary, invited Kintner to come down to his office for a personal chat. Davis asked again if there were any basis on which he would stay, and Kintner again said that the only way was if the $25 million was put back. Davis frowned and said he didn't feel it could be, and Kintner responded, "Well, there's no reason for us to talk any further." And Davis looked at him and said, "We'll be sorry to see you go."

During the Carter years, there had been times when Kintner had felt discouraged, and had talked with Fowler and others about resigning. But these had merely been his private musings, in the company of friends. He always said he was not like Reagan's Secretary of State, Alexander Haig, who used public threats of resignation to get his way. Still, within the fusion community, his earlier times of talking about resignation were well known. A number of people felt that sooner or later, he was bound to resign. When Harold Furth heard that Kintner was talking about leaving, he didn't take it too seriously; he had heard it before. MIT's Ron Davidson said that Kintner's times of being in a mood to quit were like having a revolver with a single bullet in the cylinder; if you start shooting, sooner or later it will go off. Similarly, Kintner had talked so often of resigning, sometime he was bound to do so.

Why did he do it, in the end? It was not because of Bodner, or even because of Pewitt. No one rises to direct a major research program in Washington without having a very tough hide and an ability to roll with the punches. No such senior director, holding the top grade available in Civil Service, will resign merely because a temporary OMB employee like Bodner persuades his fellow managers to reprogram $25 million within a $445 million budget. But at the root, Kintner was an engineer, who had spent his career building large nuclear projects. With all his heart he was committed to the expanding program laid down by the Buchsbaum Committee and by the Magnetic Fusion Act. He had been a protégé of Hirsch, and like Hirsch he had worked toward the day when fusion would grow into an engineering program, aimed at actually producing energy, not merely at doing more in plasma physics. He had fought for an FED, had had it within his grasp—and then he had seen it snatched away. Nor would he be likely to get it back, in the few years left before he would reach sixty-five and retire in any case. In no way would he be satisfied to manage the truncated and slowed-down program now set for the 1980s, and he knew it.

Yet his disaffection ran deeper. As he put it, "I was overwhelmed by scientists and knew that in that environment I could never recover." To him, the distinction between engineers and scientists stood at the center of what Hirsch and he had fought for, had worked to accomplish. To him, engineering meant project management. It meant hard hats, muddy boots, tight budgets, impossible deadlines—and at last winning through to success, to building something that would actually work, that would go on to do its share of the world's work. Scientists were different. They were important to have around, in fact they were essential. But they needed a good engineering manager to lead them. He knew very well what would happen if they didn't. They would piddle and diddle, lose themselves in a hundred fascinating little side issues, run experiments ad infinitum, and never get around to building anything useful.

The people he had lately been dealing with were Bodner and Palmieri from the OMB, Pewitt, Keyworth, and Trivelpiece. All had doctorates in various branches of physics. None had ever built anything like a nuclear sub or a TFTR. Together they had outmaneuvered him, had forced him to slow down when he was ready to speed up. This meant they were now the ones in control, and this was a situation not to be borne. Particularly galling was the involvement of Keyworth, who had Pewitt advising him. To Kintner, "Keyworth's view of fusion was that it wasn't ready for engineering. If you started engineering you were going to destroy the science. He was going to be a big scientist, win the Nobel Prize and all that kind of thing. He was going to take care of science, and science was the god and he didn't want to do the engineering."

In Kintner's estimate, Trivelpiece was little better. Trivelpiece could have backed him in the face-off, but for reasons of his own had elected not to do so. Kintner was sure those reasons again involved choosing science over engineering, research over building something useful. In Kintner's words, "Trivelpiece had other children to protect." These "children" were the other research programs which he was responsible for, including the nation's program in high-energy physics. Kintner was sure that Trivelpiece and Palmieri had made a bargain whereby if he did what Palmieri wanted for fusion, Palmieri would see to it that high-energy physics would be well taken care of. Such a choice was one that Kintner would not abide.

Finally, Kintner had been isolated in the debates, and he was not able to get along well with the people around him. Only three years earlier he had been sufficiently capable and resourceful to turn back John Deutch's attempt to use the Foster Committee to cut his budget. Now he was suspicious of even his closest associates. Over a year later, on reading a draft of this chapter, he wrote: "You've confirmed what I only suspected before—that Clarke also wanted to take the money out of MFTF-B. He had, as a Tokamak–Oak Ridge person, opposed MFTF from the beginning. He wrote me a personal memo recommending delaying MFTF-B. But I didn't believe before that he was working with Bodner all along. Perhaps he knew this was an issue over which I would resign and he could have my job."

Thus, when he wrote his letter of resignation to Trivelpiece, he clearly showed the bitterness he felt:

Dear Al,

This is a request for retirement from the Federal Service, effective the end of the first pay period after January 1, 1982.

I take this action reluctantly and after careful consideration. I do so for two corollary reasons.

First, it now appears that the goal-oriented magnetic fusion program which has been put together over the last five years, piece-by-piece. . . . will not be carried out, either as to the basic strategy or as to a sense of urgency. Whatever the reasons, I believe this is a national error for which a price far greater than the present savings will be paid at some future date. . . .

Second, as must be obvious to you in the short time you have been here, the ability of Program Directors like myself to maintain program direction and cohesion has been badly eroded. The Program Director . . . has been steadily reduced in position to being a messenger or clerk. . . . Emphasis now is more on procedure rather than product, on administration rather than technical matters, on preventing problems at the expense of achievement, on obtaining influence without responsibility. . . .

The ultimate example is the recent OMB-OSTP* action on the FY 1983 budget. At worst it is malicious meddling; at best it is ignorance of the broader implications of the action.

A conclusion that this is a question of who holds the power of decision underestimates both me and the problem.

I was wrong in judging that we could carry out magnetic fusion as a mission-oriented program. That attempt had been a failure. I doubt that any future attempt in this country will succeed, unless driven by crisis.

I want you to understand that this is not a problem between you and me. The factors which bring fusion to its present condition are beyond us.

Sincerely,
Edwin E. Kintner

With this he served out the last weeks of his job, drew his final paycheck, cleaned out his desk, and retired from government service for good.

---

*OSTP: Office of Science and Technology Policy: George Keyworth's office of the President's science advisor.

11

Fusion's Highest Potential

T WENTY-FIVE miles northwest of the Washington Monument, out past the edge of the suburbs, the rolling Maryland countryside gives way to the community of Germantown. Here, not far from Gaithersburg and Rockville, a five-story building of red brick spreads across some acres of ground. No other buildings are nearby, only roads and fields; it could almost have dropped out of the sky. It features corridors branching off corridors, wings angling off other wings, as if its architects had not started with any clear idea of what they were to build and so had put on further additions as they went along. If you walk down its long hallways, your strongest impression will be that you are in some huge, sprawling hospital. Indeed, you would almost expect to see patients and medical equipment, if you looked through the occasional open doors along the way. But this is no medical center. It was built in 1957 as headquarters for the Atomic Energy Commission, and was inherited twenty years later by the newly organized Department of Energy.

One particular corridor, otherwise indistinguishable from the others, conveys a distinct impression of leading into fusion country. At least that is what you would think from seeing the wall hangings, which include framed photos of tokamaks and diagrams of other plasma machines. At the end of the hallway is a two-room suite of offices. Here Kintner worked and struggled, during his turbulent months in 1981. Here he had dreamed his dreams, planned his budgets, and faced his disappointments. But today all that is past, the storms, the turbulence, the bitterness. The fusion program now is in the hands of his former deputy, John Clarke.

With a wisp of gray hair across his forehead, Clarke is everybody's idea
of what an airline pilot should look like. He is cheerful, capable, good-
humored, and quite unflappable. Then there are the suits he wears, and if
dressing for success could in itself assure the prospects of a program,
fusion would have it made. "I got them on sale," he said. "I'm glad you
like them. But no, I'm not going to give the name of my secret clothing
store."

What about the rough times during Kintner's last year?

"I don't want to imply that Ed Kintner's presence was responsible for
this rocky time," Clarke said. "It was the startup in the new administration—that's my story, and I'll stick to it. But it was a very chaotic time.
Any time an administration changes, it's chaotic. The difference now is
that we have a consistent set of people in the Department. That's not
quite true because we just lost our Under Secretary again, but that
transition went smoothly. I haven't noticed any glitch, things have just
moved ahead. So now we have a functioning Department in which these
things can be decided like gentlemen. It's much improved.

"We are now engaged in the budget preparation for the next fiscal year.
We had the 1982 budget, which was basically prepared under Carter
administration guidelines. That was whacked up by these budget cuts
when the Reagan administration took over, and it wound up at four
hundred fifty-one million. While that was going on, we were preparing
the 1983 budget, and it was the new administration setting down some
guidelines, saying that fusion was not going to grow, that caused all
those—perturbations. So 1983 was four hundred forty-seven million.
Since January we've been working on the 'eighty-four budget. And for
the first time in three budget years, we have had the opportunity to do it
properly, the way it should be done: namely, building it from the ground
up. We've had our budget exercises; we've had internal reviews—we've
been through that. We got policy guidance, and I'm pleased with it,
because it was basically an acceptance of what we wanted to do. Now,
through all of this, the thing has been smooth as silk. We've been able to
prepare our case without any interference. I can't give a budget number
for 1984; that's for the President to release.* I can say that the budget
preparations have been orderly, rational. The people in charge like Al
Trivelpiece are thorough professionals, they're experienced people.
They know what they're doing; they listen; they give you feedback; it's

---

*In January 1983, President Reagan's budget request was $467 million. The subsequent
Congressional appropriation was $471 million.

3. TOP: The 2XIIB mirror. The large tanks hold twelve neutral beams.

4. BELOW: Cutaway view of the TMX-Upgrade. The chimneylike protrusions are
orts for neutral beams. The odd angles for these beams, together with the complex
1agnet shapes shown in the cutaway, are needed to produce a thermal barrier.

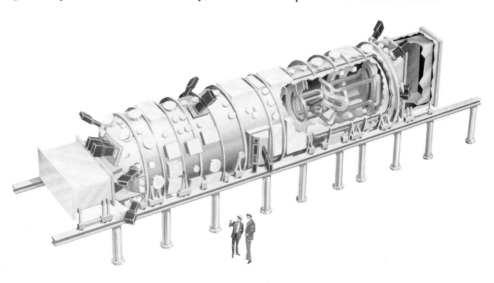

25. Coil-winding machine used in winding large magnets. The magnet protrudes through the hole in the platform at the top, and can be rotated or swung up and down

26. The completed MFTF yin-yang magnet on the sandpile where it was assembled ready to be rolled onto its sledge for transportation

7. TOP: The MFTF magnet
about to enter its 36-foot-wide
vacuum vessel. Its sledge has
been pulled across a bridge
formed from steel girders,
visible at bottom.

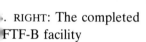

8. RIGHT: The completed
MFTF-B facility

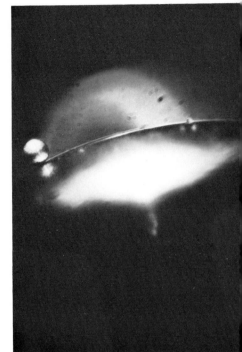

29. RIGHT: Photomicrograph of a laser-fusion pellet on the head of a pin

30. BELOW: View of the Shiva laser, showing six of its twenty arms

31. The Novette target chamber

32. The new Krypton-Fluoride Laser at Los Alamos

33. The ten-arm Nova laser

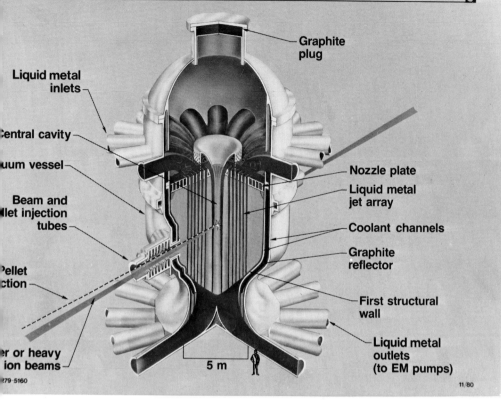

Graphite
plug

Liquid metal
inlets

Central cavity

uum vessel

Beam and
llet injection
tubes

Pellet
ction

r or heavy
ion beams

279-5160

Nozzle plate

Liquid metal
jet array

Coolant channels

Graphite
reflector

First structural
wall

Liquid metal
outlets
(to EM pumps)

5 m

11/80

54. The HYLIFE concept for a laser-fusion reaction chamber. Liquid lithium forms a massive spray as from a gigantic showerhead; the pellets and laser beam pass through he gaps in the spray. The lithium, heated by pellet microexplosions, then is used to boil water, raising steam for electric turbines in a power plant.

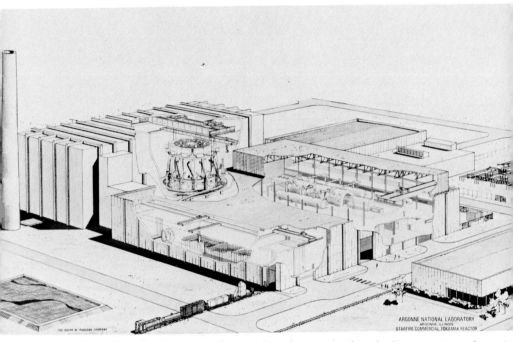

ARGONNE NATIONAL LABORATORY
ARGONNE, ILLINOIS
STARFIRE COMMERCIAL TOKAMAK REACTOR

35. The "Starfire" concept for an electric power plant built around a tokamak fusion reactor

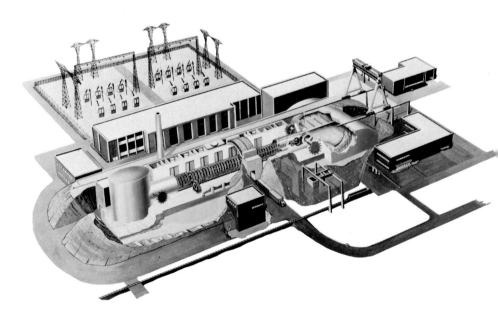

36. The "Mars" concept for a power plant based on a tandem mirror reactor

the way things ought to be. And out of that process, the fusion program has emerged smelling like a rose.

"We're not going back to the Magnetic Fusion Engineering Act, by which we were doubling the budget. We are trying to work out a program that lives with a relatively constant budget over the next five years, say. But I'll tell you what we're getting, for instance. In the beginning of the discussions with OMB, the talk was of delaying MFTF-B by three years. We are not going to delay it by three years. It's not going to be finished in 1985, but there's nothing sacred about 1985. The thing will be finished toward the end of 1986, so we have a little bit more than a year's delay. But with that delay, we have bought the ability to cover a cost overrun on TFTR, get that project finished. We have bought the ability to start another mirror program, the TARA program at MIT. TARA will underpin the MFTF-B, which is a very bold project. It's leaping forward a great deal and a lot of people are very nervous about that."

What about being able to move ahead into engineering, to look toward building the advanced fusion test reactors that will actually produce energy?

"The administration has come out and made policy statements, that it's the goal of the program to demonstrate scientific *and* engineering feasibility. That's in this administration's National Energy Plan; it went over to Congress in January, about the same time as the budget. We have seventy to eighty million dollars a year going into technology, the superconducting magnets, the plasma heaters, and so forth. That's all reactor-relevant technology, because TFTR and MFTF-B are reactor-scale devices. In fact, I was amused by something that appeared in *Nuclear News*. There was an article that said, 'Fusion abandons technology, goes back to science.' On the opposite page was a photo of TFTR, which is this gigantic assemblage of stainless steel, neutral beams, pumps—Christ, if that isn't technology, I don't know what is!

"We have decided to focus on critical issues, issues that might prevent fusion from working, ever. Tritium, there's one. We have to demonstrate over a period of time that we can handle tritium effectively, because it's radioactive. That when we spill it we can clean it up without killing anybody. It's true that you'll have only about a kilogram of tritium in a working reactor, which is a thousand times less than what you have in the core of a nuclear reactor. But tritium is a gas. It has the nasty property of diffusing through metals, so it's not easy. It's a very serious technical issue. But we're beginning to understand much more clearly exactly what the difficult problems are."

Within the OMB, David Stockman has computers programmed to keep tabs on all the various programs of federal spending. Within the federal budget, now topping $800 billion, his computers keep track of every program spending more than $50 million. There is no budget term for this amount, but it might be called a "Stockman unit." The fusion budget will be running at about half a billion dollars, or ten Stockman units. How can we be sure that the budget won't be cut drastically?

"Technical progress certainly is important; it's fundamental in keeping the program alive. But technical progress in itself doesn't tell you whether you'll be working at five Stockman units, or at ten. What will do that is demonstrating, in a very forceful way, that everything you're doing has purpose and is a necessary part of this program. You have to demonstrate that you are not running facilities just for the sake of keeping laboratories happy, that each one of these things has an essential role to play, that you've pared off the extraneous or marginal facilities. And if you can demonstrate that this program is pared down to the minimum level necessary to carry out this important federal function, that's the funding level you will get.

"It takes a certain amount of manhood, gutsiness, to choose the minimum. When you try to pare out the marginal facilities, people will say, 'Oh, you're taking too much out; you're crippling the program; you're bringing it to its knees.' What would be a marginal facility? It might be one that's still producing good solid scientific results, with a crackerjack, world-class scientific team—but which is not going to really address the next critical issue in this program. Or its results could be reproduced by one of the other facilities, which is addressing a more critical issue. So we're talking about dismembering productive, world-class scientific teams, in order to concentrate the resources in other world-class productive teams, which not only can do their work but the other people's work as well.

"Those are very hard decisions, because those teams have taken years to build up. That's where the gutsiness comes in. But the questions we are facing now demand large facilities like TFTR and MFTF-B; these are questions of reactor-scale plasmas. Those facilities are expensive. And we cannot afford to run a program in which we have our money divided up into fifteen pots. We've got to coalesce into the minimum number of large pots that are required to address these important questions."

Clarke said these things in mid-August of 1982. Three weeks later it was time for the International Atomic Energy Agency to host its biennial fusion conference, the latest in a series that had included Novosibirsk in

1968, Berchtesgaden in 1976, Innsbruck in 1978. This time the location was the new Harborplace center in Baltimore. Many of the participants stayed in the Hyatt Regency, faced with blue glass and offering sweeping views of the harborfront.

Close to a corner of the waterfront, the USS *Constellation* rides at anchor, its masts and toplines outlined in lights. This was one of the first ships in the United States navy, a frigate built in Baltimore in 1797. It was funded under the same appropriations bill that paid for the *Constitution,* Oliver Wendell Holmes' "Old Ironsides." The naval architects of the day knew how to build good ships, but their knowledge was the hard-won wisdom of artisans. Through centuries of experience, these shipwrights had learned to assemble masts and rigging that would stand up in a storm, hulls that would withstand the pounding of battle. Still, their knowledge was purely empirical. They had no theory of the action of wind on sails, no mathematical treatment of the stresses on a keel. Nearly two centuries later, with summer waning and the world's fusion leaders meeting virtually within hailing distance of that frigate, the irony was that one of their most exciting new physics results had come from just this sort of empiricism, this cutting and trying.

Ever since the advent of neutral-beam heating, tokamak physicists have been faced with a problem. The neutral beams heat the plasmas quite nicely. But in doing so, they disturb the plasmas, and set up waves within them so that they leak more rapidly. Beam-heated plasmas are less well confined than plasmas heated by passing heavy electric currents through them, as in MIT's Alcator tokamaks. During the summer of 1982, however, the world's fusion community heard news of a way around this problem. The news came from Germany, where people were working with a PLT-size tokamak called ASDEX. After several years of work, its researchers had stumbled across a way of preparing or conditioning the plasma in advance, before hitting it with the beams—and of getting much better confinement, two or even two and a half times better than anyone would have expected. Then in Baltimore in September, the prospect of getting such good stability in a beam-heated plasma had hundreds of scientists crowding into the main hall to hear the talks by the ASDEX research leaders. They had not set out deliberately or with understanding from theory, and even afterward they did not fully understand exactly what they were doing right. Once again, though, tokamak physicists had been lucky and had made an important advance.

Why was this important? Traditionally, there have been two roads toward getting good plasma confinement in a tokamak. The TFTR represents the direct approach of simply building it larger, moving its walls farther away from the plasma. The Alcators at MIT, as well as related designs such as Robert Bussard's Riggatrons, rely on the second method, to greatly increase the strength of the magnetic fields. Now, however, the ASDEX results were suggesting that there might be a third and more subtle way: conditioning the plasma to make it accept its beam heating more willingly. These tantalizing experiments thus would mark out an important frontier for tokamak physics in the 1980s, that of understanding the ASDEX results, reproducing them on other machines, and learning to use them in the fusion devices of the future.

There was good news from the United States, too. ASDEX might be something of a windfall, an unexpected bonus, but the American results represented welcome experiments that were strongly backed by theory and that had been deliberately sought. A year earlier, America's to-kamak leaders had had their meeting at Airlie House in Virginia, to talk about how tokamaks were in trouble. The issues there were steady-state operation and beta, the measure of how strong their magnetic fields would have to be to hold a given plasma pressure. At Airlie House, the fear was that if these problems could not be solved, the tokamak approach might have to be sharply redirected or even abandoned. But in Baltimore, there was news of solid progress on both these issues.

Since the late 1960s, it had been predicted from theory that beta could be increased markedly by using magnetic fields to control the shape of the plasma. All tokamak plasmas are toroidal, ring-shaped, but there are a great many ways to shape a ring. What counts is the cross-section, the shape of a slice made across this plasma ring, in the manner that we would slice a bundt cake or a doughnut the size of a pie. An inner-tube has a circular cross section, and a plasma in that shape is called a circular plasma. A tire with very wide tread would have a cross section longer than its thickness, and such a plasma is described as elongated. The prediction from theory was that an elongated plasma would have much higher beta. Near San Diego, just one or two canyons over from Bussard's own center by the sea cliffs, people at General Atomic were testing this theory. They had built the Doublet III tokamak, whose plasma had a two-lobed, or "doublet," shape, rather like two truck tires mounted side by side with the facing sidewalls removed so the tires could be glued together. In Baltimore, they announced a beta of 4.6 percent. That was some 50 percent higher than the previous record, and for the first time was close to the range of 5 to 10 percent needed for a practical tokamak reactor. What was better yet, the same theory was predicting that even higher betas

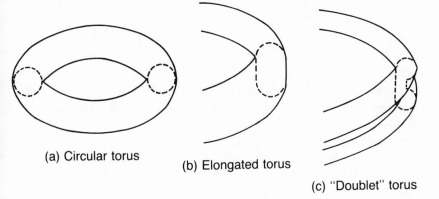

(a) Circular torus

(b) Elongated torus

(c) "Doublet" torus

FIGURE 9. Shapes of plasmas or vacuum chambers in tokamaks. Left, the circular cross section used in PLT and TFTR. Center, the elongated cross section to be used in more advanced tokamaks. Right, the "doublet" cross section used in General Atomic's Doublet III.

could be obtained with more extreme degrees of plasma elongation. If they gave the plasma cross-section the shape of a boomerang, beta might go as high as 20 percent. Even a beta of 10 percent would allow future tokamak reactors to be built much more compactly and with less powerful magnets, making them much less expensive.

The other good news came from Princeton and MIT, and again it told of experiments buttressing a theory. The theory predicted that by using powerful microwaves, it would be possible to drive currents within a tokamak plasma. Such currents are essential, both to aid in heating the plasma and to generate magnetic fields that assist in its confinement. The only way to produce such currents, however, has been with an electrical transformer. Such transformers have only limited stores of current-driving power, which is measured in volt-seconds. A tokamak reactor featuring such a transformer could run only briefly, just till the volt-seconds are all used up; then it would have to be shut down to allow the transformer to be recharged with a new store of volt-seconds. At Airlie House in 1981, this intermittent operation stood as one of the main obstacles to making tokamaks attractive as power producers.

At Princeton, during 1982, Dale Meade's experimentalists had succeeded in using microwaves to drive currents in the PLT. However, these experiments took place at low plasma density. The news from MIT was good enough to have Harold Furth saying, "Now we're nearly home free." There the Alcator people had gone to shorter microwave wavelengths. In their Alcator C, they had driven currents within plasmas the densities of which were close to what would be needed in a reactor. These experiments ran for only three seconds on PLT and a tenth of a second on Alcator C, but everyone in Baltimore knew that if you could drive currents with microwaves for a second, you could do it all day long. These experiments were giving strong hope to the idea of a steady-state tokamak. The needed plasma currents might be driven entirely by microwaves, with no need at all for a transformer. Or the transformer might be reserved for use after the plasma had first been heated with other microwaves. In hot plasma at a hundred million degrees Celsius, millions of amperes of current might be driven for hours with a fairly standard transformer. Then, with its volt-seconds running low, the tokamak would not be shut down but would be kept running with microwave current drive, while the transformer was being recharged. Then, with a fresh store of volt-seconds, the transformer would go back to work.

"Operating a tokamak may turn out to be like riding a bicycle," said Furth. "Riding a bicycle is rather a complex matter; it's hard to give a theory for it. And it's easy to get an instability, where you fall off the bike. The laws of nature allow you to fall down in a few hundred milliseconds, or to ride it all day; but it's still the same bicycle. It depends on your skill in riding it, and that's what we're doing. We're learning to ride this plasma bicycle."

"The interesting results are coming out of the big machines," said Clarke. "ASDEX, PLT, Doublet III. All of these results at this meeting are saying that TFTR will work like a house afire. The laws of nature that are coming out of this meeting say that when we understand how to effectively run TFTR, it will have spectacular results. For instance, TFTR was designed with a beta of 1.5 percent. If we can get the beta up to 3.2 percent, it would double its power output.

"We're planning an aggressive fusion program; it's getting better. We're putting full resources behind TFTR. These big machines take years of fiddling to get them to work right. For instance, ASDEX has been running now for three or four years, and they didn't find this good mode of operation the first day they turned the thing on. It took a lot of work and searching. And then there are the mirrors. Mirrors are even more aggressive, in the sense that we have commissioned these large facilities: not only MFTF-B, but TMX-Upgrade, which is a very large experiment. We've got a lot of money in that, a lot of good people, and we're doing the same thing, trying to learn how to make the mirrors work. On paper the mirrors look great, and so far, every time we've put them to an experimental test, it seems to have panned out. So I'm excited about what we're learning, I'm optimistic."

Still, all this excitement involves problems of plasma physics. What about the Buchsbaum Committee and its conclusions that the United States is ready to go forward with a Fusion Engineering Device and to embark on a path leading to real energy-producing fusion reactors? What about the issues that drove Kintner to resign? Beta and current drive may be all very well for now, but what lies beyond?

"At a half-billion dollars a year," Clarke continued, "you cannot run a program that just goes on year after year producing small scientific advances. Another kilovolt here, another heating method tested over there, another material investigated. It doesn't add up to anything, it doesn't provide any visible progress. And unless we're able to satisfy people's need for obvious progress, we'll be in trouble. The program needs a near-term goal, and even though we can't produce practical fusion power as a near-term goal, still this is what we can do. We can bring the program to a major decision point. And the decision is something we can make in five or six years: to bring the tokamak and mirror programs to the point where we can make a choice, and choose between tokamaks and mirrors.

"People don't like to hear that, but we can do that by 1988 or so. We are fielding a complete tokamak program, a complete mirror program. By 1987 we will be operating TFTR and MFTF-B, which are both reactor-size. We will be able to compare them directly. It won't be a case of comparing theoretical projections, we will have the objects there. We will have reactor-scale plasmas, we can see how they operate. And we will be able to make a choice. Now obviously, one lab is going to be awfully disappointed, Princeton or Livermore. This doesn't mean that we're going to take everybody in that lab out behind the barn and shoot them. But it means the Princeton people, who are very proud and are very good physicists, will have to acknowledge that Livermore has the better approach to fusion, and that they will have to let Livermore take the lead. Or else the Livermore people, who are also very proud of their work and are very capable, will have to accept being understudies to Princeton.

"So only one concept, mirrors or tokamaks, will survive this shootout. That's quite a thing, certainly. But frankly, if we can't do that—if we don't have the management ability and the leadership—then we're never going to get to fusion. This decision is going to terrify a lot of people, because an awful lot will be at stake. Laboratories will be on the line, people's careers will be on the line, all they've worked on. People who have spent years loving mirrors are going to have to start loving to-kamaks, or the other way round. That frightens people. It's going to be a very emotional issue. But if we face this issue, nobody can say that the fusion program hasn't made progress. That will be the ultimate test of gutsiness or manhood, to make that decision. That's big stuff.

"Then we can be in a position, in the late 'eighties, to go ahead with an Engineering Test Reactor based on the concept, tokamak or mirror, that survives this shootout. Now, what is the difference between this ETR and the FED? We changed the name, but it goes deeper than that. The FED is something that we can build today. It would produce several hundred megawatts of fusion power, and just building it would really focus our attention on engineering problems. That was Kintner's view of things, and if we had the money, that would be a good way to go. The ETR will be different because it'll be started five years later, and we'll have five years' more information to go into it. For instance, just the inclusion of current drive would reduce its cost by twenty percent. Also, the ETR will be based on the specific choice, tokamak or mirror. That means that instead of building an FED which is a generic facility simply to develop technology, we'd be doing an ETR, which for sure would be developing technology—but it would be developing the precise technology needed for that concept."

But isn't talk of an ETR simply a sop to the fusion community? Doesn't it amount to having David Stockman say, "You be good little boys and girls, and then just maybe, in 1988 or 1989, we'll let you build a machine that's even bigger and better than the one we wouldn't let you build back in 1981"?

"The machine we could build today, the FED, is a very expensive machine, two billion dollars at the most. The latest design is probably one-point-six billion. But it's still a lot of money, and no one is spending that kind of money on anything these days. It just isn't in the budget. Maybe in two or three years things will change. But I can't sit and hold my breath till federal priorities change, in planning the program. I've got to plan a fusion program that makes progress in the absence of the FED, but which is also preparing the ground so that later we can go ahead quickly. Trivelpiece and Keyworth are in complete agreement on this. No one's saying that we can't eventually have this next step in the fusion program.

"The fusion budget today is about half a billion dollars. If we were to devote all that money to one concept, tokamak or mirror, we could afford to build an FED today. We don't know enough today, to choose that concept. If you're a tokamak man, you could look on the mirror program as a reservoir of funds to build a tokamak reactor—and vice versa. The mirror people can look at the tokamak program as a nice scientific program that's keeping the money warm while they get up to the point of building a mirror reactor. But if we can bring the tokamak-mirror issue to a head, and to a head-to-head comparison, we have the resources within *this* program to start the next step, the ETR. That's why this choice is the most important decision the program has faced to date."

To Clarke and the rest of the nation's fusion leaders, then, this ETR is to achieve the goal set forth by the Foster Committee in 1978, of determining fusion's highest potential. It will realize the hopes of Hirsch who looked ahead to such a fusion reactor, of Kintner who fought to build it. Quite likely it would be the last major experiment, the last large fusion facility built by the government to do research. The next device after the ETR might well be the first facility built primarily to produce fusion power for a practical purpose. In Clarke's words, "If we pick an acceptable reactor design by 1988 or 1990, we could have it finished and operating by the year 2000. If we do that, we will not do what the

Magnetic Fusion Act calls for, which is to operate a demo plant by the year 2000. But we will have produced an information base. We will have run a prime reactor candidate at power, with all the appropriate technology. And we'll be in a very good state to begin discussions to decide where and how fusion will be commercialized. We'll have the data for it."

As the time for the tokamak-mirror choice approaches, some people no doubt will refer to it as the Super Bowl of fusion. But the stakes will be much higher than that. When the Miami Dolphins met the Washington Redskins in 1983, that was Super Bowl XVII; there would be another one a year later. Both teams would maintain their independent existence, and, although the Dolphins lost, they could go home hoping that next time it would be different. Fusion's Super Bowl, by contrast, will be more nearly as if Miami, having lost to Washington, would thereafter be relegated permanently to the status of a Redskins farm team.

When Princeton plays Livermore later in this decade, it will be Super Bowl One and Only. There will be no second chance the following year, or ever. That choice will set the direction of the federal fusion program for a long, long time; if the losing design is to be revived, that will not happen for many years. What is more, these scientific teams will not remain separate and independent. The losing lab will still have plenty of work, plenty of opportunity for its best physicists to show their stuff. But the real action will be with the winners. And just as the best players on a farm club would hope to move on up to the majors, so the best people in the losing lab will look for opportunities to leave that lab and join the other one. When a team wins a football Super Bowl, it gets trophies, TV publicity, and money. When a team wins the fusion Super Bowl, it will get official recognition, from the President and in the federal budget, that its decades of work have demonstrated fusion's highest potential and that the other team's decades of effort have fallen short.

As early as 1984, there will be a clear sign of the trend. By then the Princeton scientists will have modified the PLT and will hope to demonstrate microwave current drive for many seconds at a time, under conditions that will show that this current drive really will work for reactors of the future. At the same time, the Livermore group will have TMX-Upgrade in full operation. Thereafter, everyone will be watching closely to see if either team begins to falter or fall behind. It won't be that one day all the fusion cognoscenti will wake up and say, "My God, tokamaks won't work." It will be more that the PLT people may find themselves running behind schedule, as current drive proves trickier or less promising than they had hoped. Alternately, the scientists working with TMX-Upgrade may set up thermal barriers but be snared by plasma leakage out the sides. After all, the game for both labs will be to do the experiments that will confirm, or refute, predictions from theory. And people will be remembering the words of Lev Artsimovich: "Our relations as experimentalists with theoretical physicists should be like those with a beautiful woman. We should accept with gratitude any favors she offers, but we should not expect too much or believe all that is said."

Which team will win? With Livermore's brash upstarts from the West Coast preparing to challenge the established titleholders from the Ivy League, we can pick up part of the trend by going into the locker rooms and talking with the coaches; that is, the experiment directors. At Princeton there is Dale Meade: "Our biggest challenge is with the plasmas, with Mother Nature. The tokamak people feel challenged but confident." Will his team be nervously looking over their shoulders? "The tokamak people, or the other guys? My view is, if it weren't for the tokamaks, you'd never see an MFTF-B today. But at the moment there's not enough money to carry the tokamak program forward, beyond the TFTR. So the sooner we have this shootout, the better we're going to like it."

Then in the Bay Area, the other team has Fred Coensgen: "Whoever doesn't succeed will be disappointed. I hope the decision isn't an anticlimax. I hope it's a horse race up to the end, or else that the mirrors pull out in front early. I would hope that I wouldn't know three years ahead of time that it's hopeless. If there is some unexpected problem, we will have to be ingenious enough to solve it in reasonable time. But history has proven that we are capable. We have our plasma physics issues to address, and then we will be ready."

So the tokamaks are ahead as of now, with the mirrors playing catch-up. Moreover, the tokamaks will have two important advantages. Clarke may talk as though each will have an equal chance, but the tokamaks are more equal than the mirrors. They are getting twice as much funding. In addition, there will be far more tokamaks than mirrors, around the world. In particular, besides TFTR there will be three other world-class tokamaks, as the centerpieces of the fusion programs in Europe, Japan, and the Soviet Union.

The Soviet program was headed by Artsimovich for a number of years, but he died in 1973. Today its director is Boris Kadomtsev, the one who made the 1971 prediction that trapped-particle instabilities would bring tokamaks to grief. Kadomtsev is short and stocky, with brown skin, a gleaming bald head, and a mouthful of gold-filled teeth. He has had plenty of experience at keeping that mouth shut. At the Baltimore meeting, following a session on plans for future large tokamaks, he was asked what his country would be doing. He grinned and replied, "You capitalists are the ones who do all the planning." Still, it is known that the Soviets have had to back off considerably in their goals.

The main Soviet lab is the Kurchatov Institute in Moscow, named for the leader of their atom-bomb project in the 1940s. For a while, people there were hoping to bypass the TFTR generation of tokamaks entirely and leap directly to the T-20, a very large tokamak with superconducting coils. It would be at least as advanced as our FED, and would not only produce huge floods of neutrons; it would make them for a specific purpose—producing plutonium. However, the T-20 was just too advanced a project for them to bite off. For the 1980s, the center of their program will be the more modest T-15. It will be nearly the size of TFTR, but will use no fusion fuel, no deuterium-tritium mixtures. Unlike the TFTR, however, it will have superconducting magnet coils. Such coils can operate continuously with no flows of current, and will give the Soviets an advantage in long-pulse experiments, in which a tokamak is to be operated for many seconds at a time. Should they choose to work on microwave current drive, the T-15 will then be a fine facility for demonstrating how a tokamak can thus be kept running all day long. If Harold Furth's "tokamak bicycle" is to be ridden for long times, the rides may first take place in Moscow.

Then there is the European program. At the Culham Laboratory, near Oxford, England, there is a fusionscape fully as well developed as at Princeton. Here, too, is a massive thick-walled concrete building, a heavily instrumented control room, a set of motor-generators (which had the advantage of not being dropped during assembly), and a huge assembly of stainless steel. The focus of all this is JET, the Joint European Torus. It was completed in the summer of 1983, only a few months after TFTR, but in several ways it will go beyond the plans for TFTR. It is larger, with over four times more volume in its plasma. Also, rather than having a circular cross section as in TFTR, its plasma is elongated, in a shape that gives higher beta and will be preferred for future reactors. Because of these factors, JET will need less power to be fed into its plasma: 25 megawatts as against 30 megawatts in TFTR. Yet JET will produce much more fusion power, 100 megawatts in all. It thus will come much closer to ignition than TFTR, and may actually ignite. In the words of its director, Hans-Otto Wüster, "Ignition is not in our official program plan, but we'll take it if we can get it."

When Wüster looks across the Atlantic, he sees not only a world-class program in tokamaks, but also the world's leading mirror program. The Europeans have not had the luxury of running two such programs; they have focused on tokamaks. Still, from Wüster's viewpoint, the Yankees are putting the cart before the horse, in going for $Q = 1$ on TFTR before having the remote-handling robots in place. Our lack of these robots will limit TFTR to only its ten initial shots in DT. By contrast, the Europeans wil put their robots and other equipment in place first. Then, beginning in 1988, they expect to run off some ten thousand shots in DT, each shot lasting about ten seconds. They thus may accumulate by 1990 over twenty-four hour's worth of total DT operation, a field in which they then may well lead the world. At least those are their plans, but Wüster knows full well the Robert Burns poem about the best laid schemes o' mice an' men. Wüster himself is a tall, roly-poly man with close-cropped hair, who evidently has spent much more than an accumulated twenty-four hours in the Munich beer gardens. Rather than quote Robert Burns, however, he prefers lines from Bertolt Brecht's *Threepenny Opera:* "Machen Sie ein Plan"

> *Go ahead and make a plan*
> *And be a gr-e-a-t man about it.*
> *Then make another plan.*
> *Neither one will work!*

One place where planning is taken very seriously, however, is Japan. To invoke the name of Japan immediately conjures up visions of some future session of Congress voting to limit imports of Japanese tokamaks, as the man-made sun rises in the East. But it is not that Japan's fusion work is more advanced or broader in scope than our own. In fact, ours continues to be the more advanced. What distinguishes the Japanese is their ability to make long-range plans and carry them out. In the United States, it has proven entirely possible to set forth a bold new fusion program in October 1980, then virtually abandon it by the following September. The Japanese adopted something quite similar in June 1982: their Long-Term Program for the Development and Utilization of Atomic Energy, which sets forth their fusion plans well into the 1990s. But this program, the Japanese counterpart of our Magnetic Fusion Act, most likely will be carried out in full.

The keys to the Japanese effort are the continuity of their government and the close consultation that takes place before any new policy is announced. The Liberal Democratic Party has controlled the government there since 1955. If the Democrats had been in power in Washington all that time, then there would never have been any doubt that Frieman's 1980 fusion policy would stand. Also, in the words of Shigeru Mori, head of Japan's fusion program, "The government is a business body." The Japanese government resembles a successful group of corporations, run by professionals who have worked smoothly together for decades. Thus the very fact that we in the West can learn of their plans means that within Japan, there is little controversy and much consensus as to what is to be done.

The center of the Japanese effort is at Tokai-Mura, about 75 miles north of Tokyo and just inland from the Pacific. Tokai has long been Japan's leading nuclear-research center. Here too is a TFTR-size tokamak, the JT-60, scheduled for completion late in 1984. Like the Soviet T-15 but unlike JET and TFTR, the JT-60 will work with plasmas of hydrogen and will not burn DT. However, JT-60 has a feature not found on the others, an arrangement of special magnetic coils known as a divertor, which will serve to keep the plasma free of impurities. Significantly, the German ASDEX also features a divertor, which has been very useful in getting its good plasma confinement. The Japanese effort, then, may well be able to take the lead in making use of the good confinement first seen in ASDEX.

Close to the JT-60 site is a large vacant lot with a sign, "Future Power Reactor." This is where the Japanese are already looking to build their next big tokamak, which they call the Fusion Experimental Reactor. That will be their counterpart of our FED, or ETR. But they are not waiting to see the results of a face-off between tokamaks and mirrors. It is not that they aren't interested in mirrors; their Gamma-10, a tandem mirror, is as large as our TMX. Their big reactor will be a tokamak, however, no doubt about it; they have already decided. It will be designed to achieve ignition, and to produce some 440 megawatts of fusion power, burning DT in pulses a hundred seconds long. Their plans call for its construction to start as early as 1988 and to be finished by 1993. If they do it, Japan then may well be the first nation to have a real power-producing fusion reactor. But it won't be the last.

As the United States fusion community looks toward its own power-producing ETR, as the tokamak-mirror competition grows hotter, it will be easy to root for Livermore as its team struggles to come from behind. Everyone loves a challenger, after all. What's more, their mirrors will naturally have such nice features as steady-state operation, simplicity of assembly, high beta, and freedom from impurities—features that tokamaks will gain only through considerable ingenuity and inventiveness. Still, tokamaks have held the lead all along, and features such as microwave current drive show that they have plenty of room still for improvement. In the end, this competition may show primarily that America is a very rich country, able to carry forward with two major and independent programs in magnetic fusion, and to bring them both to the point of building reactor-size experiments before making the choice—a choice that in Japan or Europe never arose in the first place. To put it bluntly, we may well go for the tokamak just like everybody else, but unlike everyone else we first will have gone forty ways round Katie's barn before doing it.

Whichever team wins, however, the loser will have a handsome consolation prize. There will still be the extensive facilities in Livermore's Building 431, and at Princeton's Forrestal Campus. The loser's facility, be it TFTR or MFTF-B, is likely to be completely rebuilt and upgraded, with plenty of robots and well-shielded superconducting magnets. The goal would be to equip it for long-pulse DT burning, each pulse lasting for up to several minutes. It thus would produce copious streams of fusion neutrons, even if it still were to fall far short of ignition. The big machine, the ETR, will take a decade to build and will cost billions. By contrast, this consolation prize could be ready in as little as three years after the go-ahead, at a cost of a few hundred million. It would then serve such useful purposes as testing materials, studying how they stand up under neutron bombardment. Also it would give early tests of blankets for breeding tritium, as well as other systems that would go into the big machine. This consolation prize will be quite sufficient to keep the losing lab busy.

In the 1990s, then, the fusion world can expect to have grown beyond the 1980s generation of large devices, and to be busily readying the first large ignition machines, the power-producers. The United States will have its ETR, the Japanese their FER, and the Europeans will quite likely be proceeding with their NET, the Next European Torus. Come 1995 there will be plenty of old-timers who will stand in amazement at how far they will have come, compared with how it was in the 1950s and 1960s. In those days no one could hope in any serious way to build a power-producing fusion reactor. Nor were they even building experiments like PLT and TMX, to prove out the principles that would go into such future reactors. Instead they were struggling slowly and painfully to learn the most basic facts about the behavior of plasmas, hoping that this hard-won education would at least set them on a path that might someday lead them toward these principles.

The search for these principles goes back even farther than these old-timers will remember, and there is significance in this fact. It is a measure of the intrinsic difficulty of the task of achieving fusion power, of the length and twistiness of the path that had to be followed. Yet even within the fusion community, few people know where and how the road to fusion began. It is not generally appreciated that the first attempt ever made at producing atomic energy was an attempt to create controlled fusion. This work had nothing to do with the Manhattan Project, which did not even get under way until several years later. It was an effort to achieve controlled fusion in a device resembling a tokamak. It was the world's first serious experiment aimed at getting energy from the atom, and it took place in 1938.

The man who did it was Arthur Kantrowitz. Those who recognize the name may know of him for what he did in the 1950s, when he showed how to build missile nose cones that would stand up to reentering the atmosphere at 17,000 miles per hour. In the 1930s, though, he was a young physicist fresh out of school. He had graduated with a B.S. from Columbia in 1934, in the depths of the Depression. As he recalls it, "The chances were excellent that if I had gone directly for a Ph.D., I could have gotten a job clerking at Macy's for seventeen dollars a week, which many of the Columbia Ph.D.'s did at that time." Instead, he took a civil-service exam. Soon he was working at one of the few federal labs then existing, Langley Field, which was run by the National Advisory Committee for Aeronautics, the predecessor of NASA.

His work was research in aeronautics, and his boss was Eastman Jacobs, who had become quite famous within NACA for inventing a low-drag wing. At Columbia, Kantrowitz had taken courses on nuclear physics. Atomic energy was in the air, something that physicists talked about occasionally, even though no one had the foggiest idea of how to produce it. One day in 1938, Kantrowitz and Jacobs were talking over an unusual news item in a magazine: Westinghouse had bought a Van de Graaff generator. This was a huge electrical device producing sparks many feet long. It was used in atom-smashing experiments, and the conventional wisdom of the day was that if atomic energy would ever be released, it would be produced with that sort of equipment.* What did it mean, then, if Westinghouse had bought one? They concluded that the reason must be that the Westinghouse managers could see that nuclear energy was coming, and wanted to be in on it. Then how might they do it? At the time, physicists like Hans Bethe were very interested in working out the details of the nuclear reactions taking place within the sun. The question was whether atomic energy could be released in anything smaller than the sun. Kantrowitz thought it could. Within a day or two, he decided that the thing to do would be to try to heat a plasma to sunlike temperatures, using powerful radio waves, and to confine the plasma, using magnetic fields. His magnetic chamber, in turn, would have the shape of a torus.

Jacobs agreed this was something worth pursuing, even though it really had nothing to do with the lab's aircraft research. But Jacobs had a fair amount of clout in NACA's Washington office. His low-drag wing was sufficiently important that he could get anything he wanted, up to a point. By arguing that atomic energy might be important for aircraft propulsion, he got the main office to give him $5,000. That was enough to carry Kantrowitz for a year, and soon he was merrily calculating away. For starters, he had an important review article that Hans Bethe had published in *Reviews of Modern Physics* in 1936, which amounted to a menu of the nuclear reactions known at that time. Kantrowitz decided to go for the D-D reaction, fusion of deuterium nuclei in a pure deuterium plasma. Tritium had not been discovered yet and the D-T reaction, which would be much less difficult, was unknown. But even deuterium could not easily be had. It had been discovered only in 1932, and could not simply be ordered from a catalog. So work with deuterium would have to wait for a while, and Kantrowitz would do his first experiments with ordinary hydrogen.

In 1938 the very term *plasma* in physics was only ten years old. No one knew about plasma instabilities; they weren't discovered till a few years later, and it would be the mid-1950s before it would become obvious that they were important in fusion work. Kantrowitz blithely went ahead and calculated how the plasmas should leak away, by a simple process of diffusion that he could study. He also calculated how the plasma should

---

*The conventional wisdom was not far wrong. No one then could foresee today's huge tokamaks and tandem mirrors, our bridge-breaking neutral beams, or our grandstand-size lasers. But all these are large, complex, powerful electrical apparatuses, as were the Van de Graaff generators of the 1930s.

produce more energy as it got hotter, from fusion reactions. He soon realized there would be a condition in which he would get as much energy out as he was feeding in. This condition is now called breakeven, or $Q = 1$, though in 1938 these terms were decades in the future. With this understanding he could calculate, correctly, that a toroidal device, to reach breakeven, would have to be about the size of what forty years later would be built as the TFTR. A practical system for producing fusion energy would be much larger and would generate enough electricity for all of New England. Such items were not to be built for $5,000, not even in 1938, so he contented himself with a more modest experiment.

First, he would shoot for high temperatures. High-power radio-wave generators were being used in radio broadcasting, and he went ahead and built one with 150 watts of power, to heat the plasma. He calculated that with it he could get plasma temperatures of about a kilovolt, ten million Celsius degrees. At that level, the plasma would emit X rays, which he could pick up on photographic film. For his plasma chamber, he built a torus out of half-inch plates of aircraft aluminum, welded together. His torus was as large as a good-size truck tire, four feet across and eighteen inches deep. Also it had a window, so he could look in and see how the plasma was doing. Jacobs was in charge of the Langley wind tunnel, which featured a power supply of several hundred kilowatts. Kantrowitz intended to wind electric cables around the torus to make a magnetic coil. Then, with his radio oscillator and Jacobs' power supply, he would set out to prove that his theory was right and that he could produce atomic energy.

"All this business about an endless supply of energy, and all that it means, was perfectly apparent then," said Kantrowitz. "It was a very exciting thing for me, and I got very deeply involved in it. I built the apparatus with my own hands. I worked on it full-time during the day, and Jacobs worked on it evenings with me. We both had a lot of enthusiasm." The first thing was to hunt for leaks in the torus, which would spoil the vacuum. They didn't have good leak detectors in those days, so one of their main techniques was the same one an auto mechanic uses to find a leak in your tire. Kantrowitz put pressure in the torus, swabbed soapy water along the welds, and looked for bubbles. Another technique was somewhat more sophisticated; it took advantage of the fact that helium would leak more rapidly than ordinary air. He would pump the torus down to vacuum and hook up a vacuum gauge. Its needle would slowly rise as air leaked in. Then he would pass a jet of helium gas over a weld. When the gas was over the leak, the needle would go up more quickly. He sealed his leaks by painting them with enamel. After a month or two of doing this, he got the torus leak-tight, and then he could wind it with the coils and go ahead.

Going ahead meant putting little wisps of hydrogen gas into the chamber and turning on the power. The tests were done at night, after dinner, because Jacobs wanted to be in on them too. When Kantrowitz turned on the radio oscillator, he could look in through the window in the torus and see the blue glow of the plasma, following the magnetic field lines. There were bright streaks in the glow, marking the field lines. He had calculated how much X-ray emission he should get, and he had dental film to pick up the X rays. Jacobs had a darkroom at home, and on the memorable night when they finally had everything working, Jacobs took the films home and developed them. Then he gave Kantrowitz a call: "Nothing on the film."

Evidently the plasma was not getting hot enough to give off X rays. Still, there was one more thing they could do. They could hold in the circuit breakers in the power supply, which amounted to putting a penny in a fuse box to get more current. Doing that didn't work either. This plasma behavior was something Kantrowitz couldn't understand. The calculations were straightforward enough; he knew they couldn't be wrong. Had he pursued his experiments, he probably would have discovered plasma instabilities, and might even have started to learn what to do about them. After all, although he didn't know it, he was already well along toward inventing the tokamak.

By now the $5,000 was nearly gone and it was time to go after the next year's funding. The director of NACA was George Lewis, and he occasionally took trips to Langley to see what was going on. Naturally, Jacobs was one person he would want to visit. Jacobs took him into their lab and showed him what they had built, the cable-wrapped torus, the radio oscillator, the glass-enclosed vacuum pump, the connections to the wind-tunnel power supply. Lewis stood there by himself for about five minutes, looking long and hard at all of it. Lewis was not a very imaginative man, and his thinking was barely reaching to the level of jet planes, which were just beginning to come along. In his day-to-day work, he was busy trying to invent the aircraft technology of World War II. There he was at Langley, face to face with an experiment that might almost have come down a time-warp from the Princeton Plasma Physics Lab in the 1950s. Kantrowitz and Jacobs were twenty years ahead of their time, but what was important to Lewis was that they hadn't accomplished what they set out to do. So he cut it off. Right then and there, he stopped its funding.

"It was a heartbreaking experience," Kantrowitz recalls. "I had just built a whole future around this; I wanted to make it a career. This was just a tremendous blow to me that I wasn't allowed to go on with this." He went back to aeronautics, and soon returned to the physics department in Columbia University. There he quickly learned that some of its people were pursuing another approach to atomic energy. By early 1939, scientists in Germany and Denmark had split the uranium atom, and Enrico Fermi, one of the world's leaders in nuclear physics, had arrived at Columbia. Fermi set out to prove that he could make a chain reaction, working closely with a friend of Kantrowitz, a young graduate student named Herbert Anderson. Kantrowitz occasionally would drop in and ask Anderson how things were coming along, and Anderson would explain how they were getting closer. Then one day Anderson said, "I'm sorry, Arthur, I can't tell you any more." With this, Kantrowitz decided they had proved it and that it would be possible to make a nuclear chain reaction.

During the next few years, Kantrowitz was invited several times to join the Manhattan Project, but he never did. His fusion research had given him a taste for dealing with superhot gases, and eventually this would lead him into the research on missile nose-cone reentry that became so important in the 1950s. But he never followed up on his fusion work at Langley, never even wrote it up as a paper in a physics journal. He did not even take a photo of his experimental equipment. As for Jacobs, his low-drag wing design went into the propeller-driven aircraft of World War II, including the P-51 fighter plane and the B-29 bomber. But he too never returned to fusion.

Kantrowitz and Jacobs did file a patent application, in March 1939, which today stands as one of the most fascinating relics of their work. They called their invention a "Diffusion Inhibitor," and stated, "This invention relates to a method and means for inhibiting the transfer of heat. . . . WE CLAIM: The use of a magnetic field to reduce the diffusion of charged particles. . . ." The accompanying diagram showed a torus wound with a magnetic coil, in such a way that the heated plasma would be trapped in its interior and kept from reaching the walls. This invention thus amounted to discovering the basic method used in all fusion devices today to contain a hot plasma. The patent was not granted. In the end, all that came of this work was that Kantrowitz collected his papers and lab notes into a thick green folder labeled "Atom Buster," which was the name he had given his project. He kept that folder in his attic for the next forty years. When fusion research again got under way in the United States, early in the 1950s, there was nothing of Kantrowitz' work to serve as a guide, and such investigators as Lyman Spitzer had to start from scratch.

In a very real sense, then, the fusion machines of the next generation will do little more than achieve the goals that Kantrowitz, in his innocence, was quite prepared to pursue as early as 1938. Had plasmas been as easy to handle and work with as Kantrowitz hoped, then before World War II was well begun, there would have been crash programs under way to build reactors on the scale of the TFTR, even of the ETR. And since fusion reactions are a copious source of neutrons, which can be used to produce plutonium for nuclear weapons, that war might well have taken quite a different turn. The first atomic bombs might have been dropped during 1943 or 1944, and on Germany. Their first targets might have included Berlin, or perhaps an industrial center like Essen or Dortmund. That this did not happen is due to a single esoteric fact, far removed from everyday experience: that plasmas confined by a magnetic field are unstable, and are kept only with great difficulty from leaking out.

Yet despite this difficulty, well before century's end we will have gained the solution, however belatedly. What will happen then; how will fusion grow and develop? There are any number of predictions of fusion's future, and these will be the subject of the next chapter. What may be the most hopeful prediction of all, however, comes from President Reagan's science advisor, George Keyworth. It is not that Keyworth predicts early or easy success, not at all. But in his position he must necessarily be cautious, reluctant to commit himself, wary of going out on a limb. His statements can be taken as reflecting administration policy, may even be tantamount to policy, and even a carefully hedged and ambiguous statement favoring fusion thus would stand as of the first importance. But his views on the matter, which he presented at his Senate confirmation hearings in 1981 and then amplified in a personal interview, have gone much farther than that:

"If you examine all conceivable options today for ultimately meeting the world's energy requirements, I think that one by one the alternatives fall to the wayside, on technical grounds, on economic grounds. Ultimately I am left with the observation or perception that fusion will meet the needs of an expanding society, will meet it economically and safely, and that I know of no other competing alternative. There is no doubt in my mind that controlled fusion will work and will be the ultimate power source in the future. We are not at the threshold of moving from science to engineering. There is much work to do on materials and plasma physics, but nevertheless I believe that some time in the twenty-first century fusion will replace most of the commercial sources of energy."

12

Fusion in the World

W ITH views like those of George Keyworth, it is easy to imagine that somewhere there is a master plan, a well-conceived set of steps to lift fusion from the level of laboratory experiment to that of a predominant energy source. There are certainly plenty of plans and proposals for fusion's future, but the mastery in them is rather difficult to discern. Fusion will develop and grow amid the energy world of the next hundred years, and this is not something to be neatly laid out on a chart. The goals of the fusion leaders, then, are rather more modest.

Harold Furth of Princeton has stated, "I have absolutely no idea how the solution to plasma confinement will come out. It's really quite unpredictable, and it seems extremely unlikely that the ideas we have now will be recognized as being just right. The most I think we can hope to do is to propose and carry through some kind of a sensible first-generation fusion reactor, so as not to give fusion a bad name. If you come up with something that's very complicated and goes on the blink most of the time, and ruins the investors, then that's going to give fusion a bad name. But that is something people should be able to foresee.

"What will be the best fusion reactor two generations after that? That's not something we can foresee now. People will have much better ideas then. I don't think the Foster Committee saw it either, when they brought up the matter of fusion's highest potential. They were just saying, 'Don't rush in and build something that is going to turn out to be inferior.' I don't think they really thought that in five years one would know what the best possible solution would be. I certainly don't. I think the highest potential will be reached after ten generations. But I'm not looking to that. I'm looking at that potential which is high enough so that the world will take to fusion rather than rejecting it. That's what's needed. To be so ambitious as to think one can raise it to the highest potential right away— that's insane.

"The objective right now is to avoid botching it. We don't want to come up to 1990 and have people say, 'Hey, you've been working for forty years and look at this turkey you've come up with.' But the thought of fusion reactors is a very happy prospect. I think environmentally it'll be far more benign than solar power, which would be covering up vast tracts with stuff that ultimately will turn into junk. Certainly it'll be much more benign than any kind of fossil fuel or fission reactor. I think we must resist the temptation to praise it up too much, since we don't know how to do it. But certainly, if one did know how to do it, this would be the most attractive possible solution. I might see it in my own life, yes, if I don't smoke too much. I love to smoke cigars, you see. But if I don't smoke enough cigars, maybe we won't get there. The cigars help my thinking process a great deal. So I have to balance these things."

If fusion power is to take root and grow, it will be helped greatly if it can find a niche where it can do useful work and fill a need that people will be willing to pay for, without yet being made to face the challenges of commercial competition. This niche would take fusion past the stage of being work for experimenters in government-run labs, without yet demanding that it be attractive to electric utility companies, or able to compete with their present-day methods of generating electric power. It would be particularly useful if this niche would continue to shelter the nascent fusion industry within its cocoon of government sponsorship, while allowing it to perform a service directly useful to the utility industry. This would allow utility executives to gain experience in dealing with fusion, over a term of years, and to become familiar with it. Such a niche may well exist. The first use of fusion may be to produce fuels for use in existing nuclear reactors.

To link fusion power to the existing nuclear industry could certainly involve fusion in the controversy and trouble that beset the atom today. Yet there are sound reasons for proceeding this way. While public attention has focused on canceled reactor orders and abandoned construction plans, the worldwide nuclear industry is far more robust than is commonly appreciated. The industry may be in the doldrums in the United States, but it still counts over 80 reactors with operating licenses, with some 50 more under construction. In addition, the navy has over a hundred reactors powering its submarines and other ships. Outside this country, twenty-three nations now operate more than 200 reactors, with over 160 more under construction. Seventeen other nations have firm nuclear programs. Worldwide, nearly 200 additional reactors are planned or on order. Even if the United States were to cease entirely to build new reactors, these hundreds of existing ones would still be there. At the International Institute for Applied Systems Analysis near Vienna, the energy expert Wolf Häfele has constructed various scenarios for the future of the world's energy supply. Even his "low scenario" shows the equivalent of 1,300 large nuclear plants in the world of 2000, over 5,000 plants in 2030. His "high scenario" projects as many as 8,000 such plants by 2030. And each of them will need a fuel supply.

There are two sources for nuclear fuels: the elements uranium and thorium. Both exist as ores and are extracted as heavy metals resembling lead. Thorium is by far the more abundant element; worldwide, there is at least three times more thorium than uranium. Today's nuclear reactors use none of it, however; they rely on uranium. What is more, they use this uranium wastefully, extracting only a few percent of the energy potentially available.

The nuclear world has been aware of this for decades, and for most of that time has been pursuing a solution: the breeder reactor. The breeder is a type of nuclear reactor designed to produce particularly intense flows of neutrons. These neutrons irradiate uranium metal, which is stored in long rods surrounding the reactor core, and the irradiated uranium turns into plutonium. The plutonium, in turn, can be used as fuel for nuclear reactors. Indeed, the breeder can produce more fuel than it consumes. The breeder thus has been compared to a Coke machine where you put in two quarters, get a can of soda, and get your quarters back.

Even in the nuclear world, however, the breeder is far from universally popular. It is expensive, for one thing. The cost of electricity from the most advanced breeder, the French Super-Phénix, is over twice that from conventional reactors. It is not really a very effective way of getting new fuel; it would take several breeders to produce enough of this fuel for one such conventional reactor. Thus the world could ultimately need several thousand breeders, and in fact Wolf Häfele's scenarios project many more breeders than conventional reactors, by 2030. The breeder is more complex and costly than a conventional reactor, and its safe operation may be more difficult to assure. Indeed, to be quite blunt about it, the reactor could not only melt down; it could blow up. The 1982 "Clinch River Breeder Reactor Plant Technical Review" describes what such an accident could be like:

> The initiating phase covers the first phase of the accident. . . . During this phase an imbalance between reactor energy generation and heat removal would cause fuel to melt and move within the assembly . . . and the accident would progress into the meltout phase. . . . If pessimistic assumptions are made . . . a large-scale pool phase would occur. During this phase, a contiguous molten fuel region is assumed to form. . . . The final phase, hydrodynamic disassembly, is used to describe the core response to a sustained, superprompt critical excursion. . . . When this phase is entered, the accident is considered energetic . . . rapidly increasing fuel tempera-tures result in pressures which far exceed the mechanical restraint capabilities of the core components. . . . During this phase the reactor core generates a very high power. Fuel melts and vaporizes quickly due to the high power, generating pressures that are great enough to cause disruption of the core internal structure and ejection of materials from the core region. . . . The disassembly events occur . . . within ten milliseconds.

It is worthwhile to go into the technical terminology here, to appreciate the levels of horror that are veiled by its polysyllabic blandness. The reference to "fuel melting" means that as early as the "initiating phase" of the disaster, the damage to the reactor would already be greater than at the height of the accident at Three Mile Island. The "meltout phase" and "large-scale pool phase" refer to situations in which much of the structure of the reactor interior would melt and flow into a molten pool. "Hydrody-namic dissassembly" means that the structure would have no strength whatever to resist internal pressures, and would blow apart as if made of water. "Sustained, superprompt critical excursion" is a nuclear engineer-ing term for what in the vernacular is called a nuclear explosion. This explosion, in turn, blows the reactor core apart ("dissassembly") in a hundredth of a second.

This explosion would take place within a molten mass of plutonium and other metals, diluted by flows of melted steel. It thus would be very different from the detonation of an atomic bomb, in which shaped-charge explosives and neutron reflectors serve to contain the explosion until it can build up to full strength. The explosion of a fast-breeder reactor core would probably have the energy of no more than a few tons of TNT, and thus might be contained within the safety system of last resort: the thick ferroconcrete dome that surrounds the reactor. Still, this explosion would carry all the radioactivity of a full-size atom bomb. And if the dome were to give way, the accident would enter into yet another phase. This would be one of mass panic and hysteria, as a radioactive cloud rises over the shattered remains of what had formerly been a fast-breeder reactor plant.

In its operation, the breeder must begin by working with a set of nuclear conditions that conventional reactors are not even allowed to reach at all. In addition to its safety problems, the breeder is a prolific producer of nuclear wastes. For the long-term future, however, all these problems may be minor compared with the worst one. The fuel produced in a breeder is plutonium, which is the stuff of nuclear bombs. There are various grades of plutonium, which differ widely in their usefulness for this purpose. But any form of plutonium can be used to make a bomb.

If the world is to have thousands of nuclear plants, many of them breeders, then plutonium will become an industrial commodity in common use. It will be carried and transported in ordinary trucks and railroad cars, over the common highways and rail lines. Similarly, spent fuel from nuclear plants will have to be brought to reprocessing centers, where the valuable plutonium will be extracted. All this is at the root of a fear that cannot be dismissed, that of nuclear terrorism.

In a world full of breeder reactors, the plutonium-bearing nuclear fuels would not themselves be weapons-grade. The plutonium would be mixed with uranium, to be suitable for use in reactors only. But terrorists could hijack a shipment of such fuel, then run it through a clandestine chemical lab to separate out the plutonium and prepare it in concentrated form. One person who has spent a lot of time worrying about these matters is Theodore Taylor, a nuclear-safety expert at Princeton University. Taylor was the subject of John McPhee's *Curve of Binding Energy,* the theme of which was that under the proper circumstances it might be relatively easy for people to make their own atomic bombs. On the matter of nuclear terrorism, Taylor had this to say: "A terrorist group could produce plutonium by clandestinely reprocessing spent nuclear fuel. They would have to take the stuff to a swimming pool, literally, and do the work under water, to get protection against the radiation. They would go through a series of chemical steps, dissolving it in nitric acid and then precipitating out the plutonium. Is it credible that any group of terrorists would go that far? Are they likely to hire some physical chemist who knows how to do it? I don't know, but we do know something that is relevant.

"A large number of heroin production facilities have been busted over the years, mostly in southern France. The level of sophistication in chemical engineering, within those clandestine chemical laboratories, is at the same level that you need to reprocess plutonium. I spent over a year on a project in which we were examining exactly that, asking, 'How does this compare to reprocessing, compare to making a bomb?' We concluded that in terms of dedication, special expertise, and skill, they're at the same general level.

"You don't have radioactivity with heroin, of course, but you do have a great deal of poisonous material that you have to avoid breathing in. Airborne heroin is enormously toxic, as are some of the chemicals produced in the steps that lead up to it. Also, to make heroin, you have to carry a mixture of material to a temperature of a hundred degrees Celsius. If it's ninety degrees it doesn't work. If it's a hundred and ten, it explodes. In fact, several of these laboratories in France have blown up, for just that reason. One of them was being run by a chemist with a Ph.D. from the United States. He wasn't a terrorist; he was making heroin. But that's quite on a par with clandestinely reprocessing plutonium."

With all these problems and dangers, the worldwide nuclear industry thus will be very interested indeed in a method for making nuclear fuels that avoids or eases these difficulties. Fusion can provide such a method, and this in turn can provide fusion with its niche. The device that produces the nuclear fuels is called a fusion breeder, and it may well be fusion's first application, once it emerges from its current status as a purely experimental program.

The breeder reactor produces floods of neutrons useful in breeding nuclear fuels, but a fusion reactor is a far better producer of neutrons. Fusion reactions produce them in vast swarms. In fact, three-fourths of the energy from these reactions comes off as neutrons. To turn this energy into electricity is a somewhat roundabout and complicated procedure, but these neutrons can be extremely effective in breeding nuclear fuels. In fact, one fusion breeder might well produce enough fuel for fifteen conventional nuclear plants. The entire current United States nuclear industry could get its needed fuel from a half-dozen such fusion breeders, owned by the government and built on worthless federally owned land in the middle of Nevada.

Both in conventional and in breeder reactors, the neutron is a scarce and valuable commodity. The designers of such reactors take great care to achieve good "neutron economy," making the most of the neutrons. Neutron economy, in turn, leads the designers to not use thorium at all, and to use uranium in ways that encourage the production of weapons-grade plutonium. The fusion breeder would turn this around, and would make neutrons cheaply. Thus, thorium would become usable, and at a stroke the world's reserves of nuclear fuels would increase more than fourfold. This use of thorium would also offer important advantages in safeguarding these fuels against the terrorist.

When thorium is irradiated or exposed to neutrons, the fuel thus produced is uranium-233. It would be an excellent reactor material but an extremely poor bomb material. The reason is that it is formed along with uranium-232, which is so radioactive that it would strongly discourage any terrorist who wanted to use it for bombs. Moreover, the U-232 cannot be separated from the U-233 in anything resembling a clandestine heroin lab. To separate plutonium from uranium calls for chemical processing, which any clever chemist can do. To separate U-232 from U-233 calls for isotope processing, which demands highly sophisticated equipment, very large power supplies, and a great deal of capital investment. Even national governments have found it difficult.

A bomb made from uranium-233 and U-232 would need three feet of concrete shielding around it for anyone to approach it safely. The U-232 would be emitting hard gamma rays, and in the absence of this shielding would quickly give the bomb-builder a taste of his own medicine. In addition, for even greater safety against terrorism, this "spiked" U-233 could be mixed with still another isotope, U-234. Uranium-234 is inert, and is incapable of sustaining a nuclear reaction. When irradiated with neutrons, it does not turn into plutonium, and like U-232 it cannot be separated from the U-233 by methods available to terrorists. Its purpose would be to dilute the U-233, which then could still be used in a reactor but could not be used to make a bomb at all. To a terrorist, plutonium would be a nice, inert, easily-handled material, ready for immediate use. U-233 from a fusion breeder, with its admixtures of U-232 and U-234, would be different. It would be dangerous if not lethal to handle, and in the end it would not even blow up anyway.

The fusion breeder thus could offer the opportunity for a partnership between the government and the nuclear industry. The industry would not be faced with the prospect of building large numbers of breeders, which it would own and be responsible for. Instead, there would be only a few fusion breeders, which the government would pay for and own. The government, in turn, would use them to produce nuclear fuel, selling that fuel to the utilities. Naturally, the government would also use them to produce fuel for its submarines—as well as weapons-grade nuclear materials for its bombs and warheads. The first successful fusion plants might then be built in the early years of the next century, at government nuclear centers such as Richland, Washington, and Savannah River,

Georgia. They will be under heavy guard and tight military security; fusion would thus come into the world surrounded by barbed wire.

The fusion breeder then could establish fusion as a technology in being, capable of doing its share of the world's work. The government and not the utilities would accept the uncertainties attending its growing pains. Thus, in their early years, fusion plants may well have frequent interruptions and long downtimes. These would be a general feature of any complicated new plant. These shutdowns will be much less troublesome if they merely interrupt the production of nuclear fuels, rather than cutting off electric power just when it is needed. Still, during this era of the government-run fusion breeder, a nascent fusion industry will be taking shape, as contractors from the nuclear industry build and operate these plants. This fusion industry will be developing the operating experience as well as the improved designs that can bring it into its own. In time, the fusion industry will break free from its ties to the placental nuclear industry, to enter the realm of commercial production of electricity.

From the start, the fusion program has aimed at eventually building plants to produce electricity. This means that fusion in time is to emerge from its larval stage, sheltered within its cocoon of federal money, and metamorphose into something that can live and fly and take its chances in the world at large. An ETR will produce a few hundred megawatts of fusion power, will operate only part of the time, and will call for a small army of Ph.D.'s continually dancing attendance. Its commercial successors will have to produce ten times more fusion power for the same cost, must operate reasonably close to full-time, and will be in the hands of ordinary powerplant engineers and managers. That is rather a tall leap. It thus is little wonder that today there are fusion experts and then there are people in the electric-utility industry, and the two groups do not spend a great deal of time talking to each other.

If fusion is to make this metamorphosis, the utility industry will have to home-grow its own fusion experts, just as in recent decades it has had specialists making careers out of nuclear power. These will include executives who will say that fusion is certainly a wonderful application of physics, but is it the best way to take care of their customers? How will it be useful to these customers, who switch on lights and air conditioners and then receive their monthly electric bills? Today, when such questions come up, one name recurs again and again. He is Clinton Ashworth, of Pacific Gas and Electric. His office looks out from the PG & E building into the San Francisco financial district. He is one of the few senior engineers in the electric-power industry who have made it their business to work closely with fusion leaders such as Livermore's Ken Fowler, and to look ahead to this metamorphosis.

"The power industry is not looking to spend capital on much of anything right now," he said. "But the facts are that our power plants are wearing out and we're going to have to do something, eventually. We're counting very heavily on being able to keep old units reliable, and it isn't proving to be very effective. My feeling is, some of those old units are going to shut themselves down; they will retire themselves at some inopportune time. Have you ever owned an automobile that was thirty years old and had been run wide open for most of that time? That's what some of our power plants are like. You can't really replace much of anything without still having an old unit. The insulation falls off the control cables, you have to replace the control system, the piping, the turbines—the whole thing. You can patch things up for a time, but it's just old wine in new bottles. But you just plain can't patch up old units, once they reach the end of their design life. You get to the point where you're spending lots of money just trying to keep them from collapsing completely.

"We have to maintain a margin of electric-generating capacity, a reserve, and we rely on our hydroelectric plants for that. That depends on how much rain we get in the winter. If we have a very dry year, I'm not sure we have any margin. Two dry years in a row, and hydro just isn't there. In our plans we talk about new generation needed in the mid-1990s. Well, nobody knows what that is. Is it nuclear, is it coal? We don't know. I don't think anyone knows what a nuclear plant committed today would cost, when it's finished. The most recent plants that were committed in the West were in Washington State. Unit 5 was six and a quarter billion dollars, and at the time of its demise it was fifteen percent complete. It already had something like a billion and a half spent on it. I estimated Diablo Canyon in 1968 would cost three hundred fifty million. It cost two-point-four billion, for an eleven-hundred-megawatt plant, so I only missed it by a factor of seven. Fortunately, I don't have anything to do with Diablo Canyon right now. So when you talk about a 1995 nuclear unit committed today, you don't know what it'll cost."

What about coal?

"We had a coal unit proposed four years ago," Ashworth said, "and before we got through looking at the costs, we came to the conclusion that there was just no way we could commit it. This was in 1978. We had announced that we were going to build the thing, announced the site. Then we came up with a cost of three billion dollars for the plant itself.

Then another billion to buy out local industrial polluters in that area, so that we could trade off our emissions for theirs, use their emissions allowances. And another billion dollars to develop the coal property in Utah, so we could get the coal out of the ground. We were talking over five billion for a sixteen-hundred-megawatt plant, two eight-hundred-megawatt units. Today that looks like a bargain, because the Intermountain Project near Delta, Utah—I think they're up to nine billion for four seven-hundred-fifty-megawatt units."

Fusion plants may well have costs in that range, when they become available. But the amount may not be so important, if the costs are predictable. That will depend on whether the fusioneers can offer plants of standard, well-fixed design that can be built and installed as units. And doing that, in turn, will depend on how safe these plants will be. The problem with nuclear plants is that they have had to be repeatedly redesigned and rebuilt, to make up for safety problems that were not properly handled at the onset. That is what has driven up their costs. In looking ahead to tomorrow's fusion plants, Ashworth has come across an interesting little tidbit that has received very little attention elsewhere in the fusion world.

"All fusion reactors that are made of ordinary structural materials, including aluminum, will melt down. You'll have something that's got as much radioactivity as the cladding or casing on fission-reactor fuel." That is, a fusion reactor's interior parts will become so radioactive that, if left by themselves, they will heat up and melt. The way to handle this will be with cooling water or gas, to carry off the heat. But such emergency core cooling can fail; that is just what happened at Three Mile Island. The same could happen to a fusion plant.

This problem stems from what is called afterheat. If a fusion reactor is to be shut down, that will be easy to do. Just shut off the fuel, let air in to spoil the vacuum, or disturb the magnetic fields that confine the plasma. But even after shutdown, the reactor's structural materials may still be hot with radioactivity, which cannot be turned off at all. Until the radioactivity decays, it will give off intense heat, and the reactor core will need continuous cooling. If fusion reactors are designed this way, they will land right in the middle of the regulatory thicket that has snared nuclear power.

Ashworth finds it easy to imagine some future Three Mile Island for fusion. Think of a large plant built on the California coast. It is four o'clock in the morning and everything is going along smoothly, when an earthquake hits. The reactor building shakes violently. An automatic system shoots a puff of air into the reactor's vacuum chamber, and the reactor shuts down safely. But the quake has damaged the electric connections to an emergency pump circulating water through the reactor core. An auxiliary diesel generator, to provide standby power, also is damaged. The core heats up and, within an hour, despite frantic efforts, it begins to melt. There is tritium within the reactor, gaseous, radioactive, and as the meltdown progresses, the tritium escapes into the surrounding air. By early morning, California has heard the news, which has come right on top of the earthquake. What follows is widespread panic and hysteria. During the next year, eleven people die of cancer in the area downwind of the plant. During the previous year, while all the tritium was safely under control, twelve people died of cancer in the area downwind of the plant. No matter; the families of the eleven cancer victims file a class-action suit against Pacific Gas and Electric, charging wrongful death and demanding a billion dollars each. Meanwhile, PG & E faces added billions for the cost of the lost plant, and for cleaning up the mess left behind.

It will not be enough for such events to be highly improbable, or to say that they can be guarded against by suitably elaborate arrays of pipes, pumps, and plumbing. After all, that little adverb *suitably* is what makes nuclear regulators frown and powerplant builders spend billions. No, these things will have to be completely impossible, period, if fusion is to be attractive. Thus far, the fusion community has paid little attention, as Ashworth well knows: "You will find that the fusion community doesn't even know what I'm talking about. They'll say, 'Oh, it won't melt down.' You talk to the people who designed Starfire." This was a study made at Argonne National Laboratory, outside Chicago, to try to predict what a tokamak fusion plant might have for its magnets, current drive, and the like, when such plants begin to become commonplace. "The Starfire people will say it can't melt down—they don't know what they're talking about. You've got to watch 'em. They haven't even looked at whether the metal can cool down, and if you pin them down they'll say, 'Oh, well, it's got water going through it.' Well, that's not what I'm talking about. Fusion reactors have got to look better than the safest fission reactors, or they're not going to be advantageous."

At that point in our discussions, Ashworth walked over to a file cabinet and pulled out a report. Within it was a table concerning common structural materials proposed for use in building fusion reactors. The table gave the "adiabatic meltdown time": the time for heavily irradiated material to melt when it is heavily insulated, by being buried so deeply within a reactor that there is no way for it to cool down like a hot iron in the open air. The table also gave the "coolant cut-off time," after which the radioactivity will have decayed sufficiently that the pumps circulating the coolant can be safely shut off:

Meltdown and Coolant Cut-Off Times

After two years of operation at a neutron irradiation of 2.5 megawatts per square meter

Structural Material	Adiabatic Meltdown Time	Coolant Cut-Off Time
Aluminum alloy (Al-6063)	16 minutes	4.4 days
Stainless steel (SS-316)	1.6 hours	Over 30 years
Inconel (HT-9)	26 minutes	Over 30 years
Titanium alloy (Ti-6Al-4V)	1.1 hours	1 year
Vanadium alloy (V-20Ti)	2.5 hours	0.8 days
Silicon carbide	Infinity	Zero

Is Ashworth right? There is a thick two-volume work, *Fusion*, published in 1981 by Academic Press and edited by Edward Teller. It features lengthy technical articles written by leading fusion experts: Marshall Rosenbluth with a chapter on plasma physics, Richard Post of Livermore with one on magnetic mirrors, and so forth. Professor Robert Conn of UCLA, the reactor-design expert on the Buchsbaum Committee, also has a contribution in his own area of specialty, a chapter on design principles for fusion reactors, over two hundred pages long. Nowhere does he mention that there is this meltdown problem.

To Conn, the situation simply is not as bad as all that. He points out that in a nuclear reactor, the afterheat problem comes from the nuclear fuel itself, which is many times more severe than the afterheat from irradiated metal parts within the reactor. Thus, as at Three Mile Island, a reactor operator must respond in an emergency within as little as seconds and at most only a few minutes, to prevent an accident from damaging the reactor. In a fusion plant, the operator would be dealing with irradiated metal parts only, and would have much more time. Also, in a nuclear reactor the afterheat is so severe that the coolant must be circulated continuously for some time. If the circulating pumps fail or cannot be turned on in the first place, the reactor may melt. In a fusion plant, it may not be necessary to pump or circulate this coolant. The fusion reactor may be protected adequately if the coolant merely stands in place within the reactor. This liquid coolant then would carry away the afterheat simply by convection. This convection would carry the heat from the inner to the outer regions of the reactor, where it would dissipate in the surrounding air.

In Conn's words, "It is a matter of debate whether Ashworth is right. I cannot indicate to you what the public will be willing to accept twenty years from now. It is also a matter for debate whether it is enough of an advantage to have hours to respond in an emergency, rather than having seconds or minutes to respond. I would suspect that Ashworth would say that's not sufficient. Many other engineers would say that having hours to respond is a great improvement and is sufficient to make fusion a more acceptable energy source."

Nevertheless, if Ashworth is right, then what is to be done? The answer will lie in building reactors whose critical inner parts are made from materials other than metals. These would include composite materials such as carbon fiber, used on the nose and wing leading edges of the Space Shuttle. Also there would be ceramics such as silicon carbide, which might resemble the heat-resistant tiles that cover the rest of the Shuttle's wings and fuselage. Such materials do not become radioactive, ever. Indeed, Ashworth and his PG & E associates have proposed how a practical reactor might be built with these materials: "What our target was in this last design, and what we think we achieved, was a reactor that in theory you could walk away from, no matter what happened." You wouldn't need to worry about having to circulate cooling water through it, after an accident. If fusion reactors can be built to such a level of safety, then their builders can hope to stay out of the law courts and their costs may stay well controlled.

Ashworth also worries about whether the fusion program will come up with the kind of reactor he will need at first. It will be all very well for PG & E eventually to buy thousand-megawatt plants at a cost of billions, but in no way will they leap right in. Rather, they will want to start small, building pilot plants. Then they can gradually build up the confidence and experience that can lead them into the thousand-megawatt era. But Ashworth doesn't see the fusion program heading that way: "The power industry got involved in fission early, in small sizes. But the federal fusion program is trying to give birth to a full-grown adult technology, and I don't think that's possible. You can't get there from here.

"There just hasn't been much success, where somebody has tried to give birth to a new power-generation technology in anything other than small pilot-size. What we need is a series of learning plants, because invariably you learn a lot from the first thing you build. If it's small, you learn, and then you can do something about it. The mainstream fusion program doesn't seem to have that as an option. I have been concerned for a number of years that fusion just plain can't ever come into being, because it's focusing on concepts that can't be made into pilot plants. Tokamaks, tandem mirrors—they won't work, if they're built small. They just don't seem to come in sizes where you could build two or three pilots and learn from the first ones what improvements are necessary. There's no opportunity for learning. You've got to answer all the questions in the first big thing. If you try to go for something big, you end up with so much conservatism built into it that even if it works, it becomes a one-of-a-kind thing. It requires too much change to get from there to something somebody would want."

One of the features of the Magnetic Fusion Act was that it tried to set forth a master plan for the next twenty years. In particular, there was its plan for a big demonstration plant by the year 2000. Significantly, John Clarke today has backed away from this. His planning goes only as far as the ETR, but no farther: "After ETR we don't talk about a specific machine. Yes, a machine will be built, but exactly who builds it—we can't say. Personally, I have never put much faith in demo plants. We've seen people build demos, but what they've demoed was something that was not economic. You can look at things like the solar program. During the last administration there were a number of demonstration plants that demonstrated that certain types of solar power were uneconomical. That's a demonstration, but it isn't very useful."

The key to getting small fusion plants, at pilot-plant scale, is high beta. The higher the beta, the more compact and potentially cheaper a fusion plant can be. That is why it is so encouraging that magnetic mirrors have gotten very high values, actually exceeding 100 percent in some cases. That is why it is so significant that theory and experiment may point to betas of 20 percent and higher in tokamaks—and tokamaks, after all, offer much better confinement than do mirrors. Indeed, the 1982 Baltimore conference featured another significant straw in the wind. A group from Columbia University reported results from a small tokamak they had built, called Torus II. Their beta, measured by experiment, was 12 percent. By contrast, ten years ago there was good reason to think that tokamaks would be limited to a beta of only about 3 percent. Robert Bussard's Riggatrons are only one example of what can be done with high beta. For a number of years the federal program has been spending $10 to $20 million a year on compact-fusion experiments.

There is an enormous richness of approaches to fusion, very few of which have been tested with anything like the thoroughness that would demonstrate their potential. Two of the most interesting are called Spheromak and Reversed-Field Pinch. They have been tested only in small experiments thus far, at Princeton and Los Alamos. But ten years ago the magnetic-mirror concept was at a similar level, and had also been tested only in small experiments. The mirrors' rapid rise to prominence could be repeated for these newer concepts. Their attractive features include not only high beta but also freedom from much of the magnetic paraphernalia needed by tokamaks or mirrors. Their plasmas may actually generate part of their needed magnetic fields, reducing the need for exterior magnets.

Harold Furth, who loves to smoke cigars, describes their plasmas as resembling smoke rings. "They're off by themselves, you don't need to wind coils around them. They either hold together or they don't. And if they naturally hold together, like a smoke ring, then that opens up possibilities for reactor design which you just don't have any other way." Within the fusion community, people often say that if one approach will work, a lot of them will work. In bringing the tokamak or mirror to the level of an ETR, we will learn so much about plasmas and reactors that it will become quite possible to go ahead rapidly with these more compact approaches. All this means that Ashworth may well see his pilot plants after all.

Any fusion reactor will have to be classed as a nuclear reactor. Nevertheless, fusion may offer the nuclear enterprise that rarest of opportunities, a fresh start. Free of public prejudgment, free of the most immediate sources of nuclear risk, and with greatly eased problems of safety and radioactivity, fusion can offer us the chance to launch the nuclear enterprise anew, and this time to do it right. The other side of the coin is that fusion will offer the opportunity to repeat the same kind of mistakes as happened with fission. Indeed, it will take considerable care to keep this from happening. The basic problem with fission was that an inadequately matured technology was rushed into commercial use without proper preparation or development. Fusion offers a golden opportunity to do the same thing anew. Ironically, Kintner might actually have been leading the fusion program in such a direction.

Beginning in 1948, it was Hyman Rickover, Kintner's mentor, who pushed the development of what would become the standard American nuclear reactors. These reactors have very compact cores, which concentrate the afterheat to such a degree that in a core-cooling emergency, the reactor operators may have to respond correctly in less than a minute. These compact cores are a direct legacy of Rickover's atomic submarines, since the cramped space aboard these subs placed a premium on compact design.

The first such subs, the *Nautilus* and *Seawolf,* went to sea in 1954 and 1955. At nearly the same time, President Eisenhower announced a program of "Atoms for Peace," which was to emphasize the development of nuclear power plants. To speed their development, the Atomic Energy Commission undertook a series of Power Reactor Demonstration Programs, offering lucrative subsidies and other financial incentives. The utility industry in the 1950s regarded nuclear power as a risky and economically unattractive prospect, but with the government committed to pushing it, the utilities began to get interested. It was natural that they should turn to Rickover's submarine reactors, which were already being built and with which the navy was already gaining operating experience. Further, the compactness of these reactor cores made them appear cheaper to build, improving the economics. Thus, by the 1960s, when utilities started to order nuclear plants in large numbers, they were completely committed to those compact cores, which, with their concentrated afterheat, would ultimately prove to be a hidden flaw.

There is nothing like a fusion-driven submarine in the offing, to serve as an early but ultimately inappropriate focus for the development of power reactors. But there is plenty of opportunity for impatience. Fusion research has been soaking up government funds for more than thirty years, and is only now about to produce power from its reactions. It will be another twenty years or so before the early experimental fusion reactors will show what they can do. Following this half-century of effort, it will be all too easy for some Washington energy czar to declare that the important thing now is to start getting actual power from fusion, not to continue with additional years of research aimed at building the perfect reactor. This decision will be especially easy if the United States is then once again in the middle of one of its occasional energy crises. And if a too-hasty decision comes down, to standardize a fusion design and to start building, the nascent fusion industry could be forced into design compromises that in time would bring the same grief as has befallen nuclear power.

Against this backdrop, then, the issues over which Kintner resigned take on a different look. The experiences that formed Kintner's outlook were those of his years with Rickover. He wanted very much to be the Rickover of fusion, and was bitterly disappointed when he couldn't be.

But people must be careful what they wish for; their wishes may come true. So it may be with Kintner's approach to fusion. It would certainly have been exciting to rush ahead, under his lead, to carry out the Magnetic Fusion Act. But the long-term prospects for fusion may be better served by emphasizing other, less dramatic themes. To search for high beta, to develop approaches to small and compact fusion reactors that can be built as pilot plants, to work toward designs that are inherently safe and low in radioactivity—all these may do far more for fusion than pushing to build a demo plant by the year 2000, which was the focus of what Kintner was trying to do.

Still, fusion will not be here for some years, and there will be ample opportunity for an extended dialogue that will bring fusion's advantages to the fore. The best fusion designs may offer no afterheat, no long-lived or intense radioactive wastes, no danger of reactor meltdown, and greatly reduced links to nuclear weapons. With all these advantages, fusion should be much easier to build and sell to the public. Moreover, we have all along assumed that when fusion is ready, it will be judged according to the same controversies that today afflict nuclear power. Yet fusion is not an issue for the 1980s. It belongs to the next century. All we will be doing from now to 2000 will be to lay the groundwork. Only later will fusion come into its own.

When fusion is ready, as it will be around that time, it will not be judged by the passions and arguments of the late 1970s. In fact, unless the nuclear industry makes some really bad and deadly mistakes, by then it may be as out-of-date to talk of nuclear matters with heated antinuke rhetoric as it is to talk today of foreign policy issues in the manner of the early 1950s, with heated Cold War rhetoric. In the next century, the real value of fusion will be that we will need it, for we then will be closer to the end of the fossil-fuel era.

By the overwrought and panicky standards of the late 1970s, the fossil-fuel era will take an unconscionably long time a-dying. It may be late in the next century before such petroleum replacements as coal, shale oil, and tar sands finally prove unequal to the challenge, so that fusion and other alternatives must take up the slack. But when that future arrives, we will be glad to know that fusion will be available, and that it indeed will be ready.

In the next century, it will not be sufficient to build a hundred fusion plants, or even several hundred. If fusion is to be at all significant in the energy future of the nation and the world, there will have to be thousands of them, perhaps even ten thousand or more. Many of them will be clustered in great complexes or fusion parks. To a Rip Van Winkle of the late twentieth century, such fusion parks will appear arresting indeed, for they will feature what today is virtually a symbol of nuclear power— cooling towers. These huge structures, hundreds of feet high, will domi- nate the surrounding landscape. In their shadows, the buildings housing the fusion reactors will appear small indeed by comparison. However, these buildings will not be fortresslike domes of concrete and steel. Fusion's greatly eased safety problems will allow them to be built virtually as office buildings, three or four stories tall, with walls of tinted glass. As George Keyworth has stated, such plants may eventually rise to become the dominant energy source, as all alternatives fall by the wayside, on technical grounds, on economic grounds.

In seeking energy supplies for the long-term future, it is common for authors and futurists to call for sources that are cheap, nonpolluting, widely available, environmentally beneficial, inexhaustible, and compat- ible with small-scale use and with preservation of community values. Then after a while, sometimes, they begin to look around to see what is really available. But there are some features that successful energy technologies will share, along with other new technologies. They will be unobtrusive, generally noncontroversial, and, for the most part, just there. They will have much in common with the electric utility industry during its years of rapid growth and expansion in the 1950s and 1960s, when this industry was supplying its product at steady or even falling prices, and the desirability of more electricity was widely taken for granted. These future technologies will also resemble the oil industry during the midcentury decades, which quietly and at steady prices provided the nation with what it needed while staying out of the head- lines, usually. By offering reliable, predictable services in a straightfor- ward fashion, the most successful energy industries will fade into the background, indeed will become nearly invisible. The highest goal for those who would build such industries then would be that their work should be taken almost completely for granted.

Should fusion grow to achieve this, should fusion power prove to be the solution to the world's energy needs, there will certainly not be the ushering-in of a utopia in which the nations will achieve universal prosperity. But in solving an important and pressing concern, fusion will set people and governments free to worry about other things. People are very inventive when it comes to finding things to be unhappy about. If by some magic all economic needs could be met and all questions of wealth and production resolved, the world's peoples would quickly enough come forward with a host of racial, religious, political, and nationalistic problems. But if societies are freed from one set of worries, that is surely a valuable and important thing, even if they soon enough will be embracing a new set.

As we look ahead to fusion's long-term development, there are several trends that stand out as important. To begin, there will be advanced fuels. The least difficult reaction to make work uses deuterium plus tritium, D + T. However, it has several disadvantages. It requires tritium, which is radioactive and which mixes freely in water, thus demanding the strictest of controls. The tritium, in turn, must be produced by irradiating lithium with neutrons, adding an extra degree of complexity. What is most inconvenient of all, however, is that most of the energy from the reaction comes off as highly energetic neutrons. Thus the D + T reaction requires heavy radiation shielding, and its neutrons may quickly make the reactor radioactive. Also, the neutrons must be captured in lithium and the resulting heat transferred by appropriate circulating fluids to boil water and run turbines with the steam. The result is that a D + T fusion plant will require elaborate systems of pipes, valves, and other plumbing. It may be almost as much of a plumber's nightmare as any nuclear plant.

The road to advanced fuels lies in higher plasma temperatures. By everyday standards, the hundred million Celsius degrees of the D + T reaction is already thoroughly preposterous. A sphere only fifteen feet across, heated to that temperature and kept there, would radiate as much energy as the entire sun itself. Yet when we speak of these temperatures, we do not deal with massive material bodies. Instead, we achieve these temperatures in diffuse wisps of plasma held within a vacuum chamber. Thus, such temperatures do not mean that the plasma shines with a power brighter than a thousand suns. It means that the electrons and atomic nuclei in the plasma are moving very rapidly and with great energy, as they whirl around. The actual amount of heat contained within this plasma might be quite insufficient to boil water for your morning coffee. Its fusion reactions produce heat at a rapid rate, but the plasma is so diffuse that it retains very little of this heat.

Therefore, in looking ahead to advanced fuels, we may speak of a need for higher temperatures still; up to a billion degrees, perhaps even more. Even so, this is far from impossible. That is effectively the temperature today of atoms shot within a neutral beam. At such temperatures the next least difficult reaction, after D + T, is the pure deuterium reaction D + D. That gets rid of the tritium. But D + D still produces a copious flow of neutrons, so there still are the problems of radioactivity and radiation shielding.

At one billion degrees the next advanced reaction can light up. It employs D + ^{3}He, helium-3. Helium-3 is chemically identical to the helium in balloons, which is helium-4. The atom of helium-4 has two protons and two neutrons in its nucleus; that of helium-3 has one less neutron. The important advantage of D + ^{3}He is that it releases its energy purely as charged particles. Such particles can be tapped off on electrodes, thus opening the prospect of directly converting fusion energy to electricity. A direct converter would have no complex plumbing or heat-transfer systems, no turbines, no generators. Instead, it would be a simple arrangement of charged electrodes, which would separate, slow down, and collect the particles from the reaction. There would be no moving parts or circulating fluids, and the conversion efficiency would be extraordinarily high. At Livermore, Richard Post has done experiments to tap off particles leaking through the ends of a magnetic mirror, demonstrating direct conversion efficiencies up to 86 percent.

However, even D + ^{3}He has its problems. For one thing, such a mixture might also be running D + D as a side reaction. The deuterium would be there, as would the necessary temperatures and plasma conditions, so this reaction could be hard to avoid. Then D + D, in turn, would produce its share of neutrons, and so D + ^{3}He still would not avoid radioactivity or radiation shielding. This problem may well be solvable, since today there are some interesting ideas on how to suppress this D + D side reaction, while at the same time actually enhancing the power from the D + ^{3}He reactions. But there would still be the problem of getting ^{3}He, which is almost nonexistent in nature.

The quickest way to get it would be by burning D + D, which produces both tritium and helium-3. Moreover, if tritium is set aside and allowed to age, by itself it will produce helium-3. Radioactive tritium has a half-life of only twelve years, and if stored in casks, it decays to that helium isotope. In this "wine cellar" method, there would be stainless-steel barrels filled simply with water—not with ordinary H_2O, however, but rather with tritiated water, which is HTO. Since helium forms no chemical compounds, the helium-3 would simply bubble off to be easily collected, as the tritium decays. But this means that the D + ^{3}He reaction could not stand alone. It would call for a supporting industry to produce either tritium or helium-3, and which could not avoid its own floods of neutrons.

At two billion degrees the ultimate fusion reaction becomes possible. This is hydrogen plus boron. The hydrogen would be ordinary hydrogen, as in the sun or in common water; it would not even have to be deuterium. The boron would be boron-11, the most common isotope, 80.4 percent of all boron. There can be no doubt that we can get all the boron we would ever need from seawater. It exists there at a concentration only eight times smaller than that of deuterium, which we extract today routinely. Boiling the seawater to concentrate its salts into a heavy brine would make boron extraction even easier. Indeed, lithium in the sea is 200 times less abundant than deuterium, yet already lithium is extracted from such heavy brines. The hydrogen-boron reaction also produces only charged particles and opens the way for direct conversion. But the side reactions are virtually nil. Boron does not react with itself, and the hydrogen would do so only at a slow rate like that in the center of the sun. These hydrogen reactions, moreover, produce no neutrons. Boron, then, would allow us to have it all: direct conversion, no neutrons or radioactivity whatsoever, and no supporting industries that themselves produce neutrons. It would indeed give us the ultimate fusion reaction, if it can ever be achieved.

Another significant theme will be the development of entirely new types of fusion reactor, based upon new principles. Such reactors will be noteworthy for being simple and compact. Robert Bussard's Riggatron is certainly one, but it may be only a start. There are a whole range of approaches that, to Harold Furth, "seem to be ordered in terms of increasingly interesting economic potential and diminishing technical evidence." These include the stellarators and the Elmo Bumpy Torus, then range afield to concepts with names like Spheromak, Reversed-Field Pinch, Field-Reversed Mirror, Surmac. In thirty years one of them might be as exciting as tokamaks are today.

The field-reversed mirror is a particularly tantalizing approach. To appreciate how it works, think of a barrel partly filled with water. The water fills the barrel to some depth, but if the water were to spin rapidly, it would reshape itself into a thick hollow cylinder, a layer covering the sides of the barrel, with a hollow cavity in the middle. Field reversal works somewhat similarly. If a blob of plasma is held within a magnetic mirror and is made to spin rapidly, by being hit with neutral beams, then it sets up its own magnetic field, which helps to contain it. It also reshapes itself into a hollow cylinder, or shape of a muff, similar to the spinning

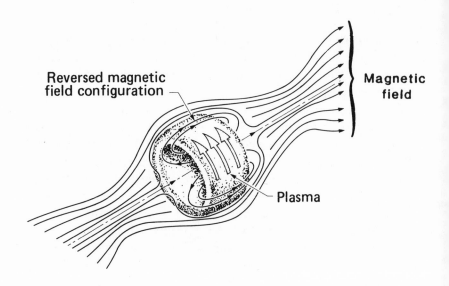

FIGURE *10*. *The field-reversed mirror concept*

water in the barrel. For a given strength of the applied magnetic fields, the plasma in a field-reversed mirror can then be up to forty times denser than in a tokamak. Indeed, the plasma pressure can be twice that of the pressure from the magnetic fields, the plasma's own field contributing the extra pressure that contains it. Then the plasma might be only the size of a beach ball, yet produce tens of megawatts of power.

Always the least costly way to produce energy is with the most compact core possible. It is the diffuse nature of sunlight, the very opposite of compactness, that limits solar energy's prospects. It was the desire for a compact core that led to today's standard designs for nuclear reactors, but this very compactness has greatly exacerbated the issues of reactor afterheat and core cooling, posing difficult safety problems. But in a fusion reactor, particularly one using advanced fuels, the trend of improving compactness by increasing plasma pressure could be pushed to the limit. Like any magnetic mirror, the field-reversed mirror would have the simple tubular shape that is so much easier to work with than any kind of toroid. It would also share the adaptability of all mirrors to direct energy conversion, and it could achieve billion-degree temperatures. What would be better still, its use of field reversal would allow it to dispense with the complex and costly arrays of magnets that the people at Livermore today believe will be necessary to seal off the ends of their mirrors. And its high plasma pressure, higher even than in a standard mirror, would make the most of the magnet arrays that it would still need.

Is there more to the field-reversed mirror than hopes and theories? The concept grew out of the success that Fred Coensgen achieved at Livermore in 1975 with the small 2XIIB mirror. That success put Livermore in line to build the very large MFTF mirror as a scaled-up version of 2XIIB. But first there was the problem of Q-enhancement, of making mirrors less leaky so they would work well enough to make good fusion reactors. In the end, Grant Logan and Ken Fowler solved that problem by coming up with the tandem mirror. But, for a while, it looked as if the solution would lie in Coensgen's invention, the field-reversed mirror. The key to making it work would be in having enough neutral-beam power on 2XIIB, which Coensgen was using to test his idea. He got close to making it work, but didn't have enough beam power. He would have needed 36 beams on 2XIIB, but had only 12. It had been hard enough to get those 12 to work, not to mention fitting them all into position, and 36 was simply too many. Nevertheless, the field-reversed mirror looked so promising that for a while Fowler was talking about carrying forward with both the tandem and field-reversed mirrors, as parallel programs. The 1978 Foster Committee's report praised both types of mirror equally as "ingenious ideas." Still, a choice had to be made, and the tandem mirror won out as being more straightforward and less risky. The field-reversed mirror went on the shelf, to await some more propitious day. But the concept is still there, and still tantalizes fusion researchers. Indeed, at Pacific Gas and Electric, it is the approach that Clinton Ashworth likes best and has emphasized in his own studies.

And beyond such ideas as these, it may be that people will invent entirely new approaches, based on principles that today receive little attention. Indeed, no less a figure than Robert Hirsch started his fusion career with such an approach. After getting his Ph.D. in 1964, he went to work in Fort Wayne, Indiana, along with Philo T. Farnsworth, the coinventor of television. They were trying to achieve controlled fusion power, but not by confining a plasma with magnetic fields as everyone else was doing. Instead, they wanted to use electric fields, which to a physicist are very different and are handled in entirely dissimilar ways. In a physics world dominated by magnetic fusion and soon to see the rise of laser fusion, they sought a third approach: electric fusion.

Farnsworth and Hirsch did calculations and predicted that this approach might work. Each of them built a series of small experimental devices to produce neutrons by burning D + T. They had funding from International Telephone and Telegraph, and from time to time someone from ITT would come round and see how they were doing. Hirsch liked to greet them with what he called "political neutrons." These visitors weren't up on the technical fine points, but they knew how to read a neutron counter. More neutrons meant that Hirsch and Farnsworth were making progress. Soon Hirsch's experiments were outstripping Farnsworth's, and he was able to get a tenfold increase in his neutron production, from one year to the next. Eventually his device was putting out 10^{10} fusion neutrons per second, enough so that he had to put radiation shielding around the device. Nor was he running his tests for a few milliseconds at a time, like his magnetic-fusion colleagues. He could keep on churning out neutrons for up to several minutes, and the only reason he couldn't run the device all day long was that after those few minutes it would start to overheat. It would be years before anyone else could match his neutron output or do better at producing lots of neutrons with relatively little power.

Three years after Hirsch got to Fort Wayne, the ITT people realized that fusion development was going to take a long time, and they asked Hirsch to try to get funding from the Atomic Energy Commission. After much debate back and forth with a review panel, the people in the fusion office invited him to come in to meet that panel and be grilled by them. To add to the impact of his written proposal, Hirsch decided to bring along a small-scale version of his device, which he would demonstrate. This was not something he could just bring in as he might wish, because it would need tritium to make neutrons. Bringing radioactive tritium into the downtown offices of the AEC was on a par with bringing a loaded rifle into the Pentagon, but eventually he got his permission. His device was about the size of a large grapefruit, and he could wheel it in on a table as if it were a cake on a dessert cart. It plugged into an ordinary 110-volt socket. It had a window in the side, and you could look in and see the bright reddish-purple glow of plasma, concentrated into a little ball at the center. Each second it produced a hundred thousand neutrons from fusion reactions. It would do this continually, for as long as Hirsch cared to keep it on.

The elders of fusiondom stroked their long gray beards and wondered what it might mean, as they looked on. Then these elders, the review committee, started asking Hirsch some questions. But he had come prepared, and he had good answers. Finally the time came to decide: would they or wouldn't they approve his getting the money? Then Tom Stix, a plasma physicist from Princeton, asked the most pertinent question of all: "If we fund this, whose budget is it going to come out of?" There was no extra funding available, and money allocated to Hirsch in Fort Wayne would have to be taken away from somebody else at Princeton or some other well-established lab. And right there and then, Hirsch's hopes for funding died.

There was simply no institutional base for electric fusion. No one else was pursuing it; no powerful lab or agency favored it. It had come out of left field from two off-the-wall inventors who worked at the wrong address, and there was no way to fit it into the established fusion program. It was not that people weren't looking for new ideas; within a year or two the tokamak would come along, and soon everyone would be hopping on its bandwagon. But electric fusion was a little too new. The response Hirsch got in 1968 was very similar to the response Bussard got, a few years later, when he sought government support for his Riggatrons.

Came the early seventies, and Hirsch was in charge of the entire fusion program. Did he now see a chance to use his position to do something for electric fusion? He did. However, he couldn't just decree, "Let there be an electric fusion program." He had to find physicists and labs who were interested, and who could prepare solid proposals to show that they might be able to get somewhere. No one was. The whole fusion world was wrapped up in magnets and lasers, and no one was out there to pick up his lead. It all was a most ironic turnabout on the cliché of the brilliant inventor, years ahead of his time, who dies penniless and alone, with no one interested in his pathbreaking invention. Farnsworth was just such a brilliant inventor, as was Hirsch, who had shared his art. Hirsch rose to lead the entire fusion program—and *still* no one was interested in what he was offering.

Even today, there is no way to know what promise may lie in electric fusion. It stands yet as one of the great might-be-somedays, like laser fusion, like the field-reversed mirror, like the ideas of direct conversion and advanced fuels, like the Riggatrons. Yet all these concepts, and their associated themes and trends, represent paths for inventiveness and ingenuity. As we pursue these paths, fusion will grow and evolve into something that today we can no more than dimly foresee.

Carried to their conclusions, then, these trends could mean that the fusion reactor of the future would fit into a household basement—maybe a large basement—yet would produce enough electricity to run Manhattan's World Trade Center. It would be quiet, simple, and efficient, and might indeed function largely without moving parts. Nor would it have radioactivity, elaborate cooling systems, turbogenerators, massive radiation shields, or any of the other baggage that today we associate with nuclear power. Such devices would fulfill the prophecy of H.G. Wells, of a world set free. They would be the universal small, neat, compact energy source. They would be flexible enough to be made to release their energy as electricity, as microwaves, as X rays or ultraviolet, as charged particles, or, when desired, as neutrons. They thus could serve to produce hydrogen from water, as an inexhaustible fuel resembling natural gas. Or their energy could be used to produce synfuels. They could power airplanes and ships. For thousands of years past, seafarers carried no fuel but relied upon the wind in the air. For thousands of years to come, mariners may again carry no fuel, but may rely upon deuterium taken from the ocean.

As Harold Furth has stated, "The prospect of adapting fusion power perfectly to the needs of society and the environment does not appear to be limited by any narrow physical constraints, only by the incompleteness of scientific knowledge and limits of technological imagination." The central problem in fusion today is plasma physics, and when we have mastered the techniques of plasma production, heating, and confinement, then the plasma will become simple. The design of new fusion devices will become straightforward. In this respect, fusion today is like aviation earlier in this century, and the rise of the fusion reactor is akin to the development of the jet engine. To enter the jet age it was necessary to understand the flow and behavior of air and hot gases within a turbojet, and to have metals that would stand up to jet heat. To enter the fusion age it is necessary to understand the flow and behavior of ionized gases within a magnetic field, and to have materials that will stand up to fusion neutrons.

In seeking insight into the future of fusion, aviation indeed provides a most useful metaphor. Within the topic of atomic energy there are fission and fusion, which will bear comparing to two approaches to aviation: lighter-than-air flight and heavier-than-air. Both date to before 1800, and like fusion and fission in the 1930s, for a time they developed almost side by side. Like fission, it was the balloon that gained the dramatic, early successes. The brothers Montgolfier had their hot-air balloon in 1783, and it was quickly followed by hydrogen balloons and cross-Channel flights.

In heavier-than-air flight, Sir George Cayley was the prophet of his day, breaking with four centuries of attempts to invent machines that would fly by flapping their wings. In 1799, to quote the historian Charles H. Gibbs-Smith, "he arrived at a correct and mature conception of the modern aeroplane." It had fixed wings, fuselage, a tail-unit of elevator and rudder, and a propulsion system separate from the rest of the aircraft as a whole. His papers of 1809–1810, "On Aerial Navigation," laid the foundations for the modern sciences of aerodynamics and flight control. But to achieve his vision was not easy, for it called for vast improvements not only in engines, but in aircraft stability and control. (Always the difficult technical problems have involved stability and control.) As a young man he had been inspired by the exploits of the balloonists. In his later years he would inspire such friends as William S. Henson, whose 1843 design for an "Aerial Steam Carriage" today looks like a Victorian's prediction of a twin-engine high-wing monoplane out of the 1930s. But true flight, like fusion, was an intrinsically recalcitrant problem, and like fusion was long delayed.

Pursuing this metaphor, we might then say that today's proud to-kamaks will be tomorrow's version of the biplane. They will serve to establish fusion as a force in being, and to demonstrate its promise for all to see. But the tokamaks may in mere decades be superseded by such advanced approaches as the field-reversed mirror. Thus far, the story of fusion has largely been one of dedicated experts persevering through the most chilling difficulties, at last winning through to the promise of success, and to their just reward. But in the next century, this history may be largely forgotten. The brief spurts of barely controlled fusion which we anticipate in the 1980s may by then loom no larger than the brief spurts of barely controlled flight that Hiram Maxim, Samuel Langley, and other steam-powered aviators achieved in the 1890s. For most people in this century, aviation began with the Wright brothers. For most people in the next century, fusion perhaps will have begun with John Clarke's Engineering Test Reactor, with Japan's Fusion Experimental Reactor—or, if Bussard is lucky, with his Riggatrons.

Pursuing the aviation metaphor even farther, we may recall the state of aviation in 1940, after a century and a half of aeronautical progress. Lindbergh had flown the Atlantic. The propeller and piston ruled the skies; this would be literally true in the coming war. Passenger service was spreading in all directions, with the DC-3 in this country, with the Clipper flying boats across the oceans. The leaders of aviation could well be proud. Yet in a few short decades, all this would be swept away.

For what would the aviation leaders of 1940 have made of the Boeing 747? Or of nonstop jet service, New York to Tokyo? (Not till 1957 did the DC-7 offer nonstop service, New York to Los Angeles!) Or airports like Dallas–Fort Worth and Narita near Tokyo, sprawling for miles and annually serving more people than all of 1940's ships out of New York harbor? Or air-traffic control systems with radar and computers? For all that existed in aviation in 1940, its true story did not even begin until the war. Similarly, the fusion advances that may unfold within our lifetimes may be no more than a prelude to fusion's true promise, which may not even begin until new and hardly-dreamed-of technologies take shape, in a century or more.

Fusion's initial growth may take place against the background of a worldwide nuclear enterprise vigorously expanding, valued and relied on by dozens of nations. Today such countries as France, Japan, Great Britain, and the Soviet Union are pushing forward with major programs of reactor construction. France in particular leads the world today in the development of the fusion breeder's rival, the breeder reactor. The energy analyst Wolf Häfele has written that "everything the fusion reactor is supposed to be able to do someday, the breeder can already do today." Yet fusion will be a vast improvement over the breeder. With all its potential advantages, it will be no surprise if fusion ultimately drives the world's nuclear industries into decline. The day may come when these industries will drift into obsolescence, unable to compete with fusion's burgeoning achievements. Beset by high costs, canceled orders, and a general sense of having passed their prime, they may come to fit the description by the energy critic Amory Lovins of the United States nuclear industry even before Three Mile Island. He said it was like a dying brontosaurus: its spine was broken but its head had not yet heard the news.

And if this is so, then another metaphor will bear on fusion's role in the coming energy transitions, as we pursue the long-drawn-out shift from nuclear and fossil fuels to energy sources which are permanent, renewable, and inexhaustible. We may compare this to another transition, which took place in the world of navigation, as seaborne commerce shifted from sail to steam.

One of the first commercially successful steamships was Robert Fulton's *Clermont* of 1803. His achievement had almost zero immediate impact. Thus, in 1805 the British fleet of Admiral Nelson smashed Napoleon's fleet at Trafalgar; but when England thus won unchallenged mastery of the sea for over a century, it was with wooden sailing ships blasting away at the enemy barely a spar's length away, with broadsides of solid iron cannonballs. As the century progressed so did the steamship, and with advancing technology it gained its place upon the oceans. But this same advancing technology also gave new life to sail. In mid-century the Yankee clippers were the wonders of the sea. And if there would soon be steel hulls for steamships, there would also be masts and hulls of steel for the clippers' successors, the great windjammers. The steamers did gain headway, but slowly, slowly; not till 1883 did Lloyd's of London register a greater tonnage of steam than sail. Even past the turn of the century such seafaring nations as France and Germany maintained a balance between the two, and with little wonder. For the best windjammers could fly at fourteen knots and more, with hulls four hundred feet long.

This state of affairs could well have persisted but for a succession of inventions, around the year 1900, which gave the advantage to steam. The turbine engine offered unprecedented power and speed. In no more than a few years it brought on the great North Atlantic liners, as well as a new generation of naval dreadnoughts. In 1916 at Jutland, for the first time since Trafalgar, men of the royal Navy found foemen worthy of their steel; but both the British and the Germans had modern battleships and battle-cruisers with armored hulls and turrets, turbine engines, and long-range guns firing twelve-inch shells. There was not a sail in sight. Elsewhere on the high seas, the U-boat took its toll, powered by another recent invention, the diesel engine. The sailing ship in commercial service did not die a natural death due to technical or economic imperatives. It was sunk en masse by the submarine.

Yet even after that war, the cargo-carrying windjammer still proudly spread its sails. Gustav Erikson of Finland owned and operated some twenty such ships in the 1920s and 1930s, as history's last great deepwater sailing fleet. His ships competed successfully with steam in the Australian grain trade, and at times they did more. In October 1934, in the South Atlantic, his *Herzogin Cecilie* outraced a British passenger liner at eighteen knots. Not till after World War II did the windjammers enter their terminal decline. The last of them in commercial trade was a four-masted iron bark that ran guano down the coast of Chile. She sank in the Pacific on June 26, 1958. By then the modern supertanker with its six-digit tonnage was already looming in the distance. On the North Atlantic the graceful luxury liners were themselves in decline, as the Boeing 707 whispered overhead. The name of that guano boat was appropriate: *Omega,* signifying the end.

This rich and redolent marine history will in time be matched by an equally compelling energy history, as its centuries-long transitions unfold. The first fusion plants will be dramatic, but in the world's immediate energy picture they will mean no more than did the *Clermont* against the backdrop of Trafalgar. They will go forward and win their place. Still, if advancing technology will give them advantage, it will also bring new advantage to tar sands, oil shale, and synthetic fuels, as well as to nuclear power. In the end, it may fall out that the capacity for further innovation in fossil fuels or nuclear energy will largely exhaust itself, while fusion, still fresh and youthful, will continue to innovate. Then, but only then, fusion may race ahead to a commanding advantage. Perhaps a few elements of a fossil-fuels industry will linger, long past their prime, still holding on in some specialized niche reminiscent of the Australian grain trade, still showing a spark of life as a fossil *Herzogin Cecilie* momentarily shows its stern to a fusion passenger liner. And in the end, when the last fossil *Omega* sinks into its Pacific, what energy technologies will still be vigorously advancing to meet new challenges? Which ones will be waning, retreating in the face of competition from alternatives that we today do not even dimly foresee?

Fusion is unquestionably one of the key technologies that will shape the coming millennium. Today we see it as a man-made sun about to rise; tomorrow we will stand in the radiance of its bright promise. Those who say we are in the sunset of our age are surely mistaken, but they can be forgiven; early dawn may look much the same as dusk. We stand today amid the landscape of the future, but we do not perceive it in detail. There are only vague forms and shadows, some of which appear as looming threats. But let us be patient; morning is at hand, and the landscape in all its intricacy will soon be disclosed. For now it is enough to know that that sun is there, its rise appears imminent, and if we look closely we can even now see its glow reflecting off the distant clouds near the horizon.

Bibliography

Allen, Oliver E. *The Windjammers*. Alexandria, Virginia: Time-Life Books, 1978.

Angier, Natalie. "Texas Recruits a Science Team," *Discover*, January 1983, pp. 24–30.

Bromberg, Joan L. *Fusion: Science, Politics, and the Invention of a New Energy Source*. Cambridge, Massachusetts: MIT Press, 1982.

Chen, Francis F. "Alternate Concepts in Magnetic Fusion," *Physics Today*, May 1979, pp. 36–42.

Cole, K.C. "Interview: Robert Bussard," *Omni*, January 1981, pp. 56–58, 90–92.

Cook, Earl. "The Flow of Energy in an Industrial Society," *Scientific American*, September 1971, pp. 134–144.

Freeman, Marsha. "Japan's Ambitious Fusion Program," *Fusion*, August 1981, pp. 24–31.

Furth, Harold P. "Progress toward a Tokamak Fusion Reactor," *Scientific American*, August 1979, pp. 50–61.

Gibbs-Smith, Charles H. *The Invention of the Airplane 1799–1909*. New York: Taplinger Press, 1965.

Gibson, Robert. "A Nuclear Bombshell That Is Killing the Industry," *Maclean's*, October 30, 1978, pp. 47–48.

Gillette, Robert. "Laser Fusion: An Energy Option, but Weapons Simulation Is First," *Science*, April 4, 1975, pp. 30–34.

Greenberg, Daniel S. "Fusion Politics," *Omni*, January 1981, pp. 52–54, 109.

Grisham, L.R. "Neutral Beam Heating in the Princeton Large Torus," *Science,* March 21, 1980, pp. 1301–1309.

Häfele, Wolf. "A Global and Long-Range Picture of Energy Developments," *Science,* July 4, 1980, pp. 174–182.

Hayes, Earl T. "Energy Resources Available to the United States, 1985 to 2000," *Science,* January 19, 1979, pp. 233–239.

Hecht, Marjorie. "The Princeton Story," *Fusion,* October 1978, pp. 18–23.

Heppenheimer, T.A. *Toward Distant Suns.* Harrisburg, Pennsylvania: Stackpole, 1979.

———. *The Real Future.* Garden City, New York: Doubleday, 1983.

Hogerton, John F. "The Arrival of Nuclear Power," *Scientific American,* February 1968, pp. 21–31.

Holdren, John P. "Fusion Energy in Context: Its Fitness for the Long Term," *Science,* April 14, 1978, pp. 168–180.

Hyde, Roderick, Lowell Wood, and John Nuckolls. "Prospects for Rocket Propulsion with Laser-Induced Fusion Microexplosions," AIAA Paper 72–1063, American Instititute of Aeronautics and Astronautics, New York.

Jefferson, Pat, and Johan Benson. "Washington Scene," *Astronautics and Aeronautics,* September 1982, pp. 10–12.

Judson, Horace F. *The Eighth Day of Creation.* New York: Simon and Schuster, 1979.

Kintner, Edwin E. "The Artsimovich Memorial Lecture," *Fusion,* October 1978, pp. 58–61.

Lubin, Moshe J., and Arthur P. Fraas. "Fusion by Laser," *Scientific American,* June 1971, pp. 21–33.

Lubkin, G.B. "Four New Tokamaks Will Each Try for a Finite Power Output," *Physics Today,* January 1978, pp. 17–20.

Metz, William D. "Nuclear Fusion: The Next Big Step Will Be a Tokamak," *Science,* February 7, 1975, pp. 421–423.

———. "MIT Chemist, Schlesinger Ally Assumes Energy Research Post," *Science,* December 16, 1977, pp. 1125–1126.

———. "Report of Fusion Breakthrough Proves to Be a Media Event," *Science,* September 1, 1978, pp. 792–794.

———. "Ambitious Energy Project Loses Luster," *Science,* May 1, 1981, pp. 517–519.

Morland, Howard. "The H-Bomb Secret," *The Progressive,* November 1979, pp. 14–23.

Nuckolls, John, Lowell Wood, Albert Thiessen, and George Zimmermann. "Laser Compression of Matter to Super-High Densities: Thermonuclear (CTR) Applications," *Nature*, September 15, 1972, pp. 139–142.

Nulty, Peter. "Good News from the Oil and Gas Hunt," *Fortune*, November 3, 1980, pp. 88–98.

"Reactor Power," *Scientific American*, July 1983, p. 60.

Robinson, Arthur L. "Fusion Energy in Our Time," *Science*, February 8, 1980, pp.622–624.

Schwarzschild, B.M. "Garching Shows Stellarators May Be Good after All," *Physics Today*, August 1980, pp. 17–19.

———. "Polarized Plasmas May Prove Useful for Fusion Reactors," *Physics Today*, August 1982, pp. 17–19.

Smith, A.C., Jr., K.E. Abreu, C.P. Ashworth, D.K. Bhadra, G.A. Carlson, E.T. Cheng, R.L. Creedon, H.H. Fleischmann, W. Grossman, Jr., S. Hanusiak, T. Kammash, C.E. Kessel, Jr., C.H. Lien, W.S. Neef, Jr., D.J. Rej, K.R. Schultz, D.M. Woodall, and C.P.C. Wong. *The Moving-Ring Field-Reversed Mirror Reactor: Annual Report Executive Summary, 1979–1980.* Research Report 922, Electric Power Research Institute. San Francisco: Pacific Gas and Electric Company, March 4, 1981.

Smith, R. Jeffrey. "Legislators Accept Fast-Paced Fusion Program," *Science*, October 17, 1980, pp. 290–291.

———. "Photovoltaics," *Science*, June 26, 1981, pp. 1472–1478.

Terkel, Studs. *Working*. New York: Pantheon, 1972.

"Up from the Elevator," *Time*, March 20, 1959, pp. 52–53.

Waldrop, M. Mitchell. "Rethinking the Future of Magnetic Fusion," *Science*, September 24, 1982, pp. 1235–1236.

———. "Princeton Physicists Meet Tokamak Deadline," *Science*, January 14, 1983, pp. 152–153.

———. "Compact Fusion: Small Is Beautiful," *Science*, January 14, 1983, pp. 154–156.

Wheaton, Bruce. "The Establishing of Television," *Science*, May 20, 1983, pp. 819–821.

Wilson, Carroll L. "Nuclear Energy: What Went Wrong," *Bulletin of the Atomic Scientists*, June 1979, pp. 13–17.

"The World Reaction to the Princeton Breakthrough," *Fusion*, October 1978, pp. 54–57.

Wyman, Robert H. "MFTF Supervisory Control System," *Energy and Technology Review*, December 1980, pp. 27–33.

Notes

This book has been written largely on the basis of interviews with principals. Wherever the text includes quotes from an individual listed among the interviewees, or refers in an obvious way to such an individual's activities, the reader should assume the source is an interview with that person. References to published sources are cited below, by the author's surname (see Bibliography), together with further information on unpublished or limited-circulation documents cited or quoted in the text.

CHAPTER 1.

Interviews: Kees Bol, Dale Meade
p.11. Oil and gas in the U.S. Cook
12. Midcentury energy prices. Hayes
12–13. This discussion of energy prospects parallels Heppenheimer, 1983
13. Photovoltaics. Smith, 1981
26–27. Space Shuttle Main Engine. Heppenheimer, 1979

CHAPTER 2.

Interviews: Stephen Dean, Harold Furth, Don Grove, Robert Hirsch, Robert Papsco, Paul Reardon

p.33. 1974 Albuquerque meeting. Metz, 1975
34–35. Inventors of television. Wheaton
36. Tokamaks at Culham in 1965. Bromberg
38. Artsimovich. Kintner
39–41. Doubts about tokamaks. Bromberg
43. Budget numbers. Fusion Power Associates, Gaithersburg, Maryland:
newsletters, June 1, 1982, and February 10, 1983
43–47. Hirsch pushes for TFTR. Bromberg

CHAPTER 3.

Interviews: Kees Bol, Don Grove, Dale Meade, Robert Papsco, Paul
Reardon, Lyman Spitzer
p.64. First plasma. Waldrop 1983a
71. Juan Perón. Bromberg

CHAPTER 4.

Interviews: Fred Coensgen, Ken Fowler, Robert Hirsch, Ted Kozman,
Grant Logan, Richard F. Post, Tom Simonen, Keith Thomasson, Robert Wyman
pp.74–75. Hirsch and Astron. Bromberg
75. Christofilos. "Up from the Elevator"
101–102. MFTF-B control system. Wyman

CHAPTER 5.

Interviews: Hal Ahlstrom, Judy Francis, Alexander Glass, Roderick
Hyde, John Nuckolls, David Solomon, Lowell Wood
Correspondence: John Holzrichter, Ray Kidder, John Nuckolls
p.111. KMS Fusion. Bromberg
117. H-bomb physics. Morland
121. Moshe Lubin's article. Lubin and Fraas
126. Hyde's New Orleans paper. Hyde, Wood, and Nuckolls
128. Cambridge University. Judson
128. *Nature* article. Nuckolls, Wood, Thiessen, and Zimmermann

CHAPTER 6.

Interviews: Hal Ahlstrom, Victor George, Alexander Glass, Roderick Hyde, Michael Monsler, John Nuckolls, Lowell Wood

Correspondence: Stephen Bodner, John Holzrichter, John Nuckolls

p.142. The cited "Laser Program Annual Report" documents are UCRL–50021–xx, where xx is the year; e.g., the 1977 report is UCRL–50021–77.

143. Nuckolls' statement on "approaching breakeven" is from the abstract of an invited review paper presented on November 8, 1977, at the 19th Annual Meeting of the Division of Plasma Physics, American Physical Society.

143–144. Metz on laser fusion. Metz 1981

152. Underground nuclear testing. Gillette

158. Budget numbers. Fusion Power Associates, Gaithersburg, Maryland: newsletters, June 1, 1982, and July 1983.

CHAPTER 7.

Interviews: Kees Bol, Robert Bussard, Stephen Dean, Bruno Coppi, Mike McCormack, Ron Parker, Don Repici, Marshall Rosenbluth, Ramy Shanny, Carl Weggel

pp.165–166. Alcator. Bromberg

172. Bussard's second review. Cole

176. Bob Guccione. "The Forbes Four Hundred," Forbes, September 13, 1982

180. University of Texas. Angier

CHAPTER 8.

Interviews: Kees Bol, Gail Bradshaw, Stephen Dean, Tony DeMeo, Harold Eubank, Harold Furth, Mel Gottlieb, Robert Hirsch, Ed Kintner

p.193. Fusion budgets. Bromberg

196. George Allen quote. Terkel

197. Graphite limiters. Furth

198. 1977 energy politics. Bromberg

198–199. John Deutch. Metz 1977

201–202. Foster Committee report. The document is DOE/ER–0008, June 1978. *See also* Greenberg

202–203. Consequences of the Foster report. Bromberg

206–208. News of PLT breakthrough. Metz 1978; Hecht

208. News conference. Metz 1978

214. World press coverage. "The World Reaction ..."

215. Technical details of the PLT achievement. Grisham

215–216. Deutch's fusion policy. Bromberg

CHAPTER 9.

Interviews: Harold Furth, Robert Hirsch, Ed Kintner, Mike McCormack, Doug Pewitt, Don Repici

Correspondence: Ed Kintner, Mike McCormack

pp.219–220. Hirsch's advisory committee. McCormack letter to Hirsch, March 8, 1979; Robinson; Greenberg

221. McCormack's "Dear Colleague" letter. Date was January 31, 1980.

222. Politics of fusion. Smith 1980

224. McCormack's letter to Eizenstat and Deutch's "Early Warning" memo were both dated December 31, 1979.

229. McCormack's "Dear Colleague" letter. Date was April 25, 1980.

230. Buchsbaum Committee report. The Buchsbaum Committee was formally known as the Fusion Review Panel of the Energy Research Advisory Board.

233. McCormack's House speech. *Congressional Record,* June 24, 1980, p. H 5563

233. McCormack's letter to O'Neill. Date was July 1, 1980.

235. The House bill was H.R. 6308. The Senate bill was S. 2926. The final legislation was Public Law 96–386, 94 Stat. 1539.

236–237. Duncan's letter to the President was drafted by Frieman and was attached as an enclosure with his Action Memorandum of Sept. 9, 1980.

CHAPTER 10.

Interviews: Stephen Bodner, John Clarke, Robert Conn, Stephen Dean, Harold Furth, Ed Kintner, Ron Parker, Doug Pewitt, Don Repici

Correspondence: Stephen Bodner, John Clarke, Ed Kintner

p.241. Elmo Bumpy Torus. Waldrop 1982

249. Romatowski's memo. Romatowski was Acting Under Secretary. The memo was his Fiscal Year 1983 Planning, Programming and Budgeting decision paper.

243–244. Michael Halbouty. Nulty
255. Stellarators. Schwarzschild 1980

CHAPTER 11.

Interviews: Charles Baker, John Clark, Fred Coensgen, Robert Conn, Catherine Fiore, Ken Fowler, Harold Furth, Arthur Kantrowitz, George Keyworth, Dale Meade, Shigeru Mori, Ron Parker, Sebastian Pease, Richard F. Post, Marshall Rosenbluth, Stephen Wolfe, Hans-Otto Wüster
p.281. Lev Artsimovich quote. Kintner
282. International tokamaks. Lubkin
283–284. Japanese fusion program. Freeman; Waldrop 1983a
291–292. Keyworth's views on fusion. Jefferson and Benson

CHAPTER 12.

Interviews: Clinton Ashworth, Hans Bethe, Fred Coensgen, Robert Conn, Harold Furth, Robert Hirsch, Ralph Moir, Richard F. Post, Marshall Rosenbluth, Theodore Taylor
Correspondence: Robert Conn
p.295. Nuclear reactor numbers. "Reactor Power"; Häfele
296. The cited Clinch River document is CRBRP–PMC–82–3.
304. Meltdown and coolant cut-off times. Smith, Jr., et al.
306–307. Compact fusion. Chen; Waldrop 1983b
307–308. Early history of nuclear power. Hogerton; Wilson
312–313. Advanced fusion reactions. Chen; Holdren; Schwarzschild 1982
313. Furth quote. Furth
319. Aviation metaphor. Gibbs-Smith
321. Amory Lovins quote. Gibson
321–322. Windjammers. Allen

Credits for Illustrations

ILLUSTRATION 1

ILLUSTRATIONS 2, 3, 4, 6, 7, 8, 23, 24, 25, 26, 27, 28, 29, 30, 31, 33, 34, and 36 and Figures 4, 5, 6, and 10

ILLUSTRATION 5

ILLUSTRATION 9 and Figure 1

ILLUSTRATIONS 10, 11, 16, 17, 18, 19, 20, 21, and 22 and Figure 3

ILLUSTRATION 12

ILLUSTRATIONS 13 and 14

ILLUSTRATION 15

ILLUSTRATION 32

ILLUSTRATION 35

Courtesy of Atlantic Richfield Company

Courtesy of Lawrence Livermore National Laboratory

Courtesy of KMS Fusion

Courtesy of Inesco, Inc.

Courtesy of Princeton Plasma Physics Laboratory

Courtesy of McCormack Associates, Inc.

Courtesy of the White House, Washington

Courtesy of the Department of Energy

Courtesy of Los Alamos National Laboratory

Courtesy of Argonne National Laboratory

Glossary

Activation. The property of being made radioactive by irradiation with neutrons.

AEC. Atomic Energy Commission; the federal agency that sponsored research on atomic energy, including fusion, prior to 1975.

Alcator. A small experimental tokamak built at MIT in the early 1970s.

Argus. A two-arm laser built at Livermore in the mid-1970s. Its arms were of the same design as the arms used in the much larger Shiva laser.

Arm. A train of laser amplifiers, mounted in tandem, together with their associated lenses and optical filters, used to step up the power of a laser pulse.

ASDEX. Axisymmetric Divertor Experiment; a large German-built technically advanced tokamak, which in 1982 demonstrated particularly good plasma confinement.

Beta. The ratio of plasma pressure to the confining pressure exerted by a magnetic field. High beta is desirable since it permits the use of less powerful magnets and leads to a smaller, more compact fusion device.

Bohm diffusion. A particularly rapid form of plasma leakage that plagued fusion experiments during the 1960s, and that resisted being understood and controlled. Named for the physicist David Bohm.

Breakeven. In laser fusion, a condition in which a pellet undergoes a microexplosion and releases as much fusion energy as there had been in the laser pulse that struck the pellet and touched off that microexplosion.

Breeder reactor. A nuclear reactor that produces a particularly intense flow of neutrons and that can use these neutrons to produce more nuclear fuel than it needs to keep itself running.

Budget examiner. In the Office of Management and Budget, a low-ranking official charged with reviewing budgets and recommending that they be cut or increased.

Caltech. California Institute of Technology.

Coil winder. A large machine used to wind electric cable around a central form so as to construct an electromagnet.

333

Confinement time. The time required for plasma to leak away from its confining magnetic fields.

Density. The number of plasma particles (ions and electrons) per cubic centimeter of volume.

Deuterium. A heavy isotope of hydrogen. Ordinary hydrogen has an atomic nucleus containing a single proton only; in deuterium the nucleus has one neutron, in addition to the proton.

Diagnostics. Instruments used to diagnose or observe the state of a plasma.

Divertor. A special set of magnetic coils used to set up an auxiliary magnetic field in a fusion device, for the purpose of removing plasma impurities.

DOE. Department of Energy; the federal cabinet-level agency responsible since 1977 for energy matters, including fusion.

Doublet III. A large tokamak at GA Technologies, Inc. (formerly General Atomic Company), near San Diego, California, used for studies on means to increase beta in tokamaks.

D + T. Deuterium plus tritium; the fusion fuels most readily used. Also, DT.

EBT. Elmo Bumpy Torus; an experimental fusion device in the shape of a torus, which may offer higher beta than a tokamak and which is favored by some researchers.

Electron. A very lightweight and mobile particle with a negative electric charge, which orbits around the nucleus within an atom.

ERDA. Energy Research and Development Administration; from 1975 to 1977 the federal agency responsible for energy matters.

ETR. Engineering Test Reactor; a very large power-producing fusion reactor which may be built at Oak Ridge National Laboratory after 1990, following a competition between tokamaks and mirrors.

FED. Fusion Engineering Device; a large power-producing tokamak, intended to be the centerpiece of the U.S. fusion program in the 1980s. It was killed by budget cuts in 1981.

FER. Fusion Experimental Reactor; a large power-producing tokamak that the Japanese propose to build in the 1990s.

Fission. The splitting of heavy atomic nuclei, resulting in release of atomic energy.

Fusion. The collision of light atomic nuclei, resulting in release of atomic energy.

Ignition. A condition in which a plasma produces so much energy from fusion reactions that it needs no auxiliary heating from outside the plasma.

Impurities. Any atoms in a plasma other than those of the fusion fuel. Impurities must be controlled because they cause the plasma to lose energy.

Ion. An atom that has lost one or more of its electrons, thus gaining a positive electric charge.

Ionization. The process of causing atoms to lose some or all of their electrons.

Isotope. A form of a chemical element. Different isotopes of the same element have virtually identical chemical properties, but differ in their nuclear properties. Their atomic nuclei have the same number of protons but differ in the number of their neutrons.

JET. Joint European Torus; a large tokamak at Culham Laboratories, United Kingdom, which is to use DT to produce fusion energy by about 1988.

JT-60. A large Japanese tokamak currently nearing completion, to be used for research in plasma physics. JT-60 will not use DT or produce fusion energy.

Kilojoule. A measure of the energy in a laser pulse. A kilojoule is the energy of a one-pound weight at 150 miles per hour.

Kilovolt. 11.6 million Celsius degrees, or 20.9 million Fahrenheit degrees. The kilovolt is a measure of plasma temperature.

KMS Fusion, Inc. A small firm in Ann Arbor, Michigan, which pioneered in the field of laser fusion.

Laser. An optical device capable of producing a highly intense, well-focused beam of light.

Laser amplifier. An optical device used to boost the power in a laser beam or in the pulse of light emitted by a laser.

Laser fusion. A method of producing fusion energy wherein a pellet of fusion fuel is hit with a powerful burst of laser light, which heats and compresses the inner regions of the pellet.

LASNEX. A computer program used to simulate the physical effects taking place within a pellet during a laser fusion experiment, when the pellet is hit with a pulse of energy from a laser.

Lawson criterion. The requirement that $n\tau$, the product of plasma density and its confinement time, must exceed 10^{14} for a fusion reaction to take place, after the plasma has first been heated to 100 million Celsius degrees. Named for the British physicist J.D. Lawson.

Lithium. A light, silvery, chemically active metal that is a source of tritium. When lithium is irradiated with neutrons, its atoms undergo nuclear reactions that transform them into tritium.

Livermore. Formally, Lawrence Livermore National Laboratory. The nation's second nuclear weapons lab, founded in 1952. Currently active in weapons design, laser fusion, and magnetic-mirror research. East of San Francisco, California

Los Alamos. Formally, Los Alamos National Laboratory. The nation's first nuclear weapons laboratory, founded during World War II. Currently a center for weapons design and laser-fusion research. Near Santa Fe, New Mexico

Magnet. As used in this book, an electromagnet, a coil of electric cable or another conductor wound around a central form. When electric current flows through the conductor, it produces a powerful magnetic field.

Magnetic bottle. An arrangement of magnets surrounding a chamber, which may be used to confine a plasma and prevent it from leaking away rapidly.

Magnetic field. The property of the space near a magnet by which iron is attracted to the magnet.

Magnetic field lines. Curving lines resembling the patterns set up when iron filings are sprinkled on a plate near a magnet. These lines are used to visualize the strength and shape of a magnetic field.

Magnetic fusion. A method of producing fusion energy by confining a plasma within a magnetic bottle and heating it to 100 million Centigrade degrees.

Magnetic mirror. A magnetic bottle consisting of a long straight tube the ends of which are plugged with magnetic fields, to prevent the plasma from escaping. The magnetic fields grow stronger near these end-plugs. Plasma particles, seeking to escape, are reflected or turned back by these stronger fields.

Microexplosion. The small explosion of a fusion pellet when it is hit with a laser pulse, in a laser fusion experiment.

Microwaves. Radio waves of short wavelength, used to heat a plasma.

MIT. Massachusetts Institute of Technology.

MFTF. Mirror Fusion Test Facility. A large mirror device built around a yin-yang magnet the size of a two-story house, built at Livermore in the late 1970s.

MFTF-B. A very large tandem mirror facility, the size of a Boeing-747 fuselage, employing two MFTF magnet sets, currently under construction at Livermore.

Nanosecond. A billionth of a second.

Neutral beam. A large boxlike device that heats plasma by shooting a powerful and intense beam of atoms into it, with high energy; also, the beam produced by such a device. The atoms in the beam carry no electric charge and hence are neutral. Therefore these atoms can cross the magnetic fields enclosing the plasma without being deflected, which they would be if they carried electric charge.

Neutron. An atomic particle having no electric charge and found in atomic nuclei.

Nova. A ten-arm, 100-kilojoule laser now under construction at Livermore for use in laser fusion experiments.

Novette. A two-arm laser currently in use at Livermore. Its arms are of the same design that will be used for Nova.

$n\tau$. The product of plasma density, n, and plasma confinement time, τ; a measure of how well the plasma is confined. $n\tau$ must exceed 10^{14} in order for fusion reactions to produce useful quantities of energy.

OMB. The U.S. Office of Management and Budget; the White House agency that prepares the federal budget to be submitted to Congress.

OSTP. The U.S. Office of Science and Technology Policy: headed by the President's science advisor, this is the White House agency that recommends policy in these areas.

PDX. Poloidal Divertor Experiment; a large tokamak at Princeton University used to study the use of a divertor in controlling impurities and improving beta.

Pellet. A small charge of fusion fuel, often contained within multiple concentric spheres of glass, plastic, and other materials, the entire arrangement being of elaborate design. It is hit with a burst of laser energy in a laser fusion experiment, with the energy heating and compressing the fuel to the point where it will undergo fusion.

Plasma. A gas heated to the point where its atoms ionize. Ordinary gases are made up of atoms, but a plasma is made up of ions and electrons. The name *plasma* was coined by the physicist Irving Langmuir to refer to rarefied, glowing, electrically conducting gases. Langmuir thought that some of the phenomena he was observing in such gases resembled phenomena observed in blood; hence he borrowed the term for this new application.

Plasma gun. A device that injects a burst of low-energy plasma. The gun is often used to produce an initial source of plasma for a fusion experiment.

PLT. Princeton Large Torus; an experimental tokamak, half the size of TFTR, built at Princeton University to test the use of neutral beams and other fusion methods.

Princeton. Formally, Princeton Plasma Physics Laboratory. The nation's leading fusion research center, located near Princeton University in New Jersey

Proton. An atomic particle having a positive electric charge and found in atomic nuclei. The nucleus of a hydrogen atom consists of a single proton only.

Q. Symbol for the ratio of fusion energy produced within a plasma to the energy fed in from neutral beams or other outside source in order to heat the plasma.

Q = 1. A condition in which the fusion energy produced by a plasma equals the fusion energy being fed in from the outside to heat that plasma.

Reprocessing. Chemical processing of nuclear fuel for the purpose of separating out the valuable plutonium within the fuel.

Riggatron. A small tokamak fusion reactor concept invented by Robert Bussard of Inesco, Inc., in La Jolla, California.

Shiva. One of the largest lasers built and operated to date, featuring twenty arms and producing laser pulses with up to fifteen kilojoules.

Space charge. The electric charge distributed within the inner volume of a mirror machine, as a result of preferential leakage of electrons from the plasma contained within that mirror.

Stellarator. A fusion device featuring a torus-shaped bottle. The bottle requires an auxiliary magnetic field to achieve good confinement, and this field is provided by special coils wrapped or twisted around the plasma chamber, in a helical pattern.

Superconductors. Electric conductors that show zero electrical resistance when they are cooled with liquid helium. Since electrical resistance dissipates an electric current, superconductors can carry powerful currents without loss. They thus can produce strong magnetic fields without needing a continuing supply of electric power.

Tandem mirror. An arrangement of several magnetic mirrors in tandem. The outer mirror cells help confine the plasma within the central mirror.

Terawatt. A trillion watts. This is the power generated momentarily by a laser delivering a kilojoule of energy in a nanosecond of time.

T-15. A large Soviet tokamak, the centerpiece of the U.S.S.R. program during the 1980s. It features superconducting magnets but will not use DT or produce fusion power, and thus will serve principally for research in plasma physics.

TFTR. Tokamak Fusion Test Reactor. A large Princeton tokamak which is to achieve Q = 1 in 1986, thereby demonstrating production of fusion power for the first time anywhere.

Thomson scattering system. A method for studying plasma properties during a fusion experiment. It uses a powerful laser, which shines through the plasma. Electrons within the plasma reflect or scatter a very small part of this laser light, but the scattered light can be picked up and used to diagnose the plasma state. Named for the British physicist J.J. Thomson.

Thermal barrier. Separation between two regions of plasma whose electrons are maintained at markedly different temperatures. Thermal barriers are expected to improve the plasma confinement in tandem mirrors.

TMX. Tandem Mirror Experiment. A magnetic bottle about the size of a railroad tank car, built at Livermore in the late 1970s and used to test the principles of tandem mirrors.

TMX-Upgrade. A new version of TMX rebuilt in the early 1980s and used to test the principles to be applied in producing thermal barriers.

Tokamak. The best-performing magnetic bottle built to date. It is torus-shaped and, like a stellarator, it requires an auxiliary magnetic field to achieve good plasma confinement. This field is provided by a powerful electric current, which is made to flow through the plasma.

Torus. The shape of a doughnut, automobile tire, or inner tube. Also, something in that shape.

Tritium. The heaviest isotope of hydrogen. Its atomic nucleus contains a proton and two neutrons. Tritium, which is produced from lithium, is essential as a fuel for fusion.

2XIIB. A moderate-size mirror device used for experiments at Livermore during the 1970s. It demonstrated how warm-plasma streaming could markedly improve plasma confinement in magnetic mirrors.

UCLA. University of California, Los Angeles.

Warm-plasma streaming. Flowing a stream of plasma through a body of much hotter plasma. This greatly reduces plasma leakage in a magnetic mirror.

Wendelstein. A large stellarator built near Munich, Germany, where it operated successfully in 1980. The name is taken from a nearby hill. It plays on the name "Matterhorn," used for the Princeton stellarator program in the 1950s, thereby suggesting that the German goals are more modest but also more achievable.

Yin-yang magnet. A large magnet having the general shape of two thick horseshoes face-to-face and crossed at right angles, with a space between them. The space is for the plasma. The name derives from the Chinese symbol for the unity of opposites, though the magnet in fact has virtually no resemblance to the shape of that symbol.

Index

AEC, 8, 35–36, 43, 46, 48, 72, 76, 110, 111–112, 117, 121, 146, 153, 169, 191, 198, 269, 308, 316–317. *See also* ERDA
afterheat, 302, 303–305, 308, 315
Ahlstrom, Harlow, 131, 151–152
Airlie House, Va., 257–258, 259, 274, 276
Albuquerque, N.M., 33, 121, 252
Alcator tokamaks, 165–166, 167, 168, 170, 173, 178, 187, 273, 274; confinement records, 165, 187, 202; current drive, 276
alcohol, fuel, 16, 175
American Physical Society, 61, 143, 187, 258
Ann Arbor, Mich., 107, 144
Apollo program, 48, 194, 224, 225, 229
Argus laser, 134, 135, 136, 142, 149–150
Artsimovich, Lev, 38, 39, 40, 61, 86, 165, 168, 205, 281
ASDEX, 273–274, 277, 284
Ashworth, Clinton, 300–301, 316; on power industry, 301–302; on fusion safety, 302–304, 305; on fusion plants, 305–306
Aspen, Colo., 71–72
Astron experiment, 74–76, 78
Atkinson, R., 26
atomic energy, 60–61, 286–288, 291, 319
atoms, 5, 14, 15, 24, 197, 299, 312
aviation, 286, 289, 290, 319–320, 322
Aviation Week, 134

Baker, Howard, 247
Baldwin, David, 84, 95

Baltimore, Md., 61, 168, 273
Baseball II mirror, 76, 77, 79, 89, 91
Basov, Nikolai, 131–132
Bell Laboratories, 45, 230
Berchtesgaden, Germany, 88–89, 205
Berk, Herbert, 84, 95
Berkeley, Calif., 51, 52, 75, 79, 82
beta, in alternate concepts, 307, 315; in mirrors, 104, 257, 306; in tokamaks, 104, 185, 257, 274–276, 277, 306
Bethe, Hans, 287
Bishop, Amasa, 35–36, 38, 40, 41, 42
Bishop, James, 210, 212
Bodner, Stephen, 158, 253, 258; on laser fusion, 158–159; joins OMB, 253; recommendations to Kintner, 254–255; and stellarators, 255–256; challenges MFTF-B budget, 256–259; strategy with Palmieri, 259–260; face-off with Kintner, 260–261, 264, 265
Bohm diffusion, 36, 39, 41, 184, 192
Bol, Kees, 183–184; on plasmas, 184–185
boron, 313
Bowen, William, 211
Bradshaw, Gail, 206, 210
breeder reactors, 191, 198, 218, 224, 238, 242, 295–297, 298, 320
Brown, Harold, 198, 199
Brueckner, Keith, 107–108, 110–112, 115, 120, 162, 172
Buchsbaum, Solomon: and Hirsch, 45–46, 110, 230; and fusion policy, *see* policy
budgets, 31, 45, 50, 59, 67, 74, 78, 93, 95, 96, 167, 198, 199, 201, 206, 210, 244, 317; laser fusion, 43, 129, 133,

339

Budgets (*continued*)
   143, 145, 146, 158, 162; magnetic fusion, 36, 43, 86, 193, 202, 215–216,
   221, 224, 232, 236, 246–247, 249–250,
   252, 256, 266, 270, 279. *See also* OMB
Bukharin, Nikolai, 26
Bussard, Robert, 163–164, 165, 166, 223,
   317, 320; and Hirsch, 164–165, 169; in-
   vents Riggatron, 166–168, 169; heads
   Inesco, 163–164, 169–170, 178, 179,
   181–182, 186–187; and Shanny, 169–
   170, 171–172, 178, 179, 181, 187; political problems, 170–173; and Litton
   Industries, 173, 175–176, 177; and
   Guccione, 176–178, 181, 182; hopes for
   future, 163, 174–175, 176, 182–183,
   185. *See also* Riggatron tokamaks

Caltech, 42, 122, 123, 124, 132, 169, 230
Cambridge University, 128
Carter, Jimmy, 13, 92, 193, 194, 198, 202,
   206, 212, 218, 223, 224–226, 228–229,
   235, 237, 238, 242, 244
cartoons, on fusion, 39, 100, 255
CBS News, 208–209
Center for Fusion Engineering, 232, 240–
   241, 250
Chicago Bridge and Iron, 50, 98
Christofilos, Nicholas, 74–76
Clarke, John, 246, 269–270; and Hirsch,
   44, 46, 47, 246; and Kintner, 246, 248–
   249, 264–265, 270; on budgets, 270,
   272; on fusion policy, 271, 277–280,
   306; management attitudes, 272, 278
Clauser, John, 93–94
coal, 8, 12, 198, 206, 301–302, 309
Coensgen, Frederic, 78–79, 85, 91, 94,
   101, 105; and 2XIIB, 81–82, 83–84, 86;
   and field-reversed mirror, 89, 315–316;
   and TMX, 89, 91; on competition with
   Princeton, 281
coil-winding machine, 97–98
Columbia University, 185, 286, 290, 306
computers, 16, 17, 18, 67, 69, 89, 96,
   101–103, 139–140, 148, 154–155, 179,
   272. *See also* LASNEX
Congress, 31, 46, 148, 150, 171, 172, 179,
   190, 193, 218–219, 222, 233–234, 237–
   238, 265; House Appropriations Committee, 74, 150, 222, 238; Joint
   Committee on Atomic energy, 41, 43,
   74, 111, 113, 218. *See also* McCormack
Conn, Robert, 219, 230, 231, 304–305
Coppi, Bruno, 165, 167–168, 170, 171,
   172, 235
cranes, lifting, 52–53, 98–99
Cutter, W. Bowman, 226, 227
Cyclops laser, 134, 135

Daedalus rocket, 155
Damm, Charles, 91
Davidson, Ronald, 253, 266
Dawson, John, 206
Dean, Stephen: on Bussard, 163; and
   mirrors, 78, 79; and PLT weekend,
   206, 208–210, 211–212, 213, 214
Del Mar, California, 179–180
DeMeo, Anthony, 208, 212–213
Department of Energy, 31, 60, 68, 92,
   150, 153, 163, 178, 182, 223; in Carter
   Administration, 198–199; Directors of
   Energy Research in, 198, 223, 244, 250;
   offices in, 200, 211–212, 251, 260, 269;
   and PLT results, 206–207, 208, 210; in
   Reagan administration, 242–244, 246–
   247; regulations of, 9, 18, 67, 68; and
   Riggatrons, 171–173. *See also* Clarke;
   Dean; Deutch; Duncan; Edwards;
   Frieman; Kintner; Pewitt; Schlesinger;
   Trivelpiece
Deutch, John, 198–199, 210–211, 230;
   and fusion policy, 199, 201, 202, 203,
   205, 215–216, 223; and McCormack,
   223–224, 227–228; and PLT, 211–212,
   213
deuterium, 8, 47, 66–67, 68–69, 187, 287,
   312–313, 318
diagnostics, 35, 36, 39, 57–58, 61, 181; in
   laser fusion, 122, 149; on MFTF-B,
   102–103; on PLT, 20–21, 196, 204,
   215; Thomson scattering, 39, 57, 58,
   61; on TFTR, 59, 64, 69–70; on 2XIIB,
   83
Dimov, G.I., 89, 96
Doublet III tokamak, 274, 275
DT, 6, 35, 41, 47, 67–69, 72–73, 117,
   283, 311, 316
Duncan, Charles, 223, 224, 226, 236–237

Ebasco-Grumman, 48–49, 51
EBT, 241–242, 247, 250, 252, 313
Edwards, James, 243–244, 246–247, 248
electric power, 11, 13–14, 19, 101, 114,
   158, 182, 198, 295, 300–302, 310, 318
electrons, 75, 86–87, 94–95, 134, 197
Ellis, William, 87
Emmett, John, 132–133, 140, 159, 160,
   161; and Shiva, 133, 134, 136, 139, 142,
   149, 162; and Nuckolls, 145, 148; and
   Nova, 145–146, 147, 148, 150
energy, major sources, 11–13, 15–16,
   295, 309–310, 322–323
Energy Daily, 206
engineers, 49–50, 133
English, Spofford, 42
ERDA, 34, 48, 84, 191, 192, 193, 198
ETR, 278–279, 284, 285, 300, 306, 320

Eubank, Harold, 195–197, 203, 215
Exxon, 13, 178, 195, 219

Farnsworth, Philo, 34–35, 316, 318
FED, 232–233, 235, 240–241, 249, 252, 257, 266, 278, 279
FER, 284, 285, 320
Fermi, Enrico, 60–61, 290
FMIT (Fusion Materials Irradiation Test), 241, 250, 252
Ford, Gerald, 34, 49, 86, 168, 193, 194, 198, 212
Fort Wayne, Ind., 34, 316
Foster, John, 199, 203; and laser fusion, 111, 146–148, 159; and magnetic fusion, 199–203, 230, 231, 279, 315–316. *See also* Policy
Fowler, T. Kenneth, 78–79, 88, 105–106, 200, 210, 301, 315; and Bodner, 258; and DT-burning, 46; and MFTF, 84–85, 86, 87; and MFTF-B, 93, 95, 96–97, 182, 233, 262; and 2XIIB, 78–82, 83–84
Frieman, Edward, 223–224, 226, 228, 229, 230–231, 234–236, 238, 239, 243
Furth, Harold, 9, 39, 64, 105, 170, 206, 253, 266; and Soviet tokamaks, 39, 40; and C-Stellarator, 40, 41; proposes TFTR, 46–47; and Foster Committee, 200, 203; and Buchsbaum Committee, 232–233; quoted, 207, 209, 242, 276–277, 293–294, 307, 313, 318
Fusion (book), 304
fusion energy, 4–5, 6, 8, 24; conditions to generate, 29, 30, 78, 115, 135, 141, 187, 204; conferences on, 36–38, 39, 61–62, 87, 88–89, 180, 205–206, 214–215, 258, 272–276; difficulty of achieving, 26, 27–28, 73, 82, 129, 285–286, 291, 319–320; engineering vs. science, controversy, 231–232, 244–245, 266–267, 271, 309; future prospects, scenarios, 15–16, 174–175, 291–292, 294, 298, 299–300, 302–303, 305, 307, 308–311, 318–323; methods for generating, 29–30, 316; news coverage of, 70, 84, 206–214; pace of development, proposals, 45, 198–199, 202, 220–221, 224–225, 229, 235, 257–259, 279–280, 285; and radioactivity, 14–15, 26; start of research in, 71–72, 286–290. *See also* plasma; policy
Fusion Energy Foundation, 207, 208
fusion reactors, 27, 232, 240–241, 293–294, 300, 302, 307, 308–309, 310, 311, 318; advanced fuels for, 311–313, 318; alternate concepts as, 241–242, 307, 313–315, 318; direct conversion in, 312,

313, 315, 318; fusion breeder, 294, 298–300, 320; laser fusion, 149, 157–158, 160, 161; materials for, 51, 54, 174, 180, 241, 302, 303–304, 305; mirrors as, 85–86, 92–93, 95, 103–105; pilot plants, 305–306, 307; safety of, 14–15, 158, 302–305; Starfire, 303; tokamaks as, 72, 86, 103–105, 165, 233, 274, 276, 284. *See also* Beta; ETR; Riggatron tokamaks

gamma rays, 14, 299
Gamow, George, 26
gasoline, 12, 15–16
Gavin, Joseph, 48, 219, 237
General Atomic Co., 41, 178, 187, 274
General Electric, 109
General Services Administration, 148, 211
glass, 108, 109, 138, 145, 148
Glass, Alexander, 144, 153
Goldston, Rob, 196, 215
Gottlieb, Melvin, 47, 193, 223; and C-Stellarator, 40–41; and DT-burning, 45, 230; and PLT, 204–205, 206, 207, 213; and PLT weekend, 210–211, 212–214; and Russian Firebird, 215
Gould, Roy, 42, 75
graphite, 166, 197
Greer, Merwyn, 150, 193
Grieger, Günter, 255
Grove, Donald, 60, 65
Grumman Corp., 48, 49, 219, 237
Guccione, Robert, 164, 176–178, 179, 187

Häfele, Wolf, 295, 320
Halbouty, Michael, 242–243
Hanford Works, Washington, 191, 218, 234, 241, 299
heat transfer, 166–167, 172, 173, 302, 303–305, 311
helium, 14–15, 58, 62, 98, 99, 100, 288, 312–313
heroin, 297–298
Hess, David, 207, 208
Hirsch, Robert: idealism of, 33–34, 194; initial interest in fusion, 33–34; with Farnsworth, 34–36, 316–317; and electric fusion, 316–318; interest in DT, 35, 43–44; joins AEC, 35–36, 37; and A. Bishop, 36–37, 38, 40, 41–42; in Novosibirsk, 37–38; pushes for tokamaks, 40–41; and Gould, 42; and laser fusion, 110, 123; leadership in Washington, 42, 216; takes over fusion program, 42–43, 74; management attitudes, 41–42, 43, 44–45, 74, 76–77,

Hirsch, Robert (*continued*)
    84–85, 202, 231; and fusion budgets,
    33, 43, 194, 216; and Trivelpiece, 168,
    250–251, 263; and Astron, 74–76, 78;
    and mirrors, 76–78, 83–85; and Bus-
    sard, 164, 168; pushes for TFTR, 45–
    46, 48, 49, 216; and Alcator, 165; and
    Kintner, 49, 191, 192, 202, 262; pro-
    moted by Ford, 34, 49, 86; resigns from
    ERDA, 194–195, 216; and Exxon, 195,
    219; and McCormack, 43, 218, 219–
    220, 225
Holifield, Chet, 41, 111, 113
Hollingsworth, Robert, 42–43
Houtermans, Fritz, 26
Holzrichter, John, 160
Hyde, Roderick, 123–127, 154, 155
hydrogen, 6, 21, 66, 68, 108, 109, 125,
    128, 166, 284, 287, 289, 313, 318

IAEA (International Atomic Energy
    Agency), 88, 205–206, 272–273
Ignitor tokamak, 167, 168
India, 131
inertial confinement fusion. *See* laser
    fusion
Inesco, Inc., 163, 168, 173, 177–180,
    186–187; facilities, 179, 180, 181, 182,
    186; financing of, 169, 177–178, 181–
    182, 186, 187; and MIT, 180–181
Innsbruck, Austria, 205, 206, 214–215
Ioffe, M.S., 80, 81, 96
Israel, 169
ITT, 35, 178, 316

Jackson, Henry, 234
Jacobs, Eastman, 286–290
Janus laser, 134
Japan, 283–284
JET tokamak, 282–283
Johnson, David, 61–62
Johnston, Tudor, 156
JT-60 tokamak, 284

Kadomtsev, Boris, 192, 215, 281–282
Kanaev, Boris, 96
Kantrowitz, Arthur, 286–291
Keeton, Kathy, 176–177
Kelley, George, 87
Kendig, Frank, 176–177
Key Biscayne, Fla., 45
Kidder, Ray, 119, 120, 162
kilojoules, 114
kilovolts, 66
Kintner, Alice, 200–201, 205
Kintner, Edwin, 89, 95, 231, 235, 266,
    270, 309; in World War II, 188–189;

and Rickover, 189–191, 232, 245, 309;
    and breeder reactors, 191–192; Hirsch's
    deputy, 49, 86, 168, 192, 245; and Q-
    enhancement, 86, 89; and Riggatrons,
    168, 170–171, 172, 226; management
    attitudes 171, 193, 199, 249, 254–256,
    261; and fusion budgets, 193, 202, 216,
    235, 246, 247–248, 249–250, 252, 261–
    262, 264; strategy for fusion, 202; and
    PLT, 192–193, 205, 206, 207, 209, 210;
    and Foster, 199–201, 202, 267; and
    Gottlieb 204–205; and Dean, 208, 209–
    210, 211–212, 220; and Schlesinger,
    210, 211, 213–214; and Deutch, 212,
    213, 242, 267 and MFTF-B, 93, 96; and
    McCormack, 220; and Buchsbaum
    Committee, 230–232; and Frieman,
    230–231, 235; fusion program of, 240–
    242, 250, 251–253, 262, 278, 309; and
    fusion engineering, 232, 240–241, 266–
    267; and Pewitt, 170, 193, 244–246,
    247, 248–249, 261; and Clarke, 248–
    249, 263, 267; and Trivelpiece, 251,
    256, 262, 263, 265, 267, 268; takes
    stand on pace of MFTF-B, 253, 259,
    261, 262, 265; and Bodner, 254–256,
    260–262, 263–265; goes to Brussels
    263–265; resigns, 265–268; significance
    of, 307–309

Kissinger, Henry, 126
KMS Fusion, 108, 110, 113, 134, 144,
    156, 162, 163, 173. See also Siegel;
    Solomon
KMS Industries, 107, 112
Knight-Ridder news service, 207
Kozman, Theodore, 99–100
Krypton-Fluoride Laser (Los Alamos),
    157

La Jolla, Calif., 163, 179–180, 243
Lance, Bert, 193, 226
Laser Focus, 207
Laser fusion: classified nature of, 115,
    117, 119, 120–121, 125, 126–128, 134–
    135, 151–152, 159, 161; competition
    with magnetic fusion, 137, 145, 162;
    computer simulations of, 111–112, 120,
    122, 123, 154; difficulty of achieving,
    119, 135–137, 144, 147, 162; lasers,
    109, 112, 113–115, 119–120, 121–122,
    129–132, 133–134, 135, 138, 145, 154,
    156–157, 159, 160, 161–162; and nu-
    clear weapons, 115, 117, 152–153, 160;
    optimistic predictions in, 108, 111, 112,
    120, 121, 128, 129, 142–143, 161, 162;
    pellets, 108–109, 111–112, 115–120,

121, 128, 134–135, 136, 141, 142, 144, 149, 150, 153, 154, 155, 156, 161; physics of, 30, 108, 113–119, 147; and politics, 137, 145, 146, 150; prospects, 128, 143–144, 149, 154, 158–159, 160–162, 318; and rocket propulsion, 124–125, 126–127, 155, 160–161; in Soviet Union, 131–132, 134–135; wavelengths for, 141–142, 145, 148, 149, 150, 151, 156. *See also* Argus; Cyclops; Emmett; Hyde; LASNEX; Nova; Novette; Nuckolls; Shiva; Wood; Zimmermann

LASNEX, 123, 128, 136, 141, 153, 154–156

La Rouche, Lyndon, 207

Lawrence Livermore National Laboratory, 8–9, 41–42, 45, 60, 74, 75, 94, 101, 117, 148, 200; Building 391, 131, 138–139, 150; Building 431, 99, 100, 101, 285; competition with Princeton, 30, 32, 84, 93, 105, 203, 257, 258, 278–281, 285; computers at, 8, 101–103, 141, 153, 154; earthquake of 1980, 97, 140; and laser fusion, 30, 111–112, 119, 133, 142–143, 146, 150, 162; and magnetic mirrors, 30, 76, 78, 81, 82, 85; and nuclear weapons, 117, 151–153; working at, 57, 79, 82–83, 102–103, 133, 139, 151–152

Lawson criterion. *See* fusion energy, conditions to generate

Lawson, J. D., 29

lead, 157, 158, 295

Levitron, 75, 78

Levitt, Morris, 207

Lewis, George, 289–290

Lidsky, Lawrence, 168

Lisberger, Steven, 140

lithium, 8, 117, 157, 186, 311, 313

Litton Industries, 175, 176

Logan, Grant, 82–83, 91–92, 102–103; and tandem mirrors, 88–89, 315; and thermal barriers, 94–95

Los Alamos National Laboratory, 45, 110, 111, 146, 149, 153, 156, 157, 164, 252, 255, 307

Los Angeles, 119, 147, 187, 320

Lovins, Amory, 321

Loweth, Hugh, 226, 227, 228, 256, 262

Lubin, Moshe, 121, 131

Lujan, Manuel, 171

machinists, 50–51

magnetic bottles, 28, 29–30, 60, 85, 104

magnetic mirrors, 30, 76, 77, 83, 86–87, 89, 92, 97, 203, 277, 307, 312, 315–316; compared with tokamaks, 103–105;

DCLC mode in, 80–82, 84, 87; field-reversed, 89, 313–316, 318, 319; n_τ in, 78, 83; space charge in, 87–88; in Soviet Union, 80, 89, 96, 106; tandem, 87–89, 92–93, 94–95, 202, 205, 284, 315–316; thermal barriers in, 95–96, 105, 258, 281; transverse leakage in, 105–106, 281. *See also* Baseball II: Coensgen; Fowler; Logan; MFTF; MFTF-B; Post; TMX; TMX-Upgrade; 2XIIB

magnets, 5–6, 17, 19, 21, 30, 76, 93, 100, 165, 166, 178; construction of, 97–99; superconducting, 27, 62, 84, 92, 97–98, 100, 282; in TFTR, 50–51, 53, 54, 62, 69; yin-yang, 84–85, 86–87, 92, 97–100. *See also* coil-winding machine

Maiman, Theodore, 119

Manhattan Project, 60–61, 62, 186, 252, 286, 290

May, Michael, 200, 201

McCarthy, Eugene, 34

McCormack, Mike, 218, 222, 234, 242; and Washington State, 217–218, 242; fusion strategy of, 221–224, 225–226, 235, 238; and Hirsch, 43, 218–220; and Magnetic Fusion Act, 220–222, 233–235; "Dear Colleague" letters, 221, 229; and Jimmy Carter, 218, 225–226, 228–229; and OMB, 226–227, 239; and Tip O'Neill, 233; and Senator Jackson, 235; significance of, 237–239. *See also* Mense

McDonnell Douglas Corp., 107, 242

McIntyre, James, 226, 227

Meade, Dale, 56, 105, 182, 183, 276; on experimental work, 59–60; on hiring Ph.D.'s, 56–57; on Livermore, 281; on TFTR operations, 64, 65, 66, 68–69

Meese, Edwin, 122, 243

Mense, Allan, 220, 221, 226

Metz, William, 143–144, 159, 199

MFTF, 84–85, 86, 92, 97, 202, 315

MFTF-B, 92–93, 95–97, 106, 140, 163, 203; construction, 97–99, 101; goals for, 101, 104, 105, 182, 257; operation of, 101–103; pace of, controversy, 253, 256–265, 271; tests of, 99–100, 101

Miami *Herald*, 207, 208

microwaves, 27, 103, 276, 318

MIT, 123–124, 125, 168, 170, 178, 180, 189, 190, 198, 230, 232; Plasma Fusion Center, 40, 41, 101, 165–166, 173, 178, 201–202, 253, 271. *See also* Alcator

Mondale, Walter, 212

Montgomery, Bruce, 165, 178

Mori, Shigeru, 284

Morland, Howard, 117
motor-generators, 18–19, 27, 52, 62, 69, 101, 181, 182, 186, 282; Princeton accident, 53, 66, 99

nanoseconds, 114
natural gas, 11–12, 153
Nature, 128, 162
Naval Research Laboratory, 131, 132, 146, 158, 169, 178, 253
neutral beams, 45, 52, 58, 167, 255, 273, 311; accident in Utah, 51–52; on magnetic mirrors, 78, 79, 81, 85, 86, 91, 93, 95, 313, 315; on PLT, 16, 20, 21, 24, 63–64, 192–193; on TFTR, 10, 47, 51–52, 62, 63–64, 65–66, 68–69
neutrons, 14–15, 18, 35, 53, 66–67, 68, 109–110, 112, 149, 152, 155–156, 157, 168, 185–186, 187, 241, 282, 285, 295, 298–299, 304, 311–313, 316–317, 318, 319
New Orleans, 61, 126
New York Times, 71, 128
nitrogen, 99, 195–196
Nixon, Richard, 45, 81, 212
Nobel Prize, 38, 131, 132, 146
notebooks, laboratory, 18, 63
Nova laser, 144–149, 154, 156, 158, 159, 162, 163; debate over design, 146–148; and Novette, 149, 150; pellets for, 149; performance of, 148–149, 156
Novette laser, 149–151, 154, 156, 157, 159; goals and performance, 150–151; and Shiva, 150
Novosibirsk, 37, 61, 205
Nuckolls, John, 121, 123, 124, 134, 135, 155, 159, 160, 199; and laser-fusion optimism, 128, 135, 142–149, 162; and Nova, 148, 149; and pellets, 117–120, 135, 136, 145, 146, 147, 153, 161
nuclear accidents, scenarios, 296, 302–303, 304
nuclear power, 13–15, 174, 185, 190, 223, 294–295, 301, 302, 307–308, 309, 315, 318, 320–321
nuclear submarines, 62, 190, 299, 308–309
nuclear terrorism, 297–298, 299
nuclear weapons, 71, 112, 115, 117, 119, 120, 134, 149, 156, 296, 299; tests, 152–153

Oak Ridge National Laboratory, 41, 44–45, 87, 185, 192, 220, 242, 247; and DT-burning, 44, 46–47
ohmic heating, 21, 167; in TFTR, 65
oil, 11–13, 153, 174–175, 243, 310

oil shale, 12–13
OMB, 31, 43, 218, 226–227, 235, 236, 248, 272; budget appeals, 193, 259–260; and 1982 budget, 246–248, 254; and 1983 budget, 250, 253–265; and Kintner, 170–171, 193, 245; and McCormack, 226–227, 228, 229; and Riggatrons, 170–171, 172. *See also* Bodner; Loweth; McIntyre; Palmieri; Repici; Stockman
Omni, 164, 176
Oregon, 123–124
oscilloscopes, 17, 18, 63, 81, 96
OSTP, 31, 228. *See also* Keyworth; Press
oxygen, 144

Pacific Intertie, 103
Parker, Ronald, 165, 170, 180–181, 187
Palmieri, Thomas, 226, 229, 246, 247; and Bodner, 253, 256, 258, 264; and MFTF-B budget, 259–260; wins Keyworth's support, 260; showdown with Kintner, 260–262; negotiates on MFTF-B, 262–263; makes deal with Trivelpiece, 265
Papsco, Robert, 48
Pasadena, Calif., 122
Pastore, John, 74, 113
patents, 110–111, 115, 168, 173, 290–291
PDX, 49, 52, 61, 66, 183, 254
Peacock, Nigel, 41
Pearlstein, Donald, 84, 97
Pease, Sebastian, 39, 41
Penthouse, 164, 174
Perón, Juan, 70–71
Pewitt, N. Douglas: and Riggatrons, 170; and TFTR, 193, 245; replaces Frieman, 243–244; views on fusion, 244–245; and Kintner, 245–246; and 1982 budget, 246–247; punishes Clarke, 248–249; in OSTP, 250, 260, 261
photovoltaics, 13
Physical Review, 127–128
physicists, 32, 40, 45; academic preparation, 55–56, 91; camaraderie of, 70, 72, 82–83, 102; on PLT, 17–20, 196, 203–204, 215; salaries of, 57; on TFTR, 55–57, 70, 72–73; wagers among, 39, 156; work of, 59–61, 79, 102–103, 133, 196
plasma, 5–7, 9, 21, 24–25, 68, 82, 184–185, 192, 287–288, 290; appearance of, 21, 24–25, 35, 69, 289, 317; computer simulations of, 91–92, 96, 169, 170, 172; confinement of, 28–29, 38, 40, 43–44, 65, 78, 81, 83, 92, 101, 104–105, 165, 184–185, 187, 202, 255, 273–274; density of, 24, 29, 30, 115, 276, 315;

heating of, 288 (*see also* microwaves; neutral beams; ohmic heating); impurities in, 6, 24, 65, 69, 70, 104, 184, 197; ignition in, 24, 62, 167, 182–183, 185, 187, 282–283; 284; injection of fuel into, 25, 65, 81–82, 187; leakage, 25, 27–28, 36, 56, 80, 85, 86–87, 88, 92, 95, 96–97, 105–106, 180, 192; mathematical theories of, 55, 184, 192, 274; operating programs in research, 254–255, 259, 264; temperatures in, 4, 5, 6, 10, 16, 21, 24, 29, 38, 64, 66, 78, 94, 187, 311–313; warm, 80, 81–82, 85, 94, 96

PLT, 11, 29, 33, 46, 49, 52, 62, 66, 254; appearance of, 20–21, 23; control room, 16–20, 63, 203–204, 215; current drive, 276, 280; motor-generators, 18–19; neutral beams, 20, 66, 192–193, 195–196, 197, 203–204; operation of, 17–19, 20, 196, 203–204; plasma, 21, 24–25, 197, 215; press coverage of, 206, 207–209, 212–213, 214; significance of, 205, 206, 212, 213, 216; temperature records in, 16, 197, 202, 204, 205, 209, 214; trapped-particle instabilities and, 193, 197, 203–205, 213

"PLT weekend," 208–214

plutonium, 218, 282, 291, 295, 296–298, 299

Pogutse, Oleg, 192, 215

policy, on fusion energy, 31–32, 75, 222–224, 227–228, 229, 239, 249; Buchsbaum Committee, 230–233, 235–237, 242, 277; Foster Committee (laser fusion), 146–148; Foster Committee (magnetic fusion) 199, 201–202, 215, 222, 232; Hirsch panel, 219–220, 222, 225, 230, 235; Magnetic Fusion Act, 220–221, 235, 271, 306; Reagan's, 271–272, 277–280

Post, Richard, 75, 80, 81, 84, 87–88, 104–105, 304, 312

Postma, Herman, 45, 46, 242

press conferences, 84, 206, 207, 210, 211, 212–213

Press, Frank, 198, 223, 226, 228, 230, 236

Princeton Club (Manhattan), 177

Princeton Plasma Physics Laboratory, 4, 8–11, 26, 36, 39, 41, 54, 85, 115, 184, 223, 307; competition with Livermore, 30, 32, 278–281, 285; and DT-burning, 44, 45–46, 60; main building, lobby, 9–10, 53, 64; status of among physicists, 60–61; and TFTR management, 48, 49–50, 51, 53; working at, 17–20, 55–57, 59–61, 70, 72–73, 196, 204, 210.

See also Furth; Gottlieb; Meade; PDX; PLT; Reardon; TFTR

Princeton University, 167, 169, 180, 208, 211, 253, 285, 297; campus, 3

Progressive, 117

proposals, 35, 47, 48, 72, 85, 89, 91–92, 96, 97, 317

Q, in mirrors, 85–86, 93, 94–95, 96, 105, 315; in tokamaks, 62–63, 86, 282

Q = 1, 62, 64, 67–70, 288

quality control, 50–51, 52

radioactivity, 14–15, 157, 158, 166, 297–298, 299, 302, 304–305, 311–313; in TFTR, 67, 68, 69

Ray, Dixy Lee, 45, 46

Reagan, Ronald, 31, 122, 242–243, 250, 252, 254, 261, 266, 270

Reardon, Paul, 47, 64; TFTR design, 48, 49–50; neutral-beam accident, 52; "Santa Claus Comes to Fusion," 64–65

Reichle, Len, 48

Reliance Trucking Co., 52–53

Repici, Dominic: and Riggatrons, 170–171; and McCormack, 226–227, 228, 229; in Reagan administration, 246, 247–248, 253, 262

restaurants, 107, 112, 179–180, 258

Reversed-Field Pinch, 307, 313

Richter, Burton, 146

Richter, Ronald, 71

Rickover, Hyman, 189–191, 307–308

Riggatron tokamaks, 166, 167, 168–169, 173, 180, 186, 274, 306, 313, 318, 320; development of, 181, 182–183, 186, 187; economic prospects, 174–176, 182, 185, 186. *See also* Bussard; Inesco; Shanny

Right Stuff, The, 60

Roberts, Michael, 44, 46, 47

robots, 157, 166, 241, 283; in TFTR, 67, 73

rockets, 26–27, 123, 124–125, 126, 155, 160–161, 163–164, 166–167, 173. *See also* spacecraft; Space Shuttle

Rockwell International Corp., 50

Romatowski, Ray, 249

Rosenbluth, Marshall, 180, 185, 230, 304

Ross, Mike, 140–141, 151

Rudakov, Leonid, 134–135

rumors, 100, 135

Rusche, Ben, 244, 248

Sakharov, Andrei, 38

Sandia Laboratories, 121, 131, 146

San Diego, 80, 108, 163, 169, 274

Savannah River, Ga., 68, 299
Schapiro, Asher, 181
Schawlow, Arthur, 133
Schlesinger, James, 92, 193, 211; and
    Hirsch, 42–43, 194–195, 251; as Energy
    Secretary, 194, 198, 206, 211–212, 223;
    attitudes toward fusion, 199, 202, 206–
    207, 210; and PLT weekend, 211–212,
    213-214
Science, 143, 144, 159–160, 199
Science Applications, Inc., 178, 243
Scientific American, 94, 121
Seaborg, Glenn, 8
seafaring, 318, 321–322
Seamans, Robert, 84, 168, 191, 194
seawater, 8, 175, 313
Shanny, Ramy, 169–170, 171, 172, 178,
    179, 181, 187
Shaw, Milton, 191
Shiva, god, 131, 152
Shiva laser, 97, 131, 156, 157, 159; ap-
    pearance of, 138–139, 149; budgets,
    129, 133, 136, 140; development of,
    133–134, 139–140, 142, 145; and fre-
    quency-multiplying crystals, 141–142,
    148; goals and performance, 134, 135,
    136–137, 138, 141, 142–143, 144, 155–
    156; operation of, 139–140; shutdown
    of, 141, 143, 150; spatial filtering
    (pinholes), 133–134, 136, 145; SWAT
    (Shiva Welfare Altruistic Trust), 156;
    and *Tron,* movie, 140–141. *See also*
    Emmett; Nuckolls; Wood
Shohet, Leon, 256
"shots" (fusion experiments), 18, 21, 24–
    25, 63, 65, 67, 69–70, 72, 81, 83, 92,
    101, 102, 139, 183, 283
Siegel, Keeve, 107–108, 109, 110, 112–
    113, 120, 153, 162, 172; death of, 113,
    144
Silicon Valley, 163
Simonen, Thomas, 87, 92, 95–96
skiing, 47, 71–72, 189
solar energy, 13–14, 15–16, 198, 206,
    242, 254, 294, 306, 315. *See also*
    alcohol
Solomon, David, 108–109, 113
spacecraft, 240–241
Space Shuttle, 15, 26–27, 125, 126, 172,
    305
speed of light, 156
Spheromak, 307, 313
Spitzer, Lyman, 70, 71–72, 73, 94, 104,
    167, 184, 195, 255, 291
stainless steel, 51, 53, 54, 98, 282, 312
Stanford University, 133, 146, 230
Starbird, Alfred, 146
"Star Wars" (laser defense), 122

stellarators, 9, 39, 40, 72, 184, 255, 256,
    257, 313. *See also* Spitzer; Wendelstein
Stix, Thomas, 317
Stockman, David, 247, 253, 256, 257, 261
stocks, investments, 110, 181, 187
sun, 4–5, 25, 26, 28, 117, 219, 287, 313
synfuels, 12–13, 15, 153, 206, 318

TARA mirror, 101, 271
Taylor, Theodore, 297–298
Teller, Edward, 28, 115, 117, 122, 304
Tennessee Valley Authority, 44
terawatts, 114
T-15 tokamak, 282
TFTR, 4, 10, 21, 25, 29, 60, 92, 145, 163,
    165, 271; assembly of, 53–54, 65–66,
    67; budget, 193–194, 215–216, 245,
    262, 271; computers, 63; design, 47,
    48–50; first plasma, 64–65; and Furth,
    46–47; goals and performance, 60, 62,
    64–69, 73, 78, 86, 105, 182, 184, 202,
    205, 206, 213, 277; and Hirsch, 43–44;
    magnets, 50–51, 53, 54, 62, 69; motor-
    generators, 52–53, 62, 66, 69, 181; neu-
    tral beams, 10, 47, 51–52, 62, 63–64,
    65–66, 68–69; operation of, 69–70,
    101; power budget of, 62–63. *See also*
    Q = 1
thorium, 295, 298–299
Thorne, Robert, 199, 206, 208, 211
TMX mirror, 89–92, 95, 96, 104, 202
TMX-Upgrade, 95–96, 101, 105–106,
    258, 280–281
tokamaks, 9–10, 30, 55, 84, 162, 319;
    beyond TFTR, 202, 203, 215, 221, 225,
    231, 257, 285; current drive, 276, 278,
    280, 285; disruptions, 58, 60, 65, 103–
    104, 167; in Europe, 282–283, 285; ex-
    periments with, 19, 26, 45, 66, 69–70,
    165–168; in Japan, 283–284, 285, 320;
    limiters, 66, 197, 203; operation of, 22,
    23, 58, 65, 69, 72, 103–105, 257, 276;
    plasma current, 22, 27, 58, 65, 69, 72,
    103, 167, 276; plasma shapes, 274–276,
    282; start of U.S. program in, 40–41; in
    Soviet Union, 36, 38, 39–40, 41, 78,
    88, 281–282; transformers, 22, 69, 103,
    276; trapped-particle instabilities, 192,
    205, 213, 216. *See also* Airlie House;
    ASDEX; Doublet III; FED; FER;
    JET; JT-60; PDX; PLT; Riggatrons; T-
    15; TFTR; Torus II
tokatron, 9
Torus II tokamak, 185, 306
Tri-State College, 169
tritium, 8, 35, 44, 47, 62, 68–69, 117,
    157, 186, 271, 285, 303, 311–313, 317
Trivelpiece, Alvin, 168, 219, 250–251,

263, 270, 279; budget options, 259–
260; showdown with OMB, 260–262;
makes deal with OMB, 262–265
TRW, Inc., 199, 200
Tsongas, Paul, 235, 238
2001: A Space Odyssey, 101–102
2XII mirror, 76, 78, 79
2XIIB, 79, 81–84, 86, 88, 91, 94, 95, 96,
103, 202, 315

UCLA, 200
University of Chicago, 61
University of Rochester, 121, 131, 146
University of Texas, 41, 95, 180
University of Wisconsin, 255, 256
uranium, 166, 185, 290, 295, 297,
298–299
U.S. Navy, 169, 188–189, 190–191, 273,
295, 308

vacuums, 17–18, 19, 21, 27, 51, 68, 80,
81, 82, 151, 186, 288–289
Valbe, Larry, 100
Van de Graaff generators, 286–287
Vermont, 189, 204–205
Vietnam War, 122
viewgraphs, 83–84, 97
Virginia, 171, 179, 195

Wagner, Carl, 178, 179
Wakefield, Ken, 49
Wallace, Henry, 183

Warhol, Andy, 10
Washington, D.C., 31, 176, 200–201,
209, 212, 246, 260. *See also* Congress;
Department of Energy; OMB; OSTP;
White House
Washington *Post,* 208, 209–210
Washington State, 217–218, 234, 301. *See
also* Hanford
Weggel, Carl, 172, 178–179, 181
Wendelstein stellarator, 255, 257
Westinghouse, 47, 48, 286–287
White House, 46, 126, 194, 207, 225–226,
236, 238, 246, 260
Willis, Eric, 211–212
Wolfe, Tom, 60
Wood, Lowell, 122, 129, 135; and laser-
fusion optimism, 120, 128, 143, 162;
and Nuckolls, 120, 122, 128; and Zim-
mermann, 122–123; and Hyde, 124–
126, 127; in 1980 earthquake, 140; on
laser fusion, quoted, 131, 138, 139,
141, 144–145, 146, 148, 150, 161
World War II, 11, 139, 183, 188–189,
195, 218, 290, 291, 320
Wüster, Hans-Otto, 283

X rays, 69, 117, 119, 149, 152, 288, 289,
318
x-y plotter, 83

Zeus laser, 161–162
Zimmermann, George, 122–123, 154, 155